T0350982

DEVELOPING
SAFETY-CRITICAL
SOFTWARE

A Practical Guide for Aviation Software and DO-178C Compliance

DEVELOPING SAFETY-CRITICAL SOFTWARE

A Practical Guide for Aviation Software and DO-178C Compliance

LEANNA RIERSON

CRC Press
Taylor & Francis Group
Boca Raton London New York

CRC Press is an imprint of the
Taylor & Francis Group, an **informa** business

CRC Press
Taylor & Francis Group
6000 Broken Sound Parkway NW, Suite 300
Boca Raton, FL 33487-2742

© 2013 by Taylor & Francis Group, LLC
CRC Press is an imprint of Taylor & Francis Group, an Informa business

No claim to original U.S. Government works

Version Date: 20121016

ISBN 13: 978-1-4398-1368-3 (hbk)

I dedicate this book in memory of Cary Spitzer, who believed in its importance and my ability to write it, and to my grandmother, Charlotte Richardson, who prayed daily for my work. Both Cary and Grandma Richardson passed away in the fall of 2011 as I was finishing the first draft of this book. I miss them both immensely and hope they would be pleased with the text that they helped inspire.

Contents

Part III Developing Safety-Critical Software Using DO-178C

Part IV Tool Qualification and DO-178C Supplements

Part V Special Topics

Preface

It's with a great sense of responsibility and complete humility that I present this book. After two years of researching and writing, I feel I've only scratched the surface. I hope that what I share and have experienced will help you in your professional endeavors.

My passion is safety and making its realization practical. As early as high school, I wrote a column for the local newspaper on safety tips, extensively researching (as much as one can in rural America before the Internet): bathtub safety, tractor safety, and electrical safety, among others. In my senior year of high school, I competed in the state persuasive speech competition—the subject was seatbelt safety. My desire to promote safety led me to the Federal Aviation Administration (FAA) and is the primary motivation behind this book.

Software is only one link in the overall safety chain. Yet, it is an important link and one which grows more vital each year. As software becomes more prevalent and is used in more critical ways, its risks and contributions to safety increase. At the same time that the risk due to software is increasing, the business models of many companies seem to be shifting—more is expected with less, schedule becomes king, and software engineers are treated like pieces on a game board that can be randomly reorganized.

This book is intended to be a tool for those caught in the churn of the industry. Perhaps you are a systems engineer or manager, a software manager, a software engineer, a quality assurance engineer, or a student striving to learn from the experiences of others. You want to do an outstanding job but are overwhelmed with schedule and budget pressures. I hope that this book, based on the last 20 years of my life in the aviation industry, will help you as you pursue quality and excellence.

Acknowledgments

This book would not have been possible without the help of many colleagues and friends. The following reviewed portions of the text and provided valuable input: Patty (Bartels) Bath of AVISTA, Jim Chelini of VEROCEL, Andrew Elliott of Design Assurance, Kelly Hayhurst of NASA's Langley Research Center, Wendy Ljungren of GE Aviation, Jeff Knickerbocker of Sunrise Certification, George Romanski of VEROCEL, Andrew Kornecki of Embry-Riddle Aeronautical University, Fred Strathmann of Springton Technologies, Steven VanderLeest of DornerWorks, and Dan Veerhusen who retired from Rockwell Collins. Thanks to each of you for your time and your friendship. You challenged my thinking and made the text better.

This book would have never been conceivable without those who mentored, guided, and opened doors of opportunity for me early in my career. There are too many to name, but I'm especially indebted to the following:

- Milt Sills, former vice president of engineering at Cessna Aircraft Company. He mentored me and stoked my love of aviation. He passed away in 2003 but continues to have a daily influence on my life.

- Jim Peterson, retired FAA systems and equipment branch manager at the Wichita Aircraft Certification Office. Jim recruited me to the FAA and instilled in me a passion to stand up for what's right no matter what the cost.

- Elizabeth Erickson, retired FAA director of Aircraft Certification Services and currently with Erickson Aviation Services. Beth believed in me and mentored me in challenging times at FAA headquarters in Washington, District of Columbia.

- Cheryl Dorsey, owner of Digital Flight. Cheryl recruited me into the ranks of software engineer at FAA headquarters. My background was avionics systems and electronic hardware, but Cheryl realized my potential and ignited my desire for safe software.

- Mike DeWalt, FAA's chief scientific technical advisor for aircraft computer software. Mike guided me when I was very young and inexperienced.

I am also grateful to RTCA, Inc. I've been privileged to serve on five committees—in leadership roles on three of them. RTCA has been very supportive of this book and generous to grant permission for references and quotes. You can learn more about RTCA, Inc. or purchase their documents by contacting them at

RTCA, Inc.
1150 18th Street NW
Suite 910
Washington, DC 20036
Phone: (202) 833-9339
Fax: (202) 833-9434
Web: www.rtca.org

Thanks also to my dear friends Susan Corcoran and Denise Polansky who helped with the final stages of this effort, to the team who prayed for me through to the finish line, and to the staff at Taylor & Francis Group for their support and help. Nora Konopka, Jill Jurgensen, and Glenon Butler have been terrific.

MATLAB® and Simulink® are registered trademarks of The MathWorks, Inc. For product information, please contact:

The MathWorks, Inc.
3 Apple Hill Drive
Natick, MA, 01760-2098 USA
Tel: 508-647-7000
Fax: 508-647-7001
E-mail: info@mathworks.com
Web: www.mathworks.com

Author

Leanna Rierson is an independent consultant in software, complex electronic hardware, and integrated module avionics (IMA) development for safety-critical systems, with emphasis on civil aviation. She has over 20 years of experience in the software and aviation industry. Rierson spent nine years as a software and avionics specialist at the United States Federal Aviation Administration (FAA)—five of those years were in the position of chief scientific and technical advisor for aircraft computer software. Rierson has published numerous papers on safety-critical software, integrated modular avionics, and aviation; led many national and international engineering teams and workshops; and developed courses, policies, handbooks, and guidance material for the FAA. She served as a subgroup cochair and editorial team leader on the RTCA special committee that wrote DO-178C and six other related documents. Rierson has taught DO-178B, and now DO-178C, to hundreds of professionals. She is an FAA Designated Engineering Representative (DER) with Level A authority in the software and complex hardware technical areas. She has worked with numerous aircraft and avionics companies, including Boeing, Cessna, Learjet, Embraer, Rockwell Collins, GE Aviation, Honeywell, and numerous others. She is currently working part-time for the Rockwell Collins avionics certification team.

Rierson holds a master's degree in software engineering from Rochester Institute of Technology and a bachelor's degree in electrical engineering from Wichita State University. She also attended Ozark Christian College and received a master's degree from Johnson Bible College. She is currently pursuing a doctorate degree.

Part I

Introduction

Part I sets the foundation for the entire book. It includes an explanation of what safety-critical software is and why it is important in today's environment. An overview of the book and an explanation of the approach taken are also provided.

1

Introduction and Overview

Acronyms

DER   Designated Engineering Representative
EASA   European Aviation Safety Agency
FAA   Federal Aviation Administration
IEEE   Institute of Electrical and Electronic Engineers
IMA   Integrated Modular Avionics
NASA   National Aeronautics and Space Administration

1.1 Defining Safety-Critical Software

A general definition for *safety* is the "freedom from those conditions that can cause death, injury, illness, damage to or loss of equipment or property, or environmental harm" [1]. The definition of *safety-critical software* is more subjective. The Institute of Electrical and Electronic Engineers (IEEE) defines safety-critical software as: "software whose use in a system can result in unacceptable risk. Safety-critical software includes software whose operation or failure to operate can lead to a hazardous state, software intended to recover from hazardous states, and software intended to mitigate the severity of an accident" [2]. The *Software Safety Standard* published by the U.S. National Aeronautics and Space Administration (NASA) identifies software as *safety-critical* if at least one of the following criteria is satisfied [3,4]:

1. It resides in a safety-critical system (as determined by a hazard analysis) AND at least one of the following:
   - Causes or contributes to a hazard.
   - Provides control or mitigation for hazards.
   - Controls safety-critical functions.
   - Processes safety-critical commands or data.
   - Detects and reports, or takes corrective action, if the system reaches a specific hazardous state.

- Mitigates damage if a hazard occurs.
- Resides on the same system (processor) as safety-critical software.

2. It processes data or analyzes trends that lead directly to safety decisions.
3. It provides full or partial verification or validation of safety-critical systems, including hardware or software systems.

From these definitions, it can be concluded that software by itself is neither safe nor unsafe; however, when it is part of a safety-critical system, it can cause or contribute to unsafe conditions. Such software is considered *safety-critical* and is the primary theme of this book.

1.2 Importance of Safety Focus

In 1993, Ruth Wiener wrote in her book *Digital Woes: Why We Should Not Depend Upon Software*: "Software products—even programs of modest size—are among the most complex artifacts that humans produce, and software development projects are among our most complex undertakings. They soak up however much time or money, however many people we throw at them. The results are only modestly reliable. Even after the most thorough and rigorous testing some bugs remain. We can never test all threads through the system with all possible inputs" [5]. Since that time, society has become more and more reliant on software. We are past the point of returning to purely analog systems. Therefore, we must make every effort to ensure that software-intensive systems are reliable and safe. The aviation industry has a good track record, but as complexity and criticality increase, care must be taken.

It is not unusual to hear statements similar to: "Software has not caused any major aircraft accidents, so why all the fuss?" The contribution of software to aircraft accidents is debatable, because the software is part of a larger system and most investigations focus on the system-level aspects. However, in other domains (e.g., nuclear, medical, and space) software errors have definitely led to loss of life or mission. The Ariane Five rocket explosion, the Therac-25 radiation overdoses, and a Patriot missile system shutdown during the Gulf War are some of the more well-known examples. The historical record for safety-critical software in civil aviation has been quite respectable. However, now is not the time to sit back and marvel at our past. The future is riskier for the following reasons:

- *Increased lines of code*: The number of lines of code used in safety-critical systems is increasing. For example, the lines of code from the Boeing 777 to the recently certified Boeing 787 have increased eight-fold to tenfold, and future aircraft will have even more software.

- *Increased complexity*: The complexity of systems and software is increasing. For example, Integrated Modular Avionics (IMA) provides weight savings, easier installation, efficient maintenance, and reduced cost of change. However, IMA also increases the system's complexity and results in functions previously isolated to physically different federated hardware being combined onto a single hardware platform with the associated loss of multiple functions should the hardware fail. This increased complexity makes it more difficult to thoroughly analyze safety impact and to prove that intended functionality is met without unintended consequences.

- *Increased criticality*: At the same time that size and complexity of software are increasing, so is the criticality. For example, flight control surface interfaces were almost exclusively mechanical 10 years ago. Now, many aircraft manufacturers are transitioning to fly-by-wire software to control the flight control surfaces, since the fly-by-wire software includes algorithms to improve aircraft performance and stability.

- *Technology changes*: Electronics and software technology are changing at a rapid rate. It is challenging to ensure the maturity of a technology before it becomes obsolete. For example, safety domains require robust and proven microprocessors; however, because a new aircraft development takes around 5 years, the microprocessors are often nearly obsolete before the product is flight tested. Additionally, the changes in software technology make it challenging to hire programmers who know Assembly, C, or Ada (the most common languages used for airborne software). These real-time languages are not taught in many universities. Software developers are also getting further away from the actual machine code generated.

- *More with less*: Because of economic drivers and the pressure to be profitable, many (maybe even most) engineering organizations are being requested to do more with less. I've heard it described as follows: "First, they took away our secretaries; next, they took away our technical writers; and then, they took away our colleagues." Most good engineers are doing the work of what was previously done by two or more people. They are spread thin and exhausted. People are less effective after months of overtime. A colleague of mine puts it this way: "Overtime is for sprints, not for marathons." Yet, many engineers are working overtime for months or even years.

- *Increased outsourcing and offshoring*: Because of the market demands and the shortage of engineers, more and more safety-critical software is being outsourced and offshored. While not always an issue, oftentimes, outsourced and offshored teams do not have the systems domain knowledge and the safety background needed to effectively find and remove critical errors. In fact, without appropriate oversight, they may even inject errors.

- *Attrition of experienced engineers*: A number of engineers responsible for the current safety record are retiring. Without a rigorous training and mentoring program, younger engineers and managers do not understand why key decisions and practices were put into place. So they either don't follow them or they have them removed altogether. A colleague recently expressed his dismay after his team gave a presentation on speed brakes to their certification authority's new systems engineer who is serving as the systems *specialist*. After the 2-hour presentation, the new engineer asked: "You aren't talking about the wheels, are you?"*
- *Lack of available training*: There are very few safety-focused degree programs. Additionally, there is little formal education available for system and software validation and verification.

Because of these and other risk drivers, it is essential to focus even more on safety than ever before.

1.3 Book Purpose and Important Caveats

I am truly honored and humbled that you have decided to read this book. The purpose of this text is to equip practicing engineers and managers with the information they need to develop safety-critical software for aviation. Over the last 20 years, I've had the privilege of working as an avionics developer, aircraft developer, certification authority, Federal Aviation Administration (FAA) Designated Engineering Representative (DER), consultant, and instructor. I've evaluated safety-critical software on dozens of systems, such as flight controls, IMA, landing gear, flaps, ice protection, nose wheel steering, flight management, battery management, displays, navigation, terrain awareness and warning, traffic collision and avoidance, real-time operating systems, and many more. I've worked with teams of 500 and teams of 3. It has been and continues to be an exciting adventure!

The variety of positions and systems has allowed me to experience and observe common issues, as well as effective solutions, for developing safety-critical software. This book is written using these experiences to help practicing aircraft systems engineers, avionics and electronic systems engineers, software managers, software development and verification engineers, certification authorities and their designees, quality assurance engineers, and others who have a desire to implement and assure safe software.

As a practical guide for developing and verifying safety-critical software, the text provides concrete guidelines and recommendations based on real-life projects. However, it is important to note the following:

---

* For those of you who don't know, speed brakes are on the wings.

- The information herein represents one person's view. It is written based on personal experiences and observations, as well as research and interactions with some of the brightest and most committed people in the world. Every effort has been made to present accurate and complete advice; however, every day I learn new things that clarify and expand my thinking. Throughout the book terms like *typically, usually, normally, most of the time, many times, oftentimes,* etc. are used. These generalizations are made based on the numerous projects I've been involved in; however, there are many projects and billions of lines of code that I have not seen. Your experience may differ and I welcome that insight. If you wish to clarify or debate a topic, ask a question, or share your thoughts, feel free to send an email to both of the following addresses: LRierson1@aol.com and Digital_Safety@sbcglobal.net.

- I have used a personal and somewhat informal tone. Having taught DO-178B (and now DO-178C) to hundreds of students, this book is intended to be an interaction between me (the instructor) and you (the engineer). Since I do not know your background, I have attempted to write in such a way that the text is useful regardless of your experience. I have integrated *war stories* and some occasional humor, as I do in the classroom. At the same time, professionalism has been a constant goal.

- Because this book focuses on safety-critical software, the text is written with higher criticality software (such as DO-178C levels A and B) in mind. For lower levels of criticality, some of the activities may not be required.

- While the focus of this book is on aviation software, DO-178C compliance, and aircraft certification, many of the concepts and best practices apply to other safety-critical or mission-critical domains, such as medical, nuclear, military, automotive, and space.

- Because of my background as a certification authority and as a DER, reading this book may at times be a little like crawling into the mind of a certification authority. (Don't be afraid.) Both FAA and the European Aviation Safety Agency (EASA) policy and guidance materials are discussed; however, the FAA guidance is primarily used unless there is a significant difference. While the content of this book is intended to explain and be consistent with the certification authorities' policy and guidance as they exist at the time of this writing, this book does not constitute certification authority policy or guidance. Please consult with your local authority for the policy and guidance applicable to your specific project.

- Having traveled the globe and interacted with engineers on six continents, I have made every effort to present an international perspective on safety-critical software and certification. However, because the majority of my work has taken place in the United States and on FAA certified projects, that perspective is presented.

- Several RTCA documents, including DO-178C, are referenced throughout this document. It has been exciting to serve in leadership roles on three RTCA committees and to have a key role in developing the referenced documents. As noted in the foreword, RTCA has been gracious to allow me to reference and quote portions of their documents. However, this book is not a replacement for those documents. I've attempted to cover the highlights and guide you through some of the nuances of DO-178C and related documents. However, if you are developing software that must comply with the RTCA documents, be sure to read the full document. You can learn more about RTCA, Inc. or purchase their documents by contacting them at the following:

  RTCA, Inc.
  1150 18th Street NW
  Suite 910
  Washington, DC 20036
  Phone: (202) 833-9339
  Fax: (202) 833-9434
  Web: www.rtca.org

- This book is written so that you can read it from beginning to end, or you can read selected chapters as needed. You will find occasional repetition between some chapters. This is intentional, since some readers may choose to use this book as a reference rather than read it straight through. References to related chapters are included throughout to help those who may not read the text cover to cover.

1.4 Book Overview

This book is divided into five parts. Part I (this part) provides the introduction and sets the foundation. Part II explains the role of software in the overall system and provides a summary of the system and safety assessment processes used for aviation. Part III starts with an overview of RTCA's DO-178C, entitled *Software Considerations in Airborne Systems and Equipment Certification*, and the six other documents that were published with DO-178C. The section then goes through the DO-178C processes—providing insight into the guidance and suggestions for how to effectively apply it. Part IV explores four RTCA guidance documents that were released with DO-178C. The topics covered are software tool qualification (DO-330), model-based development and verification (DO-331), object-oriented technology and related techniques (DO-332), and formal methods (DO-333). Part V covers special topics related to DO-178C and safety-critical software development. These special topics are focused on aviation but may also be applicable to other domains and include noncovered code (extraneous, dead, and deactivated code), field-loadable software, user-modifiable software, real-time operating systems, partitioning, configuration data, aeronautical databases, software reuse, previously developed software, reverse engineering, and outsourcing and offshoring.

There are many other subjects that are not covered, including aircraft electronic hardware, electronic flight bags, and software security. These topics are related to software and are occasionally referenced in the text. However, due to space and time limitations, they are not covered. Some of these topics are, however, slated to be covered in the new edition of CRC Press's *Digital Avionics Handbook*.

References

1. K. S. Mendis, Software safety and its relation to software quality assurance, ed. G. G. Schulmeyer, *Software Quality Assurance Handbook*, 4th edn., Chapter 9 (Norwood, MA: Artech House, 2008).
2. IEEE, *IEEE Standard Glossary of Software Engineering Terminology*, IEEE Std-610-1990 (Los Alamitos, CA: IEEE Computer Society Press, 1990).
3. National Aeronautics and Space Administration, *Software Safety Standard*, NASA-STD-8719.13B (Washington, DC: NASA, July 2004).
4. B. A. O'Connell, Achieving fault tolerance via robust partitioning and N-modular redundancy, Master's thesis (Cambridge, MA: Massachusetts Institute of Technology, February 2007).
5. L. R. Wiener, *Digital Woes: Why We Should Not Depend on Software* (Reading, MA: Addison-Wesley, 1993).

Part II

Context of Safety-Critical Software Development

Part II provides an overview of how software fits into the overall system development effort. In order to successfully implement *safe software*, one must first understand its role in the system. Focusing only on the software without considering the system and its safety characteristics is like treating the symptoms of a serious illness without getting at the root cause of the illness. In my experience, there are five key factors that must be thoroughly and constantly addressed when developing safety-critical software:

1. **Well-documented systems architecture and requirements definition.** The system architecture and requirements must focus on safety and be written down. It is next to impossible to successfully implement undocumented requirements. To ensure they are the right requirements, the system requirements also need to be validated for correctness and completeness.

2. **Solid safety practices at all levels of development.** Throughout the development, potential hazards and risks must be identified and addressed. Safety should be integral to all processes at all levels of the development—not just the responsibility of the safety organization.

3. **Disciplined implementation.** The documented requirements and safety attributes must be accurately implemented. This includes both assuring that the requirements are implemented and that no unintended functionality is added. A well-defined change management

process is essential to successful implementation and iterative development.

4. **Well-qualified personnel.** Safety-critical systems are implemented by human beings. The personnel must be knowledgeable of the domain, safety, and the technology being used to implement the system. People working in the safety-critical field should be the cream of the crop—perfectionists who strive for 100% on every task. Safety-critical systems demand a commitment to *excellence*—not just *good enough*.

5. **Thorough testing at all levels**. There is definitely a role for reviews and analyses, but they cannot replace the need to prove functionality and safety claims with the integrated software and hardware.

This book concentrates on software, but always be mindful that the software is only one aspect of the overall system. The five factors mentioned earlier should be addressed throughout the entire system development. This Part (II) on systems development and safety sets the context for subsequent Parts (III to V), where the software development and integral processes, as well as special software topics, will be detailed.

2

Software in the Context of the System

Acronyms

AC	Advisory Circular
ARP	Aerospace Recommended Practice
DAL	development assurance level
EASIS	Electronic Architecture and System Engineering for Integrated Safety Systems
EUROCAE	European Organization for Civil Aviation Equipment
FAA	Federal Aviation Administration
PSSA	preliminary system safety assessment
RAM	random access memory
RBT	requirements-based test
ROM	read only memory
SAE	Society of Automotive Engineers
TCAS	traffic collision and avoidance system

2.1 Overview of System Development

Before diving into the details of how to develop safety-critical software, it is important to understand the context of software in the overall system. Software operates in the context of a system, and safety is a property of the overall system. Software, in and of itself, is neither safe nor unsafe. However, software influences safety in a variety of system types, including aircraft systems, automotive systems, nuclear systems, and medical systems. In order to develop software that enhances rather than impairs safety, one must first understand the system in which the software operates and the overall system safety process. This chapter examines the process for developing safety-critical systems. Chapter 3 examines the system safety assessment process, which is part of the overall system development life cycle. The system and safety assessment overviews are presented from an aviation perspective; however, the concepts may be transferred to other safety-critical domains.

An aircraft is a large system comprising multiple systems—it is essentially a system of systems. Aircraft systems can include navigation, communication, landing gear, flight control, displays, collision avoidance, environmental control, in-flight entertainment, electrical power, engine control, ground steering, thrust reversers, fuel, air data, and more. Each system has an effect on the overall operation of the aircraft. Obviously, some have more safety impact than others. Because of this, the civil aviation regulations and supporting advisory material classify failure categories according to risk. Five failure condition categories are used: catastrophic, hazardous/severe-major, major, minor, and no effect (these categories are discussed in Chapter 3). Each function in each system, as well as the combination of functionality, is assessed for its overall impact on safe aircraft operation. The aircraft operation includes multiple phases, such as, taxi, take-off, climb, cruise, descent, and landing. System safety is evaluated for each phase, as the effects of failure conditions can vary greatly, depending on the phase.

Developing a safe system typically requires years of domain expertise with the specific type of system, as well as a thorough understanding of how an aircraft and its systems operate and interact with each other and human operators. In recent years, I have become alarmed that some organizations seem to believe that anyone with an engineering degree can develop an aircraft system. I am not opposed to globalization and profit, but developing the core expertise and competency to build safety-critical systems requires significant time, resources, and integrity.

For civil aviation, developers are encouraged to use SAE's Aerospace Recommended Practice (ARP) 4754 revision A (referred to as *ARP4754A*), entitled *Guidelines for Development of Civil Aircraft and Systems*. ARP4754 was published by SAE (formerly known as Society of Automotive Engineers) in 1996 and developed by a team of aircraft and avionics manufacturers, with input from certification authorities. EUROCAE (European Organization for Civil Aviation Equipment) also published an equivalent document, ED-79. In December 2010, SAE and EUROCAE published an update to the recommended practice: ARP4754A and ED-79A, respectively. On September 30, 2011, the Federal Aviation Administration (FAA) published Advisory Circular (AC) 20–174, entitled *Development of civil aircraft systems*. AC 20–174 recognizes ARP4754A as "an acceptable method for establishing a development assurance process."

ARP4754A divides the system development process into six phases: planning, aircraft function development, allocation of aircraft functions to systems, development of system architecture, allocation of system requirements to items (including allocation to software and hardware), and system implementation (including the software and hardware development) [1]. In addition, the following integral processes are identified in ARP4754A: safety assessment, development assurance level assignment, requirements capture, requirements validation, implementation verification, configuration management, process assurance, and certification authority coordination [1]. ARP4754A's scope includes the system development through the certification completion; however, in addition to the system phases described in ARP4754A, system development includes activities before product launch (as in needs determination and concept exploration) and after certificate issuance (such as, production and manufacturing, delivery and deployment, operations, maintenance and support, and retirement or disposal) [2,3]. Figure 2.1 illustrates the overall system development framework with the ARP4754A processes shown in gray.

As will be discussed in Chapter 3, the safety assessment process runs parallel to the system development process. As the system progresses from

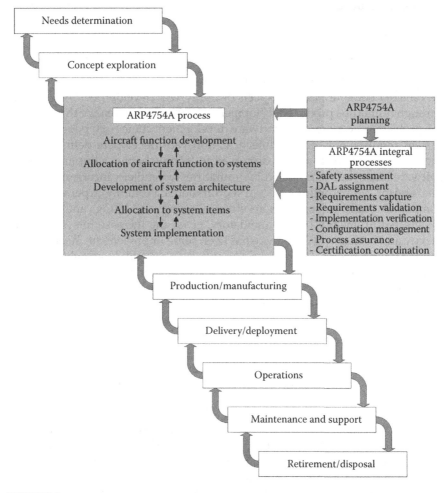

FIGURE 2.1
Overall system development process.

high-level design to implementation, the safety aspects are evaluated to ensure that the system satisfies the required safety levels. While ARP4754A proposes a top-down approach, it recognizes that most systems are developed with an iterative and concurrent approach, using both top-down and bottom-up development strategies. Most aircraft are not made of systems developed from a clean sheet. Instead, aircraft and systems tend to be *derivative projects* that build upon existing functionality and products. Therefore, it is important that the overall development life cycle consider the baseline systems and develop a realistic life cycle that meets both the customer needs (documented in the form of requirements) and the safety objectives.

2.2 System Requirements

2.2.1 Importance of System Requirements

Chapter 3 will delve into the safety assessment process. For now, however, let us consider some important aspects of the system development. After living through numerous product developments, I have come to realize that many of the problems encountered in software development come from poorly defined system requirements and architecture. Software developers definitely have their problems (which will be discussed in future chapters), but many of the software issues are exacerbated by immature, incomplete, incorrect, ambiguous, and/or poorly defined system requirements. From my experience, most companies ought to spend more effort defining, validating, and verifying system requirements. This is mandatory, given the increase in system complexity and criticality. While there has been some improvement, much more is needed if we are to continue to develop safe, reliable, and dependable systems. The certification authorities also realize this need, and with the formal recognition of ARP4754A, they are increasing their attention on the system development processes—particularly, the requirements definition, validation, and implementation verification.

2.2.2 Types of System Requirements

ARP4754A recommends that the following types of requirements be developed [1]:

- *Safety requirements* identify functions that either contribute to or directly affect the aircraft safety. The safety requirements are identified by the safety assessment process and include minimum performance constraints for both availability (continuity of function) and integrity (correctness of behavior). They should be uniquely identified and traced throughout the system life cycle.

- *Functional requirements* specify functionality of the system to obtain desired performance. Functional requirements are usually a combination of customer requirements, operational requirements, performance requirements (for example, timing, speed, range, accuracy, and resolution), physical and installation requirements, maintainability requirements, interface requirements, and constraints. Functional requirements may also be safety requirements or have safety consequences that should be evaluated for impact on safety.

- *Other requirements* include regulatory requirements and derived requirements. Derived requirements are typically the result of design decisions that are made as the system architecture matures. Since they may not trace to a higher level requirement, they need to be validated and assessed by the system safety process.

2.2.3 Characteristics of Good Requirements

Good requirements do not just happen. They require considerable effort, focus, and discipline. To write good requirements, it is important to understand the characteristics of such requirements. Requirements authors and reviewers should incorporate the following characteristics of good system requirements:

- *Atomic*—each requirement should be a single requirement.

- *Complete*—each requirement contains all of the necessary information to define the desired system functionality. Each requirement should be able to stand alone without further amplification.

- *Concise*—each requirement should simply and clearly state what must be done and only what must be done. It should be easy to read and understand—even by nontechnical users of the system. In general, a textual requirement should not contain over 30–50 words. Requirements represented using graphics should also be concise.

- *Consistent*—requirements should not contradict or duplicate other requirements. Consistent requirements also use the same terminology throughout the specification. Requirements are not the place to practice creative writing.

- *Correct*—each requirement should be the right requirement for the system being defined. It should convey accurate information. This attribute is ensured by the requirement's validation effort.

- *Implementation free*—each requirement should state *what* is required without identifying *how* to implement it. In general, requirements should not specify design or implementation. However, there may be some exceptions, such as interface or derived system requirements.

- *Necessary*—each requirement should state an essential capability, characteristic, or quality factor. If the requirement is removed, a deficiency should exist.

- *Traceable*—each requirement should be uniquely identified and easily traceable through to lower level requirements, design, and testing.

- *Unambiguous*—each requirement should only have one interpretation.

- *Verifiable*—it should be possible to confirm the implementation of each requirement. Therefore, requirements should be quantifiable and include tolerances, as appropriate. Each requirement should be written such that it can be verified by review,* analysis, or test.† Except for rare cases, if the behavior cannot be observed during verification, the requirement should be rewritten. For example, negative requirements are generally not verifiable and require a rewrite.

- *Viable*—each requirement should be able to be implemented, usable when implemented, and helpful to the overall system construction.

A requirements standard, requirements management plan, or best practices guide is recommended to provide guidance for requirements authors. Writing good requirements is not easy, even if you know for sure what you want to build. When I taught DO-178B classes for the FAA, we had a requirements writing exercise. The class was divided into teams and each team was given a LEGO® model. Each team was to write requirements for the model, tear it apart, put the pieces and the written requirements in a bag, and give it to another team to assemble using the requirements. At times, it was hilarious to see the end product, because it did not resemble the original model. Not everyone can write good requirements. In fact, from my experience, very few excel at this task. Good guidance, examples, and training based on the specific domain are invaluable. If graphical representations are used to explain the requirements, the standards should explain the rules and limitations of the graphics. For example, the standards may limit the number of symbols on a page, limit the number of pages for one graphic, limit the depth of a model, and provide detailed meaning for each symbol in the library (since various modeling tools have different interpretations for the symbols and developers have diverse levels of understanding). Chapters 5, 6, and 14 provide more discussion on requirements standards, software requirements development, and model-based development, respectively. These upcoming chapters focus on the software aspects of requirements but most of the suggestions also apply to systems requirements definition.

---

* *Reviews* are also called *inspections* in some literature.
† *Tests* are also called *demonstrations* in some literature. Tests tend to be the preferred method of verification for most certification authorities.

2.2.4 System Requirements Considerations

Requirements for safety-critical systems specify numerous functionalities. This section highlights some fundamental concepts to consider when developing systems requirements for software-intensive and safety-critical systems.*

2.2.4.1 Integrity and Availability Considerations

System requirements should address both integrity (correct functionality and operation) and availability (continuity of functionality). APR4754A defines integrity and availability as follows [1]:

- *Integrity*: "Qualitative or quantitative attribute of a system or an item indicating that it can be relied upon to work correctly. It is sometimes expressed in terms of the probability of not meeting the work correctly criteria."
- *Availability*: "Qualitative or quantitative attribute that a system or item is in a functioning state at a given point in time. It is sometimes expressed in terms of the probability of the system (item) not providing its output(s) (i.e., unavailability)."

Loss of integrity results in incorrect operation (a malfunction) of the system. Examples include misleading data on the primary flight display, incorrect resolution advisory from a traffic collision and avoidance system (TCAS), or an autopilot hardover. In order to design for integrity, architectural mitigation is often required to prevent and/or neutralize faults. This may involve the use of redundancy, partitioning, monitoring, dissimilarity, or other mitigation strategies to prevent integrity faults or to protect from such faults. Additionally, in order to address integrity, fault detection and responses or fault containment is often applied and should be captured as systems requirements.

Many systems require protection from loss of availability. Availability differs from integrity in that the function operation must be continuous and/or recoverable under all foreseeable operating conditions (including normal and abnormal conditions). Key functionality must be available during and after random hardware faults, data faults, or software or hardware upset. Systems or functions requiring high availability generally require equipment with very high reliability and/or utilize redundancy. When redundant designs are used to ensure availability, they should be able to tolerate faults and/or detect and recover from faults quickly. Systems incorporating redundant designs typically have redundancy management and fault management requirements.

---

* Please note that this section is merely an overview and does not exhaustively cover the subject.

It should be noted that some design decisions may impact both integrity and availability. Both must be considered. For example, a system may need to mitigate integrity faults without compromising availability. The original version of ARP4754 noted the relationship between integrity and availability when considering the use of dissimilarity: "It should be noted that architectural dissimilarity impacts both integrity and availability. Since an increase in integrity may be associated with a reduction in availability, and vice-versa, the specific application should be analyzed from both perspectives to ensure its suitability" [4].

2.2.4.2 Other System Requirements Considerations

In addition to integrity and availability concerns, there are other aspects to consider in the system specification, including:

1. *Safety requirements*, that is, requirements driven by the functional hazard assessment or subsequent safety analyses.

2. *Fault tolerance effects on the systems*, including the identification of fault regions. The Electronic Architecture and System Engineering for Integrated Safety Systems (EASIS) Consortium identifies a *fault region* as "a set of components whose internal disturbances are counted as exactly one fault, regardless of where the disturbances are located, how many occur, and how far they stretch within the fault regions" [5].

3. *Graceful degradation* (e.g., fail-safe mode), including the need for any containment regions. EASIS explains that "a containment region is the set of components which may be adversely affected by the respective malfunction. In other words: The containment region defines the border where fault propagation must stop" [5]. Examples include autopilot failure detection before the flight surfaces move or failed data flagged before the pilot responds.

4. *Error detection*, such as [5]:
   a. Data checks—to examine data to determine if they are valid, in range, fresh, short-circuited, open-circuited, etc.
   b. Comparison of redundant data—to compare the outputs of redundant sensors or processing units.
   c. Checking for errors in processor and memory—for example, perform built-in tests, use watchdog timers, calculate checksums for read only memory (ROM) or flash memory, use built-in processor exceptions, perform memory checks, or test random access memory (RAM) read/write.
   d. Communication monitoring—to perform checks on incoming messages in a distributed system, for example, check for message timeouts, message sequence counters, or checksums.

5. *Error handling of the detected errors.* Some examples of error handling are as follows [5]:

   a. Turn off the function.

   b. Enter a degraded mode.

   c. Provide annunciation to the user.

   d. Send a message to other systems that interact with the failed system.

   e. Reset the system and return to the previous state.

   f. Use other trustworthy data (when an input error is detected).

   g. Timeout or shut down the channel or system, when an error is continuously detected.

6. *Fault detection and handling,* which covers how the system responds to hardware faults (such as, continuous or random response).

7. *Fault avoidance,* which may be handled by selecting high quality parts (with higher specified mean time between failures), protecting against environmental effects such as electromagnetic interference, identifying certain forbidden faults in the requirements, and/or architecturally eliminating vulnerabilities.

8. *Requirements on the architecture or design* (e.g., redundancy, comparison checks, voting schemes). While this somewhat blurs the boundary between requirements and design, it is important to identify anything that requires consideration during implementation. A commentary with the requirement is a common way to identify the design considerations without defining the design too soon.

9. *Requirements on the manufacturing process.* These requirements document the key characteristics of the hardware manufacturing and production processes.

10. *Functional limitations.* Any known or desired limitations of the system need to be specified in the requirements. For example, some functions may only be allowed during certain flight phases or range of use.

11. *Critical functionality.* These requirements are needed to prevent loss of critical functionality caused by a failure.

12. *Partitioning or protection,* to provide isolation from or containment of faults. Partitioning or protection might also be used to reduce validation and verification effort, since less critical functions require less validation and verification rigor. Partitioning and protection are discussed in Chapter 21.

13. *Diversity,* which may be used to avoid common faults in two or more items. Diversity examples include different designs, algorithms, tools (e.g., compilers), or technologies [5].

14. *Redundancy,* which may be used to achieve fault tolerance or to avoid a single point of failure. Redundant components include multiple processors, sensors, etc.

15. *Timing-related requirements,* to address concerns like cockpit response times, fault detections and responses, power-up times, restart times, etc.

16. *Sequences,* including initialization, cold start, warm start, and power down requirements.

17. *Mode and state transitions,* such as flight phases, engage and disengage decisions, mode selections, etc.

18. *Environmental qualification levels,* to reflect the environmental robustness needed. The levels depend on the system criticality, expected installation environment, and regulatory requirements. RTCA/DO-160[ ]* explains the environmental qualification process for equipment installed on civil aircraft.

19. *Test provisions,* which may be needed to ensure the safety considerations are properly implemented. Examples include stimulus inputs, data recording, or other equipment interfaces to assist the testing effort.

20. *Single and multiple event upset protections* needed for some aircraft projects to protect the aircraft systems from radiation effects that occur at higher altitudes. Examples of common solutions include continuous built-in tests, error correction codes, voting schemes, and hardened hardware.

21. *Reliability and maintainability considerations,* especially to support safety analysis assumptions; for example, fault and associated data logging, initiated tests, and rigging.

22. *Latency requirements* to reduce the likelihood of faults remaining latent (hidden). Examples include alerts for fault identification and maintenance or built-in-test intervals.

23. *User and human interface requirements,* to ensure proper system use. Extensive certification guidance and standards exist to address human factors needs and to minimize operational issues.

24. *Maintenance requirements or limitations,* to ensure proper use and maintenance.

25. *Safety monitoring requirements,* when the system architecture requires safety monitors.

26. *Security considerations,* including protections such as encryptions, configuration and load checks, and interface barriers.

27. *Future growth and flexibility needs,* such as, timing and memory margins, minimized technology obsolescence, etc.

---

* [ ] indicates the latest revision or the revision required by the certification basis.

2.2.5 Requirements Assumptions

When developing the systems requirements and design, assumptions need to be documented. An assumption is a statement, principle, or premise that is not certain at the time the requirements are written [1]. There are numerous ways to document assumptions. Some commonly used methods are to: (1) include them as a commentary with the requirements or (2) create a separate document or database with links to the requirements (for instance, a separate requirements module or field). As the requirements are validated, the assumptions are also validated. Over time, the assumptions may become part of the requirements or design, rather than a separate category. For example, one may initially assume a certain maintenance period and document it as an assumption. During requirements validation, the maintenance period may be documented as a requirement (perhaps at a higher hierarchical level) so it is no longer just an assumption.

2.2.6 Allocation to Items

There are usually multiple hierarchical levels of systems requirements (such as, aircraft, system, and subsystem). The lower levels are decomposed from the higher levels, with more details (granularity) being defined at each lower level. The lowest level of system requirements is allocated to software or hardware for implementation. In ARP4754A, the software and hardware are called *items*. RTCA DO-178C is then applied to the software items and RTCA DO-254 (entitled *Design Assurance Guidance for Airborne Electronic Hardware*) to the electronic hardware items.*

2.3 System Requirements Validation and Verification

Not only are requirements developed, they are also validated and their implementation is verified. This section addresses both validation and verification.

2.3.1 Requirements Validation

Validation is the process of ensuring that requirements are correct and complete. It may include tracing, analyses, tests, reviews, modeling, and/or

---

* At this time, it is not completely clear what guidance applies to the hardware items where DO-254 is not applied (such as, non-electronic hardware or simple hardware). This area of ambiguity is expected to be cleared up in the future when DO-254 is updated to DO-254A. At this point in time ARP4754A, DO-254, or additional certification authority guidance (for example, FAA Order 8110.105, European Aviation Safety Agency Certification Memo CM-SWCEH-001, or project-specific Issue Papers) are used to cover the ambiguous areas.

similarity claims [1]. Per ARP4754A, the higher the development assurance level (DAL), the more validation methods need to be applied. Ideally, system requirements allocated to software are validated before the software team begins formal implementation. Otherwise, the software may need to be rearchitected or rewritten once the valid systems requirements are identified. Unfortunately, it is often difficult to validate certain system requirements without some working software. Prototyping the software is a common way to support requirements validation, but as Chapter 6 discusses, prototyping should be handled carefully. System requirements that have not been fully validated prior to handing off to the software team need to be clearly identified so that stakeholders can manage the risk.

2.3.2 Implementation Verification

Verification is the process of ensuring that the system has been implemented as specified. It primarily involves requirements-based tests (RBTs); however, analyses, reviews or inspections, and demonstrations may also be used. As with requirements validation, ARP4754A requires more verification methods for functions with higher DALs. Verification is an ongoing activity; however, formal* verification of implementation occurs after the hardware and software have been developed and integrated.

2.3.3 Validation and Verification Recommendations

Below are some recommendations to effectively validate and verify system requirements.

Recommendation 1: Develop a validation and verification plan.† The validation and verification plan explains the processes, standards, environment, etc. to be used. The plan should explain roles and responsibilities (who will do what), ordering of events (when it will be done), and processes and standards (how it will be accomplished). The plan should also include the content specified by ARP4754A and be understandable and implementable by the engineers performing the validation and verification.

Recommendation 2: Train all validation and verification team members. All engineers responsible for performing validation and verification activities should be trained on the plans, standards, processes, techniques, expectations, etc. Most engineers do not intentionally fail; oftentimes, their shortcomings are because of lack of understanding. Training helps validation and verification engineers understand why certain tasks are important and how to carry them out effectively.

---

* Formal verification is used for certification credit.
† The validation and verification planning may be in separate plans.

Recommendation 3: Apply appropriate methods. Validation methods focus on requirements correctness and completeness and include traceability, analysis, test, review, modeling, and similarity [1]. Verification methods focus on correctness of implementation to satisfy a given set of requirements; verification methods include reviews or inspections, analyses, tests, demonstrations, and service experience [1]. Rationale should be provided for areas where testing is not the selected method.

Recommendation 4: Develop checklists to be used by the teams. Validation and verification checklists help to ensure that engineers do not inadvertently forget something. Checklists do not replace the need to think but serve as memory joggers. The completed checklists also provide objective evidence that validation and verification efforts were performed.

Recommendation 5: Design the system and requirements to be testable. Testability is an important attribute of safety-critical systems and should be considered during the requirements capture and design. A case in point might be the implementation of test provisions that can detect errors during development and testing (e.g., bus babbler detector, partitioning violation monitor, slack checker, execution time monitor, out-of-bounds check, out-of-configuration check, built-in test, exception logs, monitors, etc.).

Recommendation 6: Develop and implement an integrated, cross-functional test philosophy. Involve as many disciplines as needed to get the right people involved at the right time, including software engineers.

Recommendation 7: Develop and run tests early and often. Proactive test development and execution is important. Early test efforts support both requirements validation and implementation verification. When writing the requirements, authors should think of how testing will be carried out. A couple of practices that help are: (1) to involve testers in requirements reviews and decisions and (2) to build an operational systems test station as soon as possible.

Recommendation 8: Plan test labs and stations early. The test facilities should be available as early as possible for multiple disciplines (e.g., software, systems, flight test, and hardware teams) to use. During planning phases, the number of users and the testing schedule should be considered to ensure adequate test facilities (number and capability). Since the customers may have some special needs or expectations, it is recommended to include them in the planning process as well.

Recommendation 9: Develop effective functional or acceptance tests and clearly identify pass or fail criteria. Since many engineers will need to use the test station and tests throughout the system development, it is important to spend time making the test environment effective, repeatable, and clear. It is ideal to automate as much as possible; however, there are normally at least a few tests that cannot be cost effectively automated.

Recommendation 10: Develop an integrated lab with accurate models for simulated inputs. The more complete the integration facility, the less the team relies on the aircraft or customer's facility. The more realistic the simulated input the better. The goal is to sort out as many problems in the lab as possible to avoid the need to troubleshoot in the aircraft or customer's facility.

Recommendation 11: Use validation and verification matrices. Matrices are a simple way to ensure that all requirements are validated and verified and to track the completion of the validation and verification effort.

Recommendation 12: Perform robustness testing. Robustness testing ensures that the system responds as expected when it receives unexpected or invalid input or when an abnormal combination or sequence of events occur. Robustness testing is often undervalued and underutilized at the systems level. When done effectively, it can mature the system and its underlying software before reaching the customer. For higher level systems (DAL A and B), ARP4754A requires that unintended functionality be considered during verification. Robustness testing provides confidence that the system will perform properly, even when events do not happen as expected.

Recommendation 13: Value the validation and verification effort. Both validation and verification are important for safety and certification. They are not just check marks or milestones on the master plan. Safety depends on how well validation and verification are done.

Recommendation 14: Keep the focus on safety. The team should be well-versed on the safety-related requirements and should spend a disproportionate amount of time on them during the validation and verification efforts. They should also recognize the signs of unintended safety compromise when testing other functions; for example, testers should not ignore excessive values, monitor flags, exceptions, or resets.

Recommendation 15: Flag issues as they arise. Issues rarely go away on their own, and frequently they get worse over time. If an issue is observed, it should not be ignored or hidden. Instead, it should be noted and evaluated. Once it is confirmed to be a valid issue, it should be documented for further investigation and resolution. An issues log (prior to baselining) and problem reports (after baselining) are typically used. Issues in the issues log that are not fixed prior to the baseline should be transferred to the problem reporting system to avoid being overlooked.

Recommendation 16: Do not sacrifice systems testing in order to compensate for schedule slips in other areas. It is a sad fact of life that verification is the last step in the chain and is under constant pressure to compensate for other slips in the schedule. This can lead to late discoveries, which are costly to fix and can severely stress the relationship with the customer.

Recommendation 17: Coordinate with software and hardware teams. Coordination between systems, software, and hardware tests can lead to earlier problem discovery (e.g., a software error might be found that also impacts systems) and optimized use of resources (e.g., there may be some systems tests that also verify software requirements and vice versa).

Recommendation 18: Gather meaningful metrics. Metrics can be used to track effectiveness, strengths, weaknesses, schedule, etc. Gathering the right metrics is important to effectively managing the validation and verification effort. Because validation and verification can consume half of the project budget* and is crucial to certification, it needs to be well-managed. Metrics can help by indicating where things are going well and where additional attention is needed.

2.4 Best Practices for Systems Engineers

Chapter 1 notes that well-qualified engineers are essential for building safe systems. Engineers are the masterminds behind the system. Without them, the system will not happen. Below are some suggested practices for the systems engineering team to implement in order to best use their talents and skills. These practices build upon the validation and verification recommendations in Section 2.3.3.

Best Practice 1: Plan ahead and put it in writing. Written plans are vital for successful certification. Develop plans for project management, safety, development, validation and verification, certification, etc. The plans should be practical, realistic, and visible for the people using them.

Best Practice 2: Plan for change. Develop an efficient change process for all levels of the system development and implementation. The change process should be clearly explained in the plans and should include a system-level change impact analysis process.

Best Practice 3: Foster a safety-focused culture. Educate the team on safety and their role in it. Safety should be embedded in organizational processes and part of team and management performance reviews. Additionally, it can be helpful to put up posters on safety, invite special speakers to inspire a safety consciousness, and remind engineers that their families' safety may depend on their actions. Safety should color all decisions and actions at all levels of the enterprise.

Best Practice 4: Document best practices. There is great value in capturing recommended practices and augmenting them with actual examples, based on years of experience. Once documented, the best practices should become

---

* Numerous project managers indicate that over half of their budget and time are spent on validation and verification.

mandatory reading for all new engineers or team members. Oftentimes, the best practices evolve into the company standards and processes.

Best Practice 5: Develop a top-level, enterprise-wide design philosophy. Organization-level practices such as, fail-safe approaches, maintenance boundaries, message approaches (such as alerts, warnings, color, and occurrence), etc. should be documented. Once documented, teams should be trained on the philosophy so that their daily decisions are consistent with the company's preferences.

Best Practice 6: Ensure the system architecture addresses known hazards from the aircraft and system hazard assessments. Oftentimes, multiple architectures are proposed and evaluated to determine which one best meets safety expectations, certification requirements, and customer needs.

Best Practice 7: Develop the system architecture with functional breakdowns early. Organize the architecture into logical groups as soon as possible, and use it to drive plans, hazard assessments, requirements, standards, program tracking, etc.

Best Practice 8: Perform an early safety assessment. The safety assessment determines the viability of the system architecture. In the early stages it does not need to be a full preliminary system safety assessment (PSSA); instead, it is used to confirm the concepts.

Best Practice 9: Capture explicit safety requirements. Safety requirements need to be documented so that they can be flowed down to software and hardware teams. Normally, safety requirements are identified with a *shall* (to show the mandatory nature) and a safety attribute.

Best Practice 10: Explain the rationale behind the requirements. It is helpful to explain where each requirement came from and how it relates to other requirements, standards, guidance, etc.

Best Practice 11: Document all assumptions and justification for derived requirements. Assumptions and justifications should be documented as the requirements are written. Do not put it off for later, since by that time the thought process may be forgotten.

Best Practice 12: Develop a proof-of-concept or prototype. A proof-of-concept or prototype helps mature the requirements and design. However, teams should be willing to throw away the prototype. Prototyping is a means to an end, not the end. Trying to upgrade a prototype to meet certification standards usually takes more time than starting again. Prototyping is discussed more in Chapter 6.

Best Practice 13: Involve hardware and software engineers in the requirements development process. The earlier hardware and software engineers understand the system, the more informed their design decisions will be, and the faster and more effectively they can implement the system.

Best Practice 14: Do not assume the requirements are right until they have been validated. It is best to keep an open mind and look for ways to improve and optimize requirements until validation. Despite some marketing claims, the customer is not always right. If something does not seem right or is unclear, raise it to the customer early.

Best Practice 15: Validate requirements as early as possible. Requirements validation is an ongoing process, but it is important to start it early using the right engineers with the right experience. Section 2.3.3 provides more suggestions for the validation effort.

Best Practice 16: Orchestrate early and frequent evaluations. Early evaluations by the customer, human factors personnel, pilots, maintenance crews, systems and aircraft integrators, certification authorities, etc. help to ensure that all stakeholders' expectations are met.

Best Practice 17: Update requirements based on input. Use input from hardware, software, safety engineers, and other stakeholders to improve and mature the requirements.

Best Practice 18: Hold early and ongoing informal reviews by peers. Peer reviews help to find issues early and ensure consistency among team members.

Best Practice 19: Use a requirements management tool. Since requirements are foundational to safe and successful implementation, it is important to manage them well. A requirements management tool can help to effectively manage requirements, by allowing faster and higher quality implementation of system requirements. Some manufacturers use a Word/textual document to house their requirements; however, unless carefully handled, such an approach can lead to requirements that are not uniquely or clearly identified and difficulties can arise in tracing to or from the requirements. This can force the software and hardware developers to interpret and make assumptions. Sometimes the interpretations and assumptions are correct but too frequently they are not. If used properly, a requirements management tool can help avoid some of these challenges. However, care must be taken not to lose the overall context and relationship of the requirements when they are captured in the tool.

Best Practice 20: Document and enforce company-wide standards and templates for systems requirements and design. Requirements standards and templates should promote the qualities that were discussed earlier in this chapter. Design standards and templates should address these kinds of topics: design rationale, constraints, allocation, limitations, safety paths, initialization, modes and controls, margins, modeling standards, etc.

Best Practice 21: Use graphical representations as appropriate. Graphics (e.g., user interfaces, display graphics, and control systems models) help to effectively communicate concepts; however, they should be supported with appropriate text.

Best Practice 22: Conduct formal reviews. Formal reviews are performed at key transition points throughout the system development. Chapter 6 discusses the peer review process.

Best Practice 23: Provide training to the systems team(s). Ongoing training is important for the systems engineers. Suggested training topics include: requirements writing, company standards, domain or context (for instance, aircraft design or avionics), architecture, safety attributes, safety techniques, safety philosophy, and best practices. Training should include concrete examples.

2.5 Software's Relationship to the System

Software is part of the system implementation. The system architecture and requirements drive the software development. Therefore, it is critical for systems, software, safety, and hardware engineers to communicate. The earlier and more frequently that communication occurs throughout the development process, the fewer problems will occur. The systems, safety, software, and hardware leads should develop a close working relationship throughout the project. I like to compare the four-disciplined team to the four legs of a chair. Without one of the legs, the chair is severely limited. Without two legs, it is useless. Communication failures among the four disciplines lead to unnecessary challenges—potentially causing wrong interpretation and implementation of system requirements, missed dependencies, inadequate tests, and wasted time and money. During a recent assessment of test data for a safety-critical avionics application, our team noticed that some of the software requirements were not tested. The software test lead said that the systems test team was exercising those requirements. When we approached the system test lead, he responded: "We aren't testing those, because the software guys are." Failure to communicate and to share test matrices led to a situation where no one was testing some of the system functions. Simple communication was all that was needed to avoid this problem.

Below are some suggestions to foster an integrated relationship between systems, safety, software, and hardware teams.

- Colocate the teams.
- Develop a core group involving each discipline's key stakeholders. This group provides leadership, makes key decisions, maintains consistency, evaluates and approves changes, etc.
- Work together as a core group to identify and document the overall design philosophy. Then share that philosophy with the entire team.

- Coordinate handoffs and transitions between the disciplines (i.e., ensure that data are mature enough for the next team to use).
- Encourage an open and backlash-free environment to identify, escalate, and resolve issues.
- Do not be afraid to involve the customer and certification authority when needed. Getting clarification on uncertain issues can prevent wrong decisions.
- Encourage extracurricular activities to improve team dynamics (go to lunch, dinners, parties, off-site meetings, sporting events, etc., together).
- Share lessons learned frequently and throughout the program.
- Promote cross-training between teams in key discipline areas (e.g., arrange for weekly or monthly presentations).
- Have safety representatives provide training on the system safety assessment results, so that everyone is familiar with the safety drivers.
- Use incremental and iterative modeling; this helps to build up the system and encourage frequent coordination.
- Encourage teams to build safety and quality into the product. It is next to impossible to add them in as afterthoughts.
- Track metrics that encourage teams to consider and assist each other (these might include measuring software functions successfully integrated on the hardware—not just the completion of hardware or software).

References

1. SAE ARP4754A, *Guidelines for Development of Civil Aircraft and Systems* (Warrendale, PA: SAE Aerospace, December 2010).
2. ISO/IEC 15288, *Systems and Software Engineering—System Life Cycle Processes*, 2nd edn. (Geneva, Switzerland: International Standard, February 2008).
3. ISO/IEC 12207, *Systems and Software Engineering—Software Life Cycle Processes*, 2nd edn. (Geneva, Switzerland: International Standard, February 2008).
4. SAE ARP4754, *Certification Considerations for Highly-Integrated or Complex Aircraft Systems* (Warrendale, PA: SAE Aerospace, November 1996).
5. Electronic Architecture and System Engineering for Integrated Safety Systems (EASIS) Consortium, *Guidelines for Establishing Dependability Requirements and Performing Hazard Analysis*, Deliverable D3.2 Part 1 (Version 2.0, November 2006).

3

Software in the Context of the System Safety Assessment

Acronyms

ASA	aircraft safety assessment
CCA	common cause analysis
CMA	common mode analysis
CNS/ATM	communication, navigation, surveillance, and air traffic management
FDAL	functional development assurance level
FHA	functional hazard assessment
HIRF	high intensity radiated fields
IDAL	item development assurance level
PASA	preliminary aircraft safety assessment
PRA	particular risk analysis
PSSA	preliminary system safety assessment
SFHA	system functional hazard assessment
SSA	system safety assessment
ZSA	zonal safety analysis

3.1 Overview of the Aircraft and System Safety Assessment Process

As discussed in Chapter 2, safety is assessed in parallel with the aircraft and system development. This chapter provides an overview of the civil aviation safety assessment process, as well as an explanation of how software fits into the system and safety framework. Figure 3.1 shows the iterative nature of the system and safety development; as the system develops, the safety aspects are identified and addressed by the design. SAE's ARP4761, entitled *Guidelines and Methods for Conducting the Safety Assessment Process on Civil Airborne Systems*

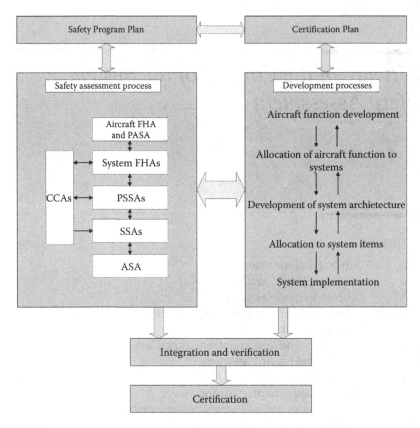

FIGURE 3.1
Overview of the system and safety assessment processes.

and Equipment, provides a detailed description of the expectations for civil aviation.* Other domains use similar approaches (e.g., Military-Standard-882, the U.S. Department of Defense standard for system safety). The main elements of the ARP4761 safety assessment process will now be described.

3.1.1 Safety Program Plan

The Safety Program Plan development is typically the first task performed in the overall assessment of the aircraft safety. The plan identifies how the aircraft safety will be ensured, how safety-related requirements will be identified, what safety standards will be applied, what data will be produced, what methods will be employed, who will perform the tasks, key milestones, and how the various tasks will be coordinated in order to ensure correctness, completeness, and consistency. ARP4574A appendix B provides a sample

---

* At this time ARP4761 is being updated. Once it is available, ARP4761A should be used instead of ARP4761.

Safety Program Plan, which may be used as a template to start an aircraft-specific safety plan. If used, the ARP4754A appendix B example will need to be tailored to fit the specific needs and organizational structure of the stakeholders involved in the specific project [1]. ARP4761 provides specific techniques for performing analyses identified in the Safety Program Plan, along with examples and sample checklists. In addition to the aircraft-level Safety Program Plan, many systems will also have a system-level safety plan for their individual system or product. These lower level safety plans may be referenced from the aircraft-level Safety Program Plan. Either the Safety Program Plan or the aircraft-level Certification Plan typically shows the relationship between the various aircraft-level and system-level plans and data.

3.1.2 Functional Hazard Assessment

The functional hazard assessment (FHA) is developed once the basic aircraft functionality and conceptual design are documented. The FHA identifies and classifies failure conditions associated with aircraft functions and combinations of functions. The effects of the failure conditions on the aircraft are assessed against the regulations and supporting guidance that establish the safety requirements and objectives for the aircraft. In the aviation world, the required safety objectives vary by aircraft type (e.g., transport category airplanes [Part 25], small airplanes [Part 23], small rotorcraft [Part 27], transport category rotorcraft [Part 29], engines [Part 33], or propellers [Part 35]).* The large transport category aircraft tend to have the most rigid requirements due to the overall risk to human life. Small aircraft have some alleviation, depending on the aircraft size and its area of operation. Table 3.1 shows the severity classifications along with the failure condition effects and required probabilities for a large transport category aircraft† (assurance levels are also shown and will be discussed in Section 3.2). Other aircraft types, engines, and propellers have similar classification; however, the probability levels for Part 23 aircraft may be lower, depending on the aircraft class. The FHA is a living document that is updated as additional aircraft functionality and failure conditions are identified during aircraft development. The final FHA identifies the functions, failure conditions, phases of operation for each function, effects of the function failure, classification of the failure, and verification methods.

Developing an FHA is an intense process. It includes the following efforts for the entire aircraft: (1) analyzing possible scenarios in which a failure could occur, (2) assessing the results if it did occur, and (3) determining if changes are needed. It uses input from past projects, accidents, and flight experience. It involves experienced engineers, pilots, human factors experts, safety personnel, and regulatory authorities. While people are usually good at understanding the consequences of a single hazard, it is the combinations

---

* Each Part is from Title 14 of the U.S. Code of Federal Regulations.
† Covered by Part 25 of Title 14 of the U.S. Code of Federal Regulations.

TABLE 3.1

Failure Condition Severity, Probabilities, and Levels

Severity Classification	Potential Failure Condition Effect	Likelihood of Occurrence	Exposure Per Flight Hour (Part 25)	Assurance Level
Catastrophic	Failure conditions, which would result in multiple fatalities, usually with the loss of the airplane	Extremely improbable	1E–9	A
Hazardous/ Severe major	Failure conditions, which would reduce the capability of the airplane or the ability of the flight crew to cope with adverse operating conditions to the extent that there would be • A large reduction in safety margins or functional capabilities • Physical distress or excessive workload such that the flight crew cannot be relied upon to perform their tasks accurately or completely • Serious or fatal injury to a relatively small number of the occupants other than the flight crew	Extremely remote	1E–7	B
Major	Failure conditions, which would reduce the capability of the airplane or the ability of the crew to cope with adverse operating conditions to the extent that there would be a significant reduction in safety margins or functional capabilities, a significant increase in crew workload or in conditions impairing crew efficiency, discomfort to the flight crew, or physical distress to passengers or cabin crew, possibly including injuries	Remote	1E–5	C

TABLE 3.1 (continued)

Failure Condition Severity, Probabilities, and Levels

Severity Classification	Potential Failure Condition Effect	Likelihood of Occurrence	Exposure Per Flight Hour (Part 25)	Assurance Level
Minor	Failure conditions, which would not significantly reduce airplane safety and which involve crew actions that are well within their capabilities. Minor failure conditions may include a slight reduction in safety margins or functional capabilities; a slight increase in crew workload, such as routine flight plan changes; or some physical discomfort to passengers or cabin crew	Reasonably probable	$1E-3$	D
No safety effect	Failure conditions that would have no effect on safety; e.g., failure conditions that would not affect the operational capability of the airplane or increase crew workload	Probable	1.0	E

of multiple realistic failures that make the FHA particularly challenging. Each aircraft has its own peculiarities, but historical evidence (e.g., accidents and incidents) of similar platforms plays a role.

3.1.3 System Functional Hazard Assessment

The system functional hazard assessment (SFHA) is similar to the FHA, except it goes into more detail for each of the systems on the aircraft.* It analyzes system architecture to identify and classify failure conditions and combinations of failure conditions. This may lead to an update of the FHA and the overall aircraft or system design.

3.1.4 Preliminary Aircraft Safety Assessment

The preliminary aircraft safety assessment (PASA) "is a systematic examination of a proposed architecture(s) to determine how failures could cause the Failure Conditions identified by the FHA" [1]. The PASA is the highest

---

* There is typically an SFHA for each system on the aircraft.

level preliminary safety analysis and is developed from the aircraft FHA. The PASA is typically performed by the aircraft manufacturer and considers combinations of systems and interfaces between them. The PASA inputs include the aircraft-level and system-level FHAs, preliminary common cause analyses (CCAs), and proposed system architectures and interfaces. The PASA may identify updates to the FHA and SFHA and will typically generate lower level safety requirements.

3.1.5 Preliminary System Safety Assessment

Like the PASA, the preliminary system safety assessment (PSSA) is a top-down examination of how failures can lead to functional hazards identified in the FHA and SFHAs. While the PASA focuses on the aircraft architecture and integration of systems, the PSSA focuses on the proposed architecture of a specific system, subsystem, or product. Each major aircraft system will have a PSSA. There may be multiple hierarchical levels of PSSAs for systems that have multiple subsystems.

The PSSAs are typically performed by the system developers and go into detail for each system on the aircraft. As the aircraft systems mature, the PSSAs provide feedback to the PASA.

The PASA and PSSA processes may lead to additional protection or architectural mitigation (e.g., monitoring, partitioning, redundancy, or built-in-test). The PASA and PSSA outputs serve as feedback or input to the FHA and SFHAs, system architecture, and system requirements. The PASA and PSSA also impact software and hardware requirements.

Overall, the PASA and PSSAs are intended to show that the proposed design will meet the safety requirements of the regulations and the FHA and SFHAs. The PASA and PSSA serve as inputs to the as-built final aircraft safety assessment (ASA) and system safety assessments (SSAs). The PASA and PSSA are the assessments that typically provide the development assurance levels for the system (functional development assurance levels— FDALs) and the software and electronic hardware (item development assurance levels—IDALs).

3.1.6 Common Cause Analysis

The CCA evaluates system architecture and aircraft design to determine sensitivity to any common cause events. This helps to ensure that (1) independence is achieved where needed and (2) any dependencies are acceptable to support the required safety level. The CCA verifies that the independence needed for safety and regulatory compliance exists, or that the lack of independence is acceptable. ARP4754A explains that the CCA "establishes and verifies physical and functional separation, isolation and independence requirements between systems and items and verifies that these requirements have been met" [1]. The output of the CCA is input to the PASA and/or PSSA and the ASA and/or SSA. Three separate analyses are typically performed to evaluate common cause; each is explained in the following [2]:

1. A *particular risk analysis (PRA)* evaluates "events or influences which are outside the system(s) and items(s) concerned, but which may violate failure independence claims" [2]. ARP4761 provides several examples of events or influences external to the aircraft or external to the system or item being evaluated. An example of something external to the aircraft is bird strike. Some business jets install their avionics in the aircraft nose bay, which is susceptible to bird strikes at lower altitudes. Therefore, redundant systems (such as navigation and communication) are physically located on opposite sides of the nose bay, preventing the common cause. Other examples of risks external to the aircraft include ice, high intensity radiated fields (HIRF), or lightning [2]. An example of something external to the system is an engine rotor burst that might occur, cutting through the hull of the aircraft. In order to reduce the susceptibility to such an event, physical separation of electrical power, flight controls, hydraulic systems, etc. is typically required. Other risks external to the system include fire, bulkhead rupture, leaking fluids, security tampering, and tire tread separation [2].

2. A *zonal safety analysis (ZSA)* is performed for each *zone* on the aircraft. Example zones include nose bay, cockpit, cabin, left wing, right wing, tail cone, fuel tanks, cargo bay, and avionics bay. A ZSA ensures that the installation "meets the safety requirements with respect to basic installation, interference between systems, or maintenance errors" [1]. The ZSA is carried out for each identified zone of the aircraft and includes design/installation guidelines, examination to ensure conformance to the guidelines, and inspection to identify any interference (based on the failure modes and effects analysis and summary).

3. The *common mode analysis (CMA)* considers any kind of common effect during development, implementation, test, manufacturing, installation, maintenance, or crew operation that may lead to a common mode failure. CMAs are performed at multiple hierarchical levels of the aircraft development—all the way from the top aircraft level down to the individual circuit cards. CMAs verify that events ANDed together in the fault tree analysis, dependence diagrams, and/or Markov analysis are truly independent [2]. This analysis has an impact on software development because it considers common modes such as software development errors, systems requirements errors, functions and their monitors, and interfaces. The CMA helps

determine what development assurance levels are assigned for the system, software, and hardware. *Development assurance* is explained in Section 3.2.

3.1.7 Aircraft and System Safety Assessments

The ASA and SSAs verify that the implemented aircraft and system design meets the safety requirements defined in the FHA, SFHAs, PASA, and PSSAs. The ASA and SSAs are performed on the as-built, implemented design, whereas the PASA and PSSAs are based on the proposed design. The ASA and SSAs are the final evidence that the aircraft meets the regulatory safety requirements.

As with the PASA and PSSAs, the ASA is usually performed by the aircraft manufacturer, and the SSAs are performed by the systems developers. The ASA considers combinations and integration of systems and is closely coordinated with the multiple SSAs.

3.2 Development Assurance

In complex and highly integrated electronic systems with software and/ or programmable hardware, it is not feasible to test all combinations of inputs and outputs to assign a probability of failure. This limitation, combined with the fact that computer programs do not deteriorate or fail over time like physical components, leads to the need for a concept called *development assurance*. Development assurance is used to ensure confidence in the process used to develop the system and its items. Development assurance is "all of those planned and systematic actions used to substantiate, at an adequate level of confidence, that errors in requirements, design and implementation have been identified and corrected such that the system satisfies the applicable certification basis" [1]. Development assurance assumes that a more rigorous process is more likely to identify and remove errors before the product is delivered than a less rigorous process. The concept of development assurance was initially introduced to address software; however, it also applies to other domains where exhaustive testing is infeasible. Table 3.2 identifies domains where development assurance is applied in civil aviation, along with the industry documents that provide the guidance for each domain.*

Development assurance levels are established by the safety assessment process, based on the potential safety impact of the system. ARP4754A defines *development assurance level* as "the measure of rigor applied to the

---

* The equivalent EUROCAE document numbers are also included.

TABLE 3.2

Development Assurance Documents for Civil Aviation

Document Number and Title	Domain Addressed
SAE/ARP4754A (EUROCAE/ED-79A): *Guidelines for Development of Civil Aircraft and Systems*	Civil aircraft and its systems
RTCA/DO-297 (EUROCAE/ED-124): *Integrated Modular Avionics (IMA) Development Guidance and Certification Considerations*	Integrated modular avionics
RTCA/DO-178C (EUROCAE/ED-12C): *Software Considerations in Airborne Systems and Equipment Certification*	Airborne software
RTCA/DO-278A (EUROCAE/ED-109A): *Guidelines for Communication, Navigation, Surveillance, and Air Traffic Management (CNS/ATM) Systems Software Integrity Assurance*	CNS/ATM ground-based software
RTCA/DO-254 (EUROCAE/ED-80): *Design Assurance Guidance for Airborne Electronic Hardware*	Airborne electronic hardware
RTCA/DO-330 (EUROCAE/ED-215): *Software Tool Qualification Considerations*	Software tools
RTCA/DO-331 (EUROCAE/ED-218): *Model-Based Development and Verification Supplement to DO-178C and DO-278A*	Airborne and CNS/ATM ground-based software
RTCA/DO-332 (EUROCAE/ED-217): *Object-Oriented Technology and Related Techniques Supplement to DO-178C and DO-278A*	Airborne and CNS/ATM ground-based software
RTCA/DO-333 (EUROCAE/ED-216): *Formal Methods Supplement to DO-178C and DO-278A*	Airborne and CNS/ATM ground-based software
RTCA/DO-200A (EUROCAE/ED-76): *Standards for Processing Aeronautical Data*	Aeronautical databases

development process to limit, to a level acceptable for safety, the likelihood of errors occurring during the development process of aircraft/system functions and items that have an adverse safety effect if they are exposed in service" [1]. For most aircraft and engines, the assurance levels shown in Table 3.1 apply.* The probabilities for each category shown in Table 3.1 are used for hardware items where reliability is applicable.

3.2.1 Development Assurance Levels

ARP4754A identifies two phases of system development: function development phase and item development phase. The function development phase includes the development, validation, verification, configuration management, process assurance, and certification coordination for the system. At the

---

* For some small aircraft (Part 23), the levels are lower. See Federal Aviation Administration Advisory Circular 23. 1309-1[ ] (where "[ ]" represents the latest revision).

lowest level, system requirements are allocated to software or hardware, referred to as *items*. The software or hardware items have their own development phases. ARP4754A guidance applies to the system development phase; DO-178B/C applies to the software development phase; and DO-254 applies to the electronic hardware development phase.

The traditional V-model illustrated in Figure 3.2 shows the function and item development phases. At the function level, an FDAL is assigned based on the system's potential impact on safety. The software and electronic hardware are assigned IDALs.

The FDAL determines the amount of rigor required at the system level (e.g., the amount of requirements reviews, testing, and independence). The IDAL determines the amount of rigor required for the specific item's development (software or electronic hardware development). Depending on the architecture, the IDAL may be lower than the FDAL. ARP4754A explains the FDAL and IDAL assignment approach. The FDAL determines what objectives of ARP4754A apply at the system level. Likewise, the IDALs determine what DO-178C (or its supplements) objectives apply for software or DO-254 objectives for electronic hardware.

It is important to note that FDAL and IDAL are assigned at the system level and as a result of the safety assessment process. The initial IDAL assignment should be completed before the DO-178C or DO-254 planning activities conclude. The IDAL is called a *software level* in DO-178C (and its supplements) and throughout the remainder of this book. We will examine the DO-178C objectives in Part III (Chapters 4 through 12).

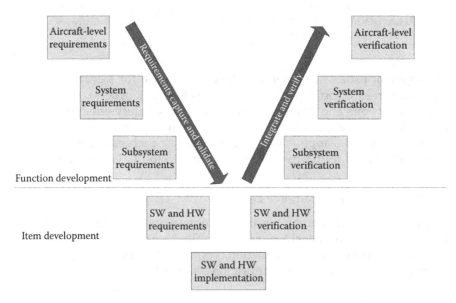

FIGURE 3.2
The V-model for system development.

3.3 How Does Software Fit into the Safety Process?

The role of software has increased over the last 20 years. As software has become more pervasive, the system and safety processes have also evolved. The traditional analog and mechanical system failure analysis and reliability techniques do not work on software because of its complexity and uniqueness. This section examines (1) software's uniqueness that drives the need for the development assurance process and (2) how software fits into the system and safety process.

3.3.1 Software's Uniqueness

Software, unlike hardware, does not follow the laws of physics—it does not wear out, break under known conditions, or fail in predictable ways. Some other unique characteristics are as follows [3]:

- Programs can be complex and difficult to understand, even if they are short.
- Software tends to improve with time because latent errors are discovered and corrected.
- Correcting an error in one area can introduce another error, which can be in a seemingly unrelated area.
- Software can interface with dozens or even hundreds of other software modules.
- Unlike hardware, software does not provide forewarning when it fails. Latent software errors can remain invisible for years and then suddenly appear.
- Software can be easily changed (which is why it's called *soft*-ware).
- Finding an error in software can be long and laborious, especially if not handled proactively.

At this point in time, it is not possible to prove if the software is perfect or judge how close to perfect it is. In fact, one might go as far as to say that all software has some errors in it because it is developed by imperfect humans. The reliability techniques applied to hardware simply do not apply to software. If we relied on traditional hardware reliability approaches, only very simple software would be allowed on the aircraft.

3.3.2 Software Development Assurance

Because of the inability to accurately apply reliability models to software, the concept of development assurance is applied by most safety-focused industries, including the aviation industry. Development assurance is defined in Section 3.2. The goal of development assurance is to apply rigorous processes to the development process in order to prevent, identify, and remove errors. The more critical the functionality, the more development and verification activities are applied. For the aviation industry, DO-178B, and now DO-178C, identify five software levels based on the safety process (DO-178 and DO-178A had three levels). Table 3.1 shows the relationship between the software levels and the failure condition categories. Level A software is the most rigorous and applies to software functionality that could cause or contribute to a catastrophic aircraft-level event. For level A, all of the objectives of DO-178C apply.* Level E software has no safety implications; therefore, none of the DO-178C objectives are required.

Table 3.3 demonstrates how the development and verification rigor and the number of objectives increase as the software level *goes up*. As the level increases, more activities and more independence are required. Each of the objectives summarized in Table 3.3 are explored in Part III. For now, it's important to understand that the higher the software level, the more verification is performed, the more independence is required, and the more errors are identified and removed.

3.3.3 Other Views

It is virtually impossible to scientifically prove the claims of the development assurance approach. Also, a good process does not necessarily mean a robust, safe product. As I often say in jest: "Not all level A software is created equally." One can have a great process and still produce buggy software; often due to use of inexperienced personnel or bad systems requirements. Likewise, a team can have a substandard process and produce robust and dependable software; primarily because of the use of highly skilled and experienced engineers. DO-178C assumes the use of qualified and well-trained people; however, it is difficult to measure the adequacy of people. Some authorities examine resumés and training history to ensure that qualified and properly trained personnel are used.

For companies without a strong safety foundation, it can be tempting to choose a software level without understanding *why* the level was needed. Divorcing the software development process from the system requirements and safety drivers can lead to major safety issues. Because of this, DO-178C, ARP4754A, and ARP4761 have attempted to improve the guidance to addressing implementation of software in the system. However, the temptation still exists and must be monitored.

Because development assurance can't be scientifically supported, there are some who question and even denounce the use of development assurance. Most of these critics promote the use of safety and reliability models at the software level. They extend the SSA process to include the software and

---

* DO-178C objectives are explained in Part III.

TABLE 3.3

Summary of DO-178C Objectives

Level	Obj Count	Obj w/ Ind	Summary of Objectives
E	0	0	No activities required
D	26	2	Plans (five plans)
			High-level requirements developed
			Architecture developed
			Executable object code developed
			Parameter data item files developed (if needed) and verified for correctness and completeness
			Some review and analysis of high-level requirements
			Only review or analyze architecture if using partitioning
			Normal and robustness testing of high-level requirements
			Requirements coverage of high-level requirements
			Testing to verify target compatibility
			Configuration management
			Quality assurance (compliance to plans and conformity review)
			Accomplishment summary and configuration index
C	62	5	Level D activities
			Development standards (three standards)
			Low-level requirements developed
			Trace data developed
			Source code developed
			Additional review and analysis of high-level requirements
			Some review and analysis of low-level requirements
			Some review and analysis of architecture
			Review and analysis of some source code
			Verification of parameter data item files
			Normal and robustness testing of low-level requirements
			Requirements coverage of low-level requirements
			Review test procedures
			Review test results
			Statement coverage analysis
			Data and control coupling analysis
			Additional quality assurance (review plans and standards, compliance to standards, and transition criteria)
B	69	18	Level C and D activities
			Additional review and analysis of high-level requirements (target compatibility)
			Additional review and analysis of low-level requirements (target compatibility and verifiability)
			Additional review and analysis of architecture (target compatibility and verifiability)

(*continued*)

TABLE 3.3 (continued)

Summary of DO-178C Objectives

Level	Obj Count	Obj w/ Ind	Summary of Objectives
			Additional review and analysis of source code (verifiability)
			Decision coverage
A	71	30	Level B, C, and D activities
			Modified condition/decision coverage analysis
			Source to object code traceability verification

Obj, objectives; Ind, independence.

minimize the reliance on development assurance. Such efforts employ safety use cases and models to demonstrate that the system is reliable under a given set of conditions expected to be encountered. Extensive effort and skill are required to develop the scenarios and exercise the system under those scenarios. The safety focus is beneficial to ensure that the software design supports the system safety. However, to date, software reliability modeling techniques are still limited and controversial.

3.3.4 Some Suggestions for Addressing Software in the System Safety Process

As was previously noted, safety is a system attribute—not a software attribute. Sound systems, safety, hardware, and software engineering are necessary for safe system implementation. Here are some suggestions for improving the tie between the disciplines:

- Assign safety attributes to the software requirements that specifically support safety protections, in order to identify which software requirements are most crucial to safety. This also helps to ensure that the safety implications of changes are fully considered.

- Involve the system and safety engineers in the review of software requirements and architectures.

- Coordinate derived software requirements with the safety team as soon as they are identified to ensure that there are no safety impacts. Include rationale in derived requirements, so that they can be understood, evaluated for safety, and maintained.

- Colocate system, safety, hardware, and software teams.

- Perform PSSA early in order to identify software levels and clearly identify what system functions have safety impact; that is, make it clear what drives the software level assignment or what mitigations (protections) are needed in the software.

- Provide a summary of the PSSA to the entire software team to ensure that they understand the rationale for the software level and the impacts of their work on safety.
- Consider carrying the safety assessment down into the software to support the overall safety assessment process. This is particularly beneficial when protection (e.g., partitioning or monitoring) is implemented in the software.
- Train the software team in the basics of system development, the SSA process, development assurance level assignment, common techniques used for developing safety-critical systems, and what they should be aware of for their system.
- Use experienced teams. If junior engineers are used (we all have to start somewhere), pair them with experienced engineers.
- Develop a safety-focused culture. If top-level management supports safety and walks the talk, it makes an incredible difference in how it is viewed and implemented down in the trenches.

References

1. SAE ARP4754A, *Guidelines for Development of Civil Aircraft and Systems* (Warrendale, PA: SAE Aerospace, December 2010).
2. SAE ARP4761, *Guidelines and Methods for Conducting the Safety Assessment Process on Civil Airborne Systems and Equipment* (Warrendale, PA: SAE Aerospace, December 1996).
3. J. P. Bowen and M. G. Hinchey, Ten commandments of formal methods… Ten years later, *IEEE Computer*, January 2006, 40–48.

Part III

Developing Safety-Critical Software Using DO-178C

RTCA/DO-178C (referred to simply as *DO-178C* throughout this book) is entitled *Software Considerations in Airborne Systems and Equipment Certification*. It was developed as a set of recommendations through consensus of the international community* sponsored by RTCA[†] and EUROCAE[‡] and was published on December 13, 2011. It is expected that the Federal Aviation Administration (FAA), European Aviation Safety Agency (EASA), and other civil aviation authorities will soon recognize this document as an acceptable means of compliance to the regulations.[§] DO-178C and its EUROCAE equivalent (ED-12C) were preceded by DO-178/ED-12, DO-178A/ED-12A, and DO-178B/ED-12B. DO-178C provides guidance for developing airborne software; however, most of the guidance is also applicable to other safety-critical domains. Part III provides an overview of DO-178C, the differences from DO-178B to DO-178C, and the recommendations for developing software

---

* The joint committee included membership from industry, regulatory authorities, and academia from the United States, Canada, Europe, South America, and Asia.
[†] RTCA (formerly known as Radio Technical Commission for Aeronautics) is an association of both government and industry aviation organizations in the United States. RTCA receives some funds from the federal government but is also supported by annual fees from members and the sales of documents.
[‡] EUROCAE (European Organization for Civil Aviation Equipment) is an international organization whose membership is open to manufacturers and users in Europe of equipment for aeronautics, national civil aviation administrations, trade association, and, under certain conditions, non-European members.
[§] The recognition typically comes in the form of an Advisory Circular (AC) for the FAA and equivalent advisory materials by other certification authorities. It is expected that FAA AC 20-115C will be published to recognize DO-178C.

to comply with DO-178C. This section will not repeat what is included in DO-178C but will provide practical information on how to apply DO-178C to real projects.

At this time, the aviation industry is transitioning from DO-178B to DO-178C. Much of the guidance is common between the two versions of the document. Therefore, most of the dialogue in Part III applies to both DO-178B and DO-178C users. In anticipation of the transition to DO-178C, the term *DO-178C* is used throughout this book. Unless specifically noted as being different from DO-178B, the information will generally also apply to DO-178B users.

Part III is summarized as follows:

- Chapter 4 provides a history of DO-178, an overview of DO-178C, a summary of the primary differences between DO-178B and DO-178C, and a summary of the six documents published with DO-178C.

- Chapter 5 examines the DO-178C planning process and recommends best practices for effective planning.

- Chapters 6, 7, and 8 discuss software requirements capture, design, and implementation, respectively. Each chapter explores what is required by DO-178C objectives and discusses recommendations for satisfying these objectives.

- Chapters 9, 10, 11, and 12 examine the integral processes of verification, software configuration management, software quality assurance, and certification liaison, respectively. Again, these chapters will briefly discuss DO-178C objectives but will focus on practical ways to satisfy the objectives.

Chapters 6 (requirements) and 9 (verification) are longer than the other chapters due to the vital importance of these two subject areas to safety and DO-178C compliance.

4

Overview of DO-178C and Supporting Documents

Acronyms

AL	assurance level
CAST	Certification Authorities Software Team
CC1	control category #1
CC2	control category #2
CNS/ATM	communication, navigation, surveillance, and air traffic management
COTS	commercial off-the-shelf
DP	discussion paper
EUROCAE	European Organization for Civil Aviation Equipment
FAQ	frequently asked question
OOT	object-oriented technology
OOT&RT	object-oriented technology and related techniques
PSAA	Plan for Software Aspects of Approval
PSAC	Plan for Software Aspects of Certification
RT	related techniques
SC-205	Special Committee #205
SW	software
TQL	tool qualification level
WG-71	Working Group #71

4.1 History of DO-178

DO-178 and its EUROCAE* equivalent (ED-12) have a 30-year history. Table 4.1 summarizes the evolution of DO-178 and related documents.

In 1982, the first version of DO-178 was published. The document was very brief and only provided a high-level framework for software development.

---

* European Organization for Civil Aviation Equipment.

TABLE 4.1

Evolution of DO-178 and Related Documents

Document	Year Published	Content
DO-178	1982	Provides very basic information for developing airborne software.
DO-178A	1985	Includes stronger software engineering principles than DO-178. Includes both verification and validation of requirements.
DO-178B	1992	Significantly longer than DO-178A. Provides guidance in form of *what* (objectives), rather than *how*. Provides visibility into life cycle processes and data. Does not include requirements validation.
DO-248B	2001	Includes errata for typographical errors in DO-178B. Also provides FAQs and DPs to clarify DO-178B. Was preceded by DO-248 in 1999 and DO-248A in 2000. Is not considered to be *guidance*—it is only clarification.
DO-278	2002	Applies DO-178B to CNS/ATM software. Adds some CNS/ATM-specific terminology and guidance.
DO-178C	2011	Content is very similar to DO-178B; however, it clarifies several areas, adds guidance for parameter data items, and references DO-330 for tool qualification.
DO-278A	2011	Stands alone from DO-178C, unlike DO-278 which made direct references to DO-178B. Very similar to DO-178C with a few terminology changes and additional guidance needed for CNS/ATM software.
DO-248C	2011	Updates DO-248B to align FAQs and DPs with DO-178C updates. Also expanded to address DO-278A topics, to clarify additional topics that came about since DO-248B, and to add rationale for DO-178C objectives and supplements.
DO-330	2011	Provides guidance on tool qualification. It is a stand-alone document. DO-178C and DO-278A reference DO-330.
DO-331	2011	A technology supplement to DO-178C and DO-278A. Provides guidance on model-based development and verification.
DO-332	2011	A technology supplement to DO-178C and DO-278A. Provides guidance on OOT&RT.
DO-333	2011	A technology supplement to DO-178C and DO-278A. Provides guidance on formal methods.

Between 1982 and 1985 the software engineering discipline matured significantly, as did DO-178. DO-178A was published in 1985. Many systems that are still in existence today used DO-178A as their means of compliance. However, there were a few challenging characteristics of DO-178A. First, DO-178A addressed both requirements validation and verification. DO-178 and DO-178A preceded the development of ARP4754 and ARP4761, which put the system and safety assessment framework into place. With the development of ARP4754, it was determined that requirements validation (ensuring that one has the right requirements) is a system activity; whereas requirements verification (ensuring that the requirements allocated to software are implemented correctly) is a software activity. Second, DO-178A was not objective-based, making it challenging to objectively assess compliance to it. Third, DO-178A allowed structural testing, rather than structural coverage analysis. As Chapter 9 will explain, structural testing is not adequate to identify unintended functionality or to measure the completeness of requirements-based testing.

DO-178B was published in late 1992. It was developed in parallel with ARP4754 and ARP4761, which were published a short time later. DO-178B contains more detail than DO-178A and became the *de facto* standard for software development in civil aviation. DO-178B is objective-based; it strives to identify *what* the developer must do but gives flexibility for *how* it is accomplished. The document also requires more insight into what is being done by requiring documented evidence (life cycle data) for all activities used for certification credit; this makes it possible to objectively assess compliance. Additionally, as mentioned earlier, DO-178B does not address requirements validation; that is the responsibility of the systems team. And, DO-178B significantly clarified the expectations for structural coverage analysis.

In 1999–2001, RTCA published DO-248, DO-248A, and DO-248B. DO-248 started out as an annual report to clarify some of the more challenging aspects of DO-178B. The final version of the report was published in 2001 as DO-248B (its EUROCAE equivalent is ED-94B); it included the contents from previous releases. DO-248B included 12 errata on DO-178B (minor typographical corrections), 76 frequently asked questions (FAQs), and 15 discussion papers (DPs). A FAQ is a short clarification of less than one page and is written in the form of a question and answer. A DP is also clarification but is longer than one page and is written in the form of a topic explanation. DO-248B was never considered guidance material, but merely clarified guidance already contained in DO-178B. It is an informational and educational document to help users better understand and apply DO-178B. As will be explained later in this chapter, DO-248B has been replaced by DO-248C.

In 2005, RTCA and EUROCAE formed a joint committee to update DO-178B (and ED-12B) to provide a path forward for addressing software technology changes.* Objectives of the committee were to continue the following [1,2]:

- Promoting safe implementation of aviation software
- Providing clear and consistent ties with the systems and safety processes
- Addressing emerging software trends and technologies
- Implementing a flexible approach to change with the technology

---

* The joint committee is RTCA Special Committee #205 (SC-205) and EUROCAE Working Group #71 (WG-71). The committee was divided into an executive committee and seven sub-groups.

The input to the committee's efforts were DO-178B, DO-278, DO-248B, certification authorities' publications (including policy, guidance, issue papers, Certification Authorities Software Team [CAST]* papers, research reports, etc.), the committee's terms of reference, an ongoing issues list, and the experience and expertise of hundreds of professionals. The output was seven documents, often referred to as *green covers*, since the hard copy of each RTCA document is published with a green cover. RTCA published the following documents at the end of 2011 (a nice Christmas gift for those of us on the committee who committed 6.5 years of our lives to the effort):

- DO-178C/ED-12C, *Software Considerations in Airborne Systems and Equipment Certification*
- DO-278A/ED-109A, *Guidelines for Communication, Navigation, Surveillance, and Air Traffic Management (CNS/ATM) Systems Software Integrity Assurance*
- DO-248C/ED-94C, *Supporting Information for DO-178C and DO-278A*
- DO-330/ED-215, *Software Tool Qualification Considerations*
- DO-331/ED-218, *Model-Based Development and Verification Supplement to DO-178C and DO-278A*

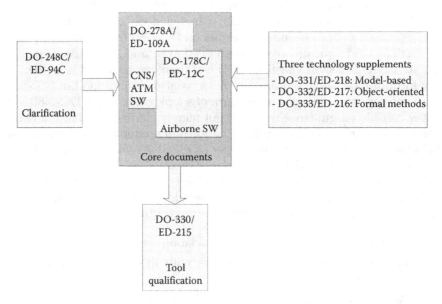

FIGURE 4.1
Relationship of DO-178C and related documents.

---

* CAST is a team of international certification authorities who strive to harmonize their positions on airborne software and aircraft electronic hardware in CAST papers.

- DO-332/ED-217, *Object-Oriented Technology and Related Techniques Supplement to DO-178C and DO-278A*
- DO-333/ED-216, *Formal Methods Supplement to DO-178C and DO-278A*

Figure 4.1 illustrates the relationship of the seven documents. Each document is briefly described later and will be more thoroughly explained in future chapters. The RTCA DO- and EUROCAE ED- documents are the same except for page layout (RTCA uses a letter-sized page, whereas EUROCAE uses size A4) and the added French translation for EUROCAE documents. Throughout the remainder of this book, only the RTCA numbers will be referenced.

4.2 DO-178C and DO-278A Core Documents

DO-178C and DO-278A are virtually the same, except DO-178C is focused on airborne software and DO-278A on CNS/ATM ground-based software. Section 4.2.1 summarizes the primary differences between the two documents. Because of the fact that the DO-178C and DO-278A are nearly the same and DO-178C has a larger user community than DO-278A, after Section 4.2.1, only DO-178C will be mentioned unless there is a notable difference.

DO-178C and DO-278A are often called *the core documents* since they provide the core concepts and organization that the other documents are built upon. These core documents are intended to be as technology-independent as possible. DO-248C and the supplements reference the two core documents and provide the technology-specific details.

DO-178C is very similar to DO-178B; however, it has been updated to address unclear aspects of DO-178B, to be consistent with the system and safety documents (ARP4754A and ARP4761), and to have a full set of objectives tables that agree with the front matter of the document. Like DO-178B, DO-178C is composed of 12 sections, an Annex A that summarizes the objectives in the main body, an Annex B that includes a glossary, and some supporting appendices. The relationship of the DO-178C sections are illustrated in Figure 4.2 and briefly described in the following:

- *DO-178C section 1* provides an introduction to the document.
- *DO-178C section 2* includes the system framework which is aligned with ARP4754A.
- *DO-178C section 3* provides a brief overview of the DO-178C software life cycle processes: planning, development (requirements, design, code, and integration), verification, software configuration management, software quality assurance, and certification liaison.

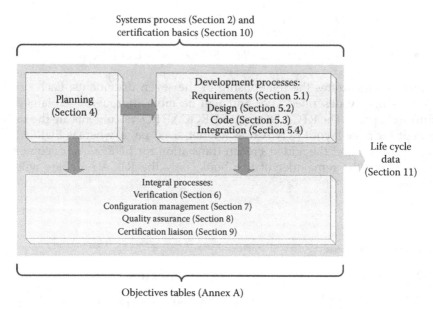

FIGURE 4.2
Overview of DO-178C.

- *DO-178C section 4* explains the planning process objectives and activities. The planning objectives are summarized in DO-178C Table A-1.

- *DO-178C section 5* provides an explanation of the objectives and activities of the development processes. The development processes includes the requirements, design, code, and integration phases. The development objectives are summarized in DO-178C Table A-2.

- *DO-178C section 6* describes the verification process. Verification is an integral process, which begins with the verification of the plans and goes all the way through the reporting and review of verification results. Verification includes reviews, analyses, and tests. Verification plays a significant role in the software development assurance; over half of the DO-178C objectives are verification objectives. The verification objectives are summarized in DO-178C Tables A-3 through A-7.

- *DO-178C section 7* explains the software configuration management process, which includes change control and problem reporting. The configuration management objectives are summarized in DO-178C Table A-8.

- *DO-178C section 8* provides guidance for the software quality assurance process. Objectives for software quality assurance process are summarized in DO-178C Table A-9.

- *DO-178C sections 9 and 10* summarize the certification liaison process and the overall aircraft or engine certification process.

The certification liaison objectives are summarized in DO-178C Table A-10.

- *DO-178C section 11* identifies the life cycle data generated while planning, developing, and verifying the software. In many situations the term *software life cycle data* is also referred to as *data items* or *artifacts*. The expected contents of each life cycle data item is briefly explained in DO-178C section 11. Each data item is discussed in the chapters ahead.

- *DO-178C section 12* includes guidance on some additional considerations that go above and beyond the previous sections, including previously developed software, tool qualification levels (TQLs), and alternative methods (such as exhaustive input testing, multiple-version dissimilar software, software reliability models, and product service history).

- *DO-178C Annex A* includes tables for each process that correlate the objectives and outputs. The DO-178B Annex A tables and the main body of DO-178C were slightly modified to ensure consistency. The DO-178C Annex A tables summarize objectives, applicability by software level, independence requirements for the objective, output that is typically generated to satisfy the objective, and the amount of configuration control required on the output. Section 4.2.2 explains more about the relationship between the DO-178C Annex A tables and the DO-178C document body, as well as how to interpret the tables.

- *DO-178C Annex B* includes the glossary of terms that are used in DO-178C. DO-178C uses common software engineering definitions but modifies some of them for the aviation-specific domain.

- *DO-178C appendix A* provides background information, a brief summary of differences from DO-178B to DO-178C, and the committee's terms of reference.

- *DO-178C appendix B* lists the committee membership. Anyone who attended at least two meetings was included in the membership list; therefore, it is a rather long list.

Table 4.2 provides a brief summary of the significant changes from DO-178B to DO-178C for each section (the DO-178C section references are included in parentheses). Pertinent details of the difference are explained in future chapters of this book.

4.2.1 DO-278A and DO-178C Differences

DO-278A is entitled *Guidelines for Communication, Navigation, Surveillance, and Air Traffic Management (CNS/ATM) Systems Software Integrity Assurance.* DO-278A focuses on ground-based CNS/ATM software, which can have a direct impact on aircraft safety. The main differences between DO-178C and DO-278A are noted here.

TABLE 4.2

Primary Differences between DO-178B and DO-178C

Section	Summary of Significant Changes from DO-178B to DO-178C
General	Aligned and clarified objectives and activities throughout
	Aligned objectives in the main body with Annex A
	Clarified objectives versus activities throughout
1 (Introduction)	Expanded the points on how to use the document (1.4)
	Added an explanation of the technology supplements and the need to describe the use of supplements in the software plans (1.4.o)
2 (Systems)	Updated to align with ARP4754A (2.1–2.5, 2.5.6, 2.6)
	Provided more details on the information exchange between systems and software (2.2)
	Added parameter data item explanation and examples (2.5.1)
4 (Planning)	Added a planning activity for parameter data items, when used (4.2.j)
	Added outsourcing and supplier oversight as a subject to explain in the plans (4.2.l)
	Clarified the need to assess the impact on airborne software when known tool problems and limitations exist for the development environment (4.4.1.f)
	Added an activity to consider robustness in the development standards (4.5.d)
5 (Development)	Added examples of derived requirements (5.0)
	Added guidance for parameter data items (5.1.2.j, 5.4.2)
	Provided some clarification on interfaces between software components in the form of data flow and control flow (5.2.2.d)
	Added deactivated code design considerations to the design guidance section (5.2.4)
	Added an activity during coding phase to ensure that the autocode generator conforms to planned constraints, if an autocode generator is used (5.3.2.d)
	Added guidance for patches (5.4.2.e and 5.4.2.f)
	Added *trace data* as an output of the development effort (5.5 and Table A-2) and removed traced items in DO-178B sections 5.1.2, 5.2.2, 5.3.2
	Clarified the need for bidirectional traceability between hierarchical levels (5.5)
6 (Verification)	Clarified that robustness is key to verification and clarified robustness testing (6.1.e and 6.4.2.b)
	Clarified the verification process activities (6.2)
	Updated the testing graphic to demonstrate how the section 6 guidance fits together (Figure 6-1)
	Updated criteria for architecture consistency review (6.3.3.b)
	Added explanation that compiler, linker, and some hardware options may impact the worst-case execution timing and should be assessed (6.3.4.f)
	Clarified integration objectives and added compiler warnings as a potential error during integration (6.3.5)

TABLE 4.2 (continued)

Primary Differences between DO-178B and DO-178C

Section	Summary of Significant Changes from DO-178B to DO-178C
	Moved review and analysis of test cases, procedures, and results from 6.3.6 (in DO-178B) to 6.4.5 (in DO-178C)
	Explained that test procedures are generated from test cases (6.4.2.c)
	Updated objectives to be consistent with Annex A (6.4, 6.4.4, 6.4.5)
	Clarified requirements-based test coverage analysis (6.4.4.1)
	Clarified that structural coverage analysis also evaluates interfaces (6.4.4.2)
	Clarified structural coverage guidance, particularly for object code coverage and relationship to requirements-based testing (6.4.4.2)
	Clarified that data and control coupling analyses are based on requirements-based tests (6.4.4.2.c)
	Added a class of non-covered code called *extraneous code*, which includes *dead code* (6.4.4.3.c)
	Clarified deactivated code guidance (6.4.4.3.d)
	Added bidirectional tracing between requirements and test cases, test cases and test procedures, and test procedures and test results (6.5)
	Added *trace data* as an output of test development effort (6.5 and Table A-6)
	Added parameter data items verification guidance (6.6)
7 (Configuration Management)	Added parameter data item configuration management, when parameter data are used (7.0.b, 7.1.h, 7.2.1.e, 7.2.7.c, 7.2.7.d)
	Reorganized the location of a few sections for better flow (7.4, 7.5)
8 (Quality Assurance)	Updated quality assurance objectives to include consistency review of plans and standards (8.1.a, Table A-9)
	Moved objective 8.1.a (in DO-178B) to 8.1.b (in DO-178C) and added evaluation of supplier processes to the objective (8.1.b)
	Added activity to ensure supplier process and output compliance (8.2.i)
9 (Certification Liaison)	Added parameter data item file as type design data, if parameter data are used (9.4.d)
11 (Life Cycle Data)	Added need to include supplier oversight information in the PSAC (11.1.h)
	Added autocode generator options and constraints (when applicable) to the development environment section of the Software Development Plan (11.2.c.3)
	Added need to include compiler, linker, and loader information in the Software Development Plan (11.2.c.4)
	Clarified that re-verification content in the Software Verification Plan also needs to address affected software, not just changed software (11.3.h)
	Clarified that compiler, linker, and loader data are used with source code rather than part of it (11.11)
	Added guidance to address software verification that supports systems processes (11.14)
	Added autocode options to list of information to include in the Software Life Cycle Environment Configuration Index (11.15.b)

(continued)

TABLE 4.2 (continued)

Primary Differences between DO-178B and DO-178C

Section	Summary of Significant Changes from DO-178B to DO-178C
	Added parameter data item files, if used, to the Software Configuration Index (11.16.b, 11.16.g)
	Added procedures, methods, and tools for modifying user-modifiable software, if used, to the Software Configuration Index (11.16.j)
	Added loading procedures and methods to the Software Configuration Index (11.16.k)
	Slightly reorganized the Software Accomplishment Summary contents (11.20)
	Clarified the reference to the Software Configuration Index in the Software Accomplishment Summary (11.20.g)
	Added section on supplier oversight in the Software Accomplishment Summary (11.20.g)
	Clarified the unresolved problem report summary information that should be included in the Software Accomplishment Summary (11.20.k)
	Added sections to explain content of trace data (11.21) and parameter data item files (11.22)
12 (Additional Considerations)	Added autocode generator and hardware changes to list of changes that could impact software (12.1.3.c and 12.1.3.f)
	Updated tool qualification guidance and added reference to DO-330 (12.2)
	Removed formal methods as an alternative method, since it is now addressed in DO-333
	Clarified product service history guidance (12.3.4)
Annex A (Objective Tables)	Updated Annex A tables to add references to activities (all tables)
	Clarified Annex A tables outputs (all tables)
	Added or modified objectives for parameter data items (Table A-2 objective 7 and Table A-5 objectives 8 and 9)
	Updated objectives wording for consistency with referenced guidance (Table A-1 objectives 2 and 7, Table A-2 objectives 2 and 5, and Table A-9 objective 5)
	Removed applicability of design objectives for level D (Table A-2 objectives 4, 5, and 6)
	Added trace data as an output (Tables A-2 and A-6)
	Added objective to address source to object code tracing (Table A-7 objective 9)
	Updated objectives for quality assurance's role in the planning process (Table A-9 objective 1)
	Separated objectives for quality assurance's role in evaluated planning and standard compliance (Table A-9 objectives 2 and 3)
Annex B (Glossary)	Clarified some terms (e.g., derived requirements)
	Added some missing terms
	Deleted a few terms that are not used in DO-178C

- DO-178C uses *airborne software*; whereas DO-278A uses *CNS/ATM software*.

- DO-178C uses *certification*; whereas DO-278A uses *approval*.

- DO-178C uses airworthiness requirements; whereas DO-278A uses applicable approval requirements.

- DO-178C uses parameter data item files; whereas DO-278A uses adaptation data item files.

- DO-178C uses Plan for Software Aspects of Certification (PSAC); whereas DO-278A uses Plan for Software Aspects of Approval (PSAA).

- DO-278A section 2 differs slightly from DO-178C, since the tie to the safety assessment process is different for CNS/ATM ground systems vs. aircraft.

- DO-278A uses assurance levels 1–6 (AL1–AL6) instead of software levels A–E. The DO-278A assurance levels are equivalent to DO-178C levels, except DO-278A has an additional level between DO-178C's levels C and D. Table 4.3 shows the mapping between DO-278A and DO-178C levels.

- DO-278A has brief sections on security requirements, adaptability, cutover (hot swapping), and post-development life cycle, which are not provided in DO-178C.

- DO-278A's section on service experience is slightly different from DO-178C's section on product service history.

- DO-278A has a lengthy section on commercial off-the-shelf (COTS) software that does not exist in DO-178C.

TABLE 4.3

Correlation of DO-178C and DO-278A Levels

DO-178C Software Level	DO-278A Assurance Level
A	AL1
B	AL2
C	AL3
No equivalent	AL4
D	AL5
E	AL6

Source: RTCA DO-278A, *Guidelines for Communications, Navigation, Surveillance, and Air Traffic Management (CNS/ATM) Systems Software Integrity Assurance,* RTCA, Inc., Washington, DC, December 2011. Based on DO-278A Table 2-2. Used with permission of RTCA, Inc.

4.2.2 Overview of the DO-178C Annex A Objectives Tables

It is often said that DO-178C is best read from the back forward. DO-178C
Annex A summarizes the objectives, applicable sections, and required
data from DO-178C sections 4 through 11. However, if one doesn't read
the front matter, the DO-178C Annex A objectives tables will not provide a complete understanding of the subject matter. The DO-178C
Annex A tables essentially provide a roadmap for applying objectives
by software level. This subsection briefly explains the subject of each
table, how the tables relate, and how to interpret the tables. Objectives
are further discussed in Chapters 5 through 12 as the technical matter
is more deeply explored.

Table 4.4 contains the title of the 10 DO-178C Annex A tables, as well as
the number of objectives included in each table. Table 4.5 shows the number
of objectives applicable by software level. Figure 4.3 shows the relationship
between the 10 DO-178C objectives tables. The planning process drives

TABLE 4.4

Summary of DO-178C Annex A Tables

Table #	No. of Objectives	DO-178C Annex A Table Title
A-1	7	Software planning process
A-2	7	Software development processes
A-3	7	Verification of outputs of software requirements process
A-4	13	Verification of outputs of software design process
A-5	9	Verification of outputs of software coding and integration process
A-6	5	Testing of outputs of integration process
A-7	9	Verification of verification process results
A-8	6	Software configuration management process
A-9	5	Software quality assurance process
A-10	3	Certification liaison process

Source: RTCA DO-178C, *Software Considerations in Airborne Systems
            and Equipment Certification*, RTCA, Inc., Washington, DC,
            December 2011.

TABLE 4.5

Number of DO-178C Objectives by Software Level

Software level	A	B	C	D	E
Number of objectives	71	69	62	26	0

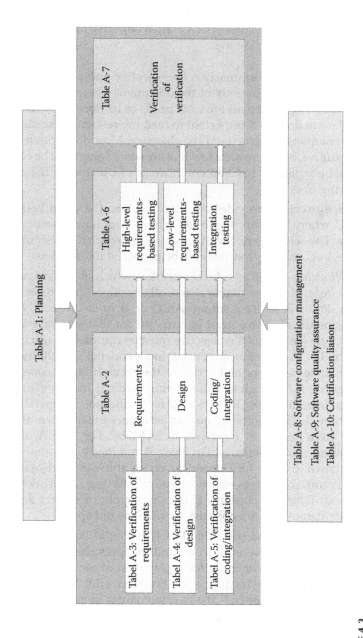

FIGURE 4.3
Relationship of DO-178C Annex A tables.

drives the other processes, and the configuration management, quality assurance, and certification liaison processes take place throughout the entire project.

Table 4.6 shows a representative objective from DO-178C (Table A-6 objective 3). There are five major sections of each objective table. Each section is briefly described in the following:

1. The *Objective* section summarizes the objective and provides a reference back to the main body of the document where the objective is explained. The objectives are *what* needs to be done. They do not explain *how* to do it. It is important to read the referenced section in order to completely understand the objective. Most descriptions are rather intuitive, but there are a few that are more complex. The objectives will be explained in Chapters 5 through 12, including the more challenging objectives.

2. The *Activity* section provides reference back to the main body of DO-178C, where there are suggested activities for how to satisfy the objective. The activities provide the typical approach to satisfying an objective, but other approaches may be proposed in the plans.

3. The *Applicability by Software Level* identifies which objectives apply by software level. As previously noted, all objectives apply to level A; however, levels B–D are a subset. An ○ or ● in columns A, B, C, or D means the objective must be satisfied for that software level. A filled circle (●) means the objective must be completed with independence. For verification objectives, independence simply means that a separate person or tool performs the verification (i.e., one that did not generate the original artifact being verified). Independence is explained later in this book (Section 9.3 for verification independence and Section 11.1.3 for software quality assurance independence).

   There are a total of 71 objectives in DO-178C. For level A, all 71 objectives need to be satisfied. For levels B and lower, the number is less, as shown in Table 4.5. Depending on the software development approach, some objectives may not be applicable (e.g., if partitioning is not used, Table A-4 objective 13 does not apply, or if there are no parameter data, Table A-5 objectives 8 and 9 do not apply).

4. The *Output* columns identify the data that should be generated to prove that the objective is satisfied and its applicable reference to DO-178C section 11, where a brief explanation of each data item is provided. There is a saying in the certification world: "If it isn't written down, it didn't happen." The output column identifies the data used to prove compliance to the objectives. Table 4.7 provides a summary of all the DO-178C section 11 data items that are referenced in the Annex A tables. The items shown with an asterisk are always

TABLE 4.6

Example DO-178C Objective

	Objective		Activity	Applicability by Software Level				Output		Control Category by Software Level			
	Description	Ref	Ref	A	B	C	D	Data Item	Ref	A	B	C	D
3	Executable object code complies with low-level requirements	6.4.c	6.4.2 6.4.2.1 6.4.3 6.5	●	●	○		Software verification cases and procedures	11.13	①	①	②	
								Software verification results	11.14	②	②	②	
								Trace data	11.21	①	①	②	

Source: RTCA DO-178C, *Software Considerations in Airborne Systems and Equipment Certification*, RTCA, Inc., Washington, DC, December 2011. Used with permission of RTCA, Inc.

TABLE 4.7

Overview of Software Life Cycle Data

DO-178C Section	Life Cycle Data Item	Description
*11.1	PSAC	Top-level software plan used to document agreements with the certification authority
11.2	Software development plan	Describes software development procedures and life cycle to guide the development team and ensure compliance to DO-178C development objectives
11.3	Software verification plan	Describes software verification procedures to guide the verifiers and to ensure compliance to DO-178C verification objectives
11.4	Software configuration management plan	Establishes the software configuration management environment, procedures, activities, and processes to be used throughout the software development and verification effort
11.5	Software quality assurance plan	Establishes the plan for software quality assurance's oversight of the project to ensure compliance to DO-178C objectives and the plans and standards
11.6	Software requirements standards	Provides guidelines, methods, rules, and tools for the requirements authors
11.7	Software design standards	Provides guidelines, methods, rules, and tools for the designers
11.8	Software code standards	Provides guidelines, methods, rules, and tools for using the programming language effectively
11.9	Software requirements data	Defines the high-level and derived high-level software requirements
11.10	Software design description	Defines the software architecture, low-level requirements, and derived low-level requirements
11.11	Source code	Consists of code files that are used together with the compile, link, and load data to create the executable object code and integrate it into the target computer
11.12	Executable object code	The code that is directly read by the target computer's processor
11.13	Software verification cases and procedures	Details how the software verification processes are implemented
11.14	Software verification results	The output of the verification processes
11.15	Software life cycle environment configuration index	Identifies the software environment, including any tools used to develop, control, build, verify, and load the software

TABLE 4.7 (continued)

Overview of Software Life Cycle Data

DO-178C Section	Life Cycle Data Item	Description
*11.16	Software configuration index	Identifies the configuration of the software product, including source code, executable object code, and supporting life cycle data. Also includes build and load instructions
11.17	Problem reports	Identifies product and process issues in order to ensure resolution
11.18	Software configuration management records	Includes results of the various software configuration management activities
11.19	Software quality assurance records	Includes results of the software quality assurance activities, including the software conformity review
*11.20	Software accomplishment summary	Summarizes the compliance to DO-178C, any deviations from the PSAC, software characteristics, and open problem reports
11.21	Trace data	Provides evidence of traces between requirements, design, code, and verification data
11.22	Parameter data item file	Includes data (such as configuration data) directly usable by the target computer's processor

Source: RTCA DO-178C, *Software Considerations in Airborne Systems and Equipment Certification*, RTCA, Inc., Washington, DC, December 2011.
* These documents are submitted to the certification authority, as a minimum.

5. The *Control Category by Software Level* indicates the level of configuration management that applies to the output data item. A ① means control category #1 (CC1) and requires a more rigorous configuration management process. A ② indicates control category #2 (CC2) and is less rigorous. CC1 and CC2 are explained in Chapter 11.

4.3 DO-330: Software Tool Qualification Considerations

DO-178C defines a software tool as: "A computer program used to help develop, test, analyze, produce, or modify another program or its documentation. Examples are an automated design tool, a compiler, test tools, and modification tools" [1]. Because of the increased volume of tools being used and the amount of third-party tool manufacturers, tool qualification guidance was expanded and separated from the DO-178C core into a stand-alone document, DO-330. DO-178C section 12.2 explains how to determine if a tool needs to be qualified and to what level the tool needs to be qualified. There are five TQLs defined; TQL-1 is the most rigorous. DO-330 explains what needs to be done to qualify the tool. DO-330 is organized similar to DO-178C and has its own set of objectives. DO-330 is a stand-alone document that allows tool developers to develop qualifiable tools without having detailed knowledge of DO-178C itself. DO-330 was created to be as domain independent as possible so it may be used by other domains (such as aeronautical database developers, electronic hardware developers, system safety assessment tool developers, system developers). The DO-330 guidance applies to tools implemented in software (i.e., it does not apply to tools implemented in hardware). Some adaptation may be needed for each specific domain that references the guidance to identify the appropriate TQLs and usage of the tool in the life cycle for that domain. However, the overall qualification process is relatively independent of the domain in which the tool is used. DO-330 appendices C and D provide several FAQs on tool qualification. DO-330 and tool qualification are examined in Chapter 13.

4.4 DO-178C Technology Supplements

To develop an approach that can change with technology and trends, the RTCA and EUROCAE committee opted to develop technology supplements. The vision is that as technology changes, the main body of DO-178C can remain unchanged and technology supplements can be modified or added, as needed. Three technology supplements were published at the same time as DO-178C. Each technology supplement is objective-based and expands on the objectives in the DO-178C core for its specific technology. Each supplement explains how objectives from the DO-178C core apply, are modified, or are replaced. Each supplement is briefly summarized in the following and will be explained in more detail in subsequent chapters. If multiple supplements are used together, the PSAC needs to explain what supplement and objectives apply to each portion of the software life cycle and resulting data.

4.4.1 DO-331: Model-Based Development Supplement

DO-331, entitled *Model-Based Development and Verification Supplement to DO-178C and DO-278A*, defines a model as [3]:*

---

* Brackets added for clarification.

An abstract representation of a set of software aspects of a system that is used to support the software development process or the software verification process. This supplement [DO-331] addresses model(s) that have the following characteristics:

a. The model is completely described using an explicitly identified modeling notation. The modeling notation may be graphical and/or textual.

b. The model contains software requirements and/or software architecture definition.

c. The model is of a form and type that is used for direct analysis or behavioral evaluation as supported by the software development process or the software verification process.

DO-331 is based on the DO-178C core and modifies or adds guidance to the core. Both model-based development and model-based verification are addressed in the DO-331 supplement. DO-331 provides guidance for *specification models* and *design models*, as well as model coverage analysis and model simulation. Most of the variances between DO-178C and DO-331 are in sections 5 and 6 (the development and verification processes). There are also impacts on planning (section 4), software life cycle data (section 11), and the Annex A objectives tables. Model-based development and verification and DO-331 are discussed more in Chapter 14.

4.4.2 DO-332: Object-Oriented Technology Supplement

DO-332, entitled *Object-Oriented Technology and Related Techniques Supplement to DO-178C and DO-278A*, explains that: "Object-oriented technology is a paradigm for system analysis, design, modeling, and programming centered on objects. There are concepts and techniques commonly found in object-oriented languages that need to be taken into account when considering software safety" [4]. DO-332 identifies the most pertinent issues surrounding object-oriented technology (OOT) and provides guidance on what needs to be done to address them. Additionally, DO-332 addresses six *related techniques* commonly associated with, but not limited to, OOT: parametric polymorphism, overloading, type conversion, exception management, dynamic memory management, and virtualization. As Chapter 15 will explain, the guidance for these *related techniques* may be applicable to non-OOT projects, when the languages utilize these advanced techniques.

Like the other supplements, the DO-332 supplement is based on the DO-178C core and identifies where guidance needs to be modified or enhanced. Some changes from DO-178C are identified in DO-332 sections 4 through 6, and 11. In general, however, the changes are relatively minor. In addition to the guidance provided in DO-332 sections 1 through 12, the supplement includes an Annex D that identifies vulnerabilities in object-oriented technology and related techniques (OOT&RT) and provides guidance on what needs to be done to address these vulnerabilities. The Annex D with vulnerabilities is unique for DO-332; other supplements do not have such a section. Like DO-330 and the other supplements, DO-332 has an appendix, which summarizes FAQs. The OOT&RT FAQs are divided into the following categories: (1) general, (2) requirements, (3) design, (4) programming, and (5) verification. DO-332 and OOT are discussed more in Chapter 15.

4.4.3 DO-333: Formal Methods Supplement

DO-333, entitled *Formal Methods Supplement to DO-178C and DO-278A*, defines formal methods as: "Descriptive notations and analytical methods used to construct, develop and reason about mathematical models of system behavior. A formal method is a formal analysis carried out on a formal model" [5]. DO-178B identified formal methods as an alternative method; however, the guidance was vague and not widely applied. Since DO-178B's publication advances have been made in the area of formal methods, and tools and techniques have been improved to make the technology more feasible. Therefore, the formal methods section has been removed from DO-178C and addressed in a supplement (DO-333). DO-333 uses the DO-178C core as the foundation and supplements the core with formal methods-specific guidance. The main sections that differ between DO-178C and DO-333 are sections 5, 6, and 11, as well as the Annex A objectives tables. Most other sections remain relatively unchanged. Chapter 16 provides more information on formal methods and DO-333.

4.5 DO-248C: Supporting Material

DO-248C, entitled *Supporting Information for DO-178C and DO-278A*, is a collection of FAQs and DPs on DO-178C topics. It replaces DO-248B and is very similar to DO-248B, with the following differences:

- As can be deduced from the title, DO-248C addresses DO-278A as well as DO-178C.*
- DO-248C does not have any errata, since no DO-178C errata were identified at the time when DO-248C was published.
- Some FAQs and DPs in DO-248B were removed from DO-248C since the information now exists elsewhere (e.g., in DO-178C, DO-330, the supplements, or the system and safety documents [ARP4754A and ARP4761]).

---

* For the remainder of this section, whenever DO-178C is mentioned, the same also applies for DO-278A.

- Some FAQs and DPs were updated to be consistent with DO-178C (e.g., FAQ #42 on structural coverage at object code level and FAQ #67 on data and control coupling analysis).
- Some new FAQs and DPs were added to clarify pertinent topics. Some of the new FAQs or DPs were added to clarify new DO-178C topics (e.g., DP #20 on parameter data items). Other FAQs or DPs were added to clarify topics that have become more pertinent since DO-248B's publication (e.g., DP #16 on cache management, DP #17 on floating-point arithmetic, and DP #21 on single event upset). Some FAQs or DPs were added to address some common certification-related subjects (e.g., FAQ #81 on merging high-level and low-level requirements, FAQ #83 on low-level requirements in the form of pseudocode, and DP #19 on independence).
- Section 5 was added to DO-248C; it includes rationale for DO-178C, DO-278A, DO-330, and the supplements.

DO-248C is not written to be read from front to back. Instead, it is a series of short papers that explain commonly misunderstood or troublesome aspects of DO-178C. The key to using DO-248C is its appendices C and D. DO-248C appendix C provides an index of keywords in the document. DO-248C appendix D provides a mapping of the DO-248C papers to the DO-178C sections. When researching a topic or a DO-178C section, one can use appendix C or D to find the relevant DO-248C sections. Hyperlinks are included in the electronic version to enhance the usability. Several DO-248C FAQs and DPs will be mentioned throughout this book as germane to the subject.

It should be noted that FAQs and DPs on tool qualification or the technology supplements are included in those documents, rather than in DO-248C.

It is expected that as the aviation community gains experience with DO-178C, DO-248C will be updated to capture errata and provide additional clarification.

References

1. RTCA DO-178C, *Software Considerations in Airborne Systems and Equipment Certification* (Washington, DC: RTCA, Inc., December 2011).
2. RTCA DO-248C, *Supporting Information for DO-178C and DO-278A* (Washington, DC: RTCA, Inc., December 2011).
3. RTCA DO-331, *Model-Based Development and Verification Supplement to DO-178C and DO-278A* (Washington, DC: RTCA, Inc., December 2011).
4. RTCA DO-332, *Object-Oriented Technology and Related Techniques Supplement to DO-178C and DO-278A* (Washington, DC: RTCA, Inc., December 2011).

5. RTCA DO-333, *Formal Methods Supplement to DO-178C and DO-278A* (Washington, DC: RTCA, Inc., December 2011).
6. RTCA DO-278A, *Guidelines for Communications, Navigation, Surveillance, and Air Traffic Management (CNS/ATM) Systems Software Integrity Assurance* (Washington, DC: RTCA, Inc., December 2011).

5

Software Planning

Acronyms

CC1	control category #1
CC2	control category #2
COTS	commercial off-the-shelf
CRC	cyclic redundancy check
FAA	Federal Aviation Administration
MISRA	Motor Industry Software Reliability Association
PQA	product quality assurance
PQE	product quality engineer
PR	problem report
PSAC	Plan for Software Aspects of Certification
RTOS	real-time operating system
SAS	Software Accomplishment Summary
SCI	Software Configuration Index
SCM	software configuration management
SCMP	Software Configuration Management Plan
SDP	Software Development Plan
SLECI	Software Life Cycle Environment Configuration Index
SOI	stage of involvement
SQA	software quality assurance
SQAP	Software Quality Assurance Plan
SVP	Software Verification Plan

5.1 Introduction

You've probably heard the saying: "If you fail to plan, you plan to fail." Or, maybe you've heard another one of my favorites: "If you aim at nothing, you'll hit it every time." Good software doesn't just happen—it takes extensive planning, as well as well-qualified personnel to execute and adapt the plan.

The DO-178C planning process involves the development of project-specific plans and standards. DO-178C compliance (and development assurance in general) requires that plans are both documented and followed. If the plans are written to address the applicable DO-178C objectives, then following them will ensure compliance to the objectives. Hence, DO-178C compliance and certification authority approval hinges on a successful planning effort.

DO-178C identifies five plans and three standards and explains the expected contents of each document. While most organizations follow the five-plans-and-three-standards suggestion, it is acceptable to package the plans and standards as best fits the needs of the organization. Although not recommended, the developer could combine all the plans and standards in a single document (keep in mind that in this case, the entire document would need to be submitted to the certification authority). Using a nontraditional planning structure can create some challenges on the first exposure to the certification authority and may require additional submittals; however, it is acceptable as long as all of the necessary topics identified in DO-178C are adequately discussed. Regardless of how the plans and standards are packaged, the recommended contents from DO-178C sections 11.1 through 11.8 need to be addressed, and the documents must be consistent.

5.2 General Planning Recommendations

Before explaining each of the five plans, let's consider some general planning recommendations.

*Recommendation 1: Ensure that the plans cover all applicable DO-178C objectives and any applicable supplement objectives.** The plans should be written so that the team will comply with all of the applicable objectives when they execute the plans. In order to ensure this, a mapping between the DO-178C (and applicable supplements) objectives and the plans is helpful. When completed, the mapping is often included in an appendix of the Plan for Software Aspects of Certification (PSAC).† I recommend presenting the mapping as a four-column table—each column is described here:

1. *Table/Objective #*: Identifies the DO-178C (and applicable supplements) Annex A table and objective number.

2. *Objective Summary*: Includes the objective summary, as it appears in Annex A of DO-178C and/or the applicable supplements.

---

* Chapter 4 provides an overview of the technology supplements. Chapters 14 through 16 provide more details on each of the technology supplements.
† The PSAC is the top-level software plan that is submitted to the certification authority. It is discussed later in this chapter.

3. *PSAC Reference*: Identifies sections in the PSAC that explain how the objective will be satisfied.

4. *Other Plans Reference*: Identifies sections in the team's plans (e.g., Software Development Plan [SDP], Software Verification Plan [SVP], Software Configuration Management Plan [SCMP], and Software Quality Assurance Plan [SQAP]) that explain the detailed activities to satisfy the objective.

Recommendation 2: Write the plans and standards in such a way that they can be followed by the teams implementing them. The target audience for the PSAC is the certification authority; however, the intended audience for the other plans and the standards is the teams (e.g., development team, verification team, and quality assurance team) that will execute the plans. They should be written at the appropriate level for the team members to understand and properly execute them. If it is an experienced team, the plans may not require as much detail. However, a less experienced team or a project involving extensive outsourcing typically requires more detailed plans.

Recommendation 3: Specify what, how, when, and who. In each plan, explain what will be done, how it will be done, when it will be done (not the specific date, but when in the overall progression of activities), and who will do it (not necessarily the specific names of engineers, but the teams who will perform the tasks).

Recommendation 4: Complete the plans and put them under configuration management before the software development begins. Unless plans are finalized early in the project, it becomes difficult to establish the intended direction, find the time to write them later, and convince the certification authority that they were actually followed. Even if plans are not released before the development begins, they should be drafted and put under configuration management. The formalization (i.e., review and release) should occur as soon as possible. Any changes from the draft versions to the released versions of the plans should be communicated to the teams to ensure they update data and practices accordingly.

Recommendation 5: Ensure that each plan is internally consistent and that there is consistency between the plans. This seems like common sense, but inconsistency is one of the most common problems I discover when reviewing plans. When plans are inconsistent, the efforts of one team may undermine the efforts of another team. If the SDP says one thing and the SVP says another, the teams may make their own decisions on which to follow. There are several ways to ensure consistency among the plans, including using common authors and reviewers. One of the most effect ways I've found to create consistent plans is to gather the technical leads and spend some time collectively determining the software life cycle from beginning to end, including the activities, entry criteria, exit criteria, and output for each phase. A large white board or a computer projection to capture the common vision works great. Once the life cycle is documented and agreed upon, then the plans can be written.

Recommendation 6: Identify consistent transition criteria for processes between plans. Careful attention should be paid to defining consistent transition criteria between the processes in all of the plans, particularly the development and verification plans.

Recommendation 7: If developing multiple software products, company-wide planning templates may be helpful. The templates can provide a starting point for project-specific plans. It is best to create the templates using a set of plans that have been used on at least one successful project and that have implemented lessons learned. I recommend that the templates be regularly updated based on lessons learned and feedback from certification authorities, customers, designees, other teams, etc.

Recommendation 8: Involve certification experts (such as authorized designees*) as early in the planning process as possible. Getting input from a designee or someone with certification experience can save time and prevent unnecessary troubles later in the project.

Recommendation 9: Obtain management buy-in on the plans. Effective project execution is dependent on management's understanding and support of the plans. There are many ways to encourage management buy-in; one of the most effective is to involve management in the planning process, as authors or reviewers. Once the plans are written, management will need to establish a detailed strategy to implement the plans (e.g., identify the necessary resources); the success of this strategy depends on their understanding of the plans.

Recommendation 10: Include appropriate level of details in the plans. Rather than including detailed procedures in the plans, the procedures can be located in an engineering manual or in work instructions. The plans should identify the procedures and provide a summary of the procedures, but the details do not have to be in the plans themselves. For example, structural coverage analysis and data and control coupling analyses normally require some specific procedures and instructions; these procedures can be packaged somewhere besides the SVP. This approach allows some flexibility to choose the best solution when it comes time to actually execute the SVP. It is important that the plans clearly explain what procedures apply, how the procedures will be configuration controlled, how modifications will occur, and how any modifications will be communicated to the team.

---

* Designees are representatives of the certification authority or delegated organization who are authorized to review data and either approve or recommend approval. Projects typically coordinate with designees prior to interaction with the certification authority.

Recommendation 11: Brief the certification authority prior to submitting the PSAC. In addition to submitting the PSAC and possibly the other plans to the certification authority, it is valuable to provide a presentation to the certification authority on the planned approach. This is particularly recommended for new or novel products or for companies who may not have a strong working relationship with their local certification authority. Oftentimes, the verbal discussion will identify issues that need to be resolved prior to submitting the plans to the certification authority.

Recommendation 12: Disperse plans to the teams. Once the plans are mature, make sure the teams understand their content. I am amazed at how many projects I audit where the engineers don't know where to find the plans or what the plans say. Since engineers are not always thrilled with documentation, I recommend that training sessions be mandatory for all team members, as well as reading assignments for their specific roles. Be sure to keep accurate training records, since some certification authorities may ask to see such evidence.

Recommendation 13: Develop the plans to be changeable. A forward-thinking project realizes that the plans may change as the project progresses; therefore, it is wise to document the process for changing the plans in the plans themselves. DO-178C section 4.2.e indicates that the PSAC should include an explanation of the means to revise software plans throughout the project.

Sometimes the planned processes do not work as expected. Improvements or a complete overhaul to the process may be needed. The plans need to be updated to reflect the modifications. During the planning process, there should be thought given to the process for updating plans in a timely manner. This will make it easier *when* it actually happens.

Sometimes it is tempting to simply document changes in the Software Accomplishment Summary (SAS)* rather than updating the plans. For changes that happen late in the process, this might be acceptable. However, there are some risks. First, the certification authority may not allow it, since many authorities prefer to always have the plans current. Second, the next team that uses the plans (either for updating the software or for a starting point for the next project) may not know about the changes. Therefore, if plans are not updated, the changes should be clearly documented and agreed upon with the certification authority. The changes should also be formally communicated to the engineering teams and documented in problem report(s) against the plans to ensure they are not overlooked.

Recommendation 14: Plan for postcertification process. In addition to changes during the initial certification effort, it is recommended that the planning structure accommodate the postcertification process. Questions to consider

---

* The SAS is the compliance report for the DO-178C compliant software and is submitted to the certification authority. It is completed at the end of the software project and identifies deviations from the plans. The SAS is discussed in Chapter 12.

are the following: Will postcertification changes require a new PSAC? Or will the change be described in a change impact analysis that is an addendum to the PSAC? Or will changes be described in an appendix of the PSAC? Or will some other packaging approach be used?

If the postcertification process is not considered during the initial certification, it may take more time later and will likely result in an updated PSAC being required for the change, no matter how insignificant the change may seem. It's worth noting that if the plans are kept up to date during the initial development effort, it makes postcertification changes much easier.

The remainder of this chapter explains each of the five plans and three standards mentioned in DO-178C and includes some recommendations to consider while developing them.

5.3 Five Software Plans

This section explains the expected contents of the five plans and provides suggestions to consider. Several technical topics that may be unfamiliar are introduced in this chapter and will be further explained in future chapters.

5.3.1 Plan for Software Aspects of Certification

The PSAC is the one plan that is always submitted to the certification authority. It is like a contract between the applicant and the certification authority; therefore, the earlier it is prepared and agreed upon, the better. A PSAC that is not submitted until late in the project introduces risk to the project. The risk is that a team may get to the end of a project only to find out that the processes and/or data are not compliant. This can cause schedule and budget slips, considerable rework, and increased scrutiny by the certification authority.

The PSAC provides a high-level description of the overall project and explains how the DO-178C (and applicable supplements) objectives will be satisfied. It also provides a summary of the other four plans. Since the PSAC is often the only plan submitted to the certification authority, it should stand alone. It doesn't need to repeat everything in the development, verification, configuration management, and quality assurance plans, but it should provide an accurate and consistent summary of those plans. Occasionally, I come across a PSAC that merely points to other documents (I call it a *pointer PSAC*). Instead of summarizing the development, verification, configuration management, and software quality assurance (SQA) processes, it simply points to the other plans. This certainly reduces redundant text. However, it also means that the other plans will most likely need to be submitted to the certification authority in order for the authority to fully understand the processes. Ideally, the PSAC includes enough detail that the certification authority understands the processes and is able to make an accurate judgment about the compliance, but not so much detail that it repeats the contents of the other plans.

The PSAC should be written clearly and concisely. I have reviewed PSACs that repeated the same thing multiple times and that actually copied sections of DO-178B. The clearer the document is, the less confusing it will be and the more likely it will be approved by the certification authority.

DO-178C section 11.1 provides a summary of the expected PSAC contents and is often used as the outline for the PSAC. Following are the typical sections of the PSAC, along with a brief summary of the contents of each section [1]:

- *System overview*: This section explains the overall system and how the software fits into the system. It is normally a couple of pages long.

- *Software overview*: This section explains the intended functionality of the software, as well as any architectural concerns. It, too, consists of a couple of pages. Since the plan is for software, it is important to explain the software that will be developed, not just the system.

- *Certification considerations*: This section explains the means of compliance. If the DO-178C supplements are being used, this is a good place to explain what supplement will apply to what part of the software. Additionally, this section usually summarizes the safety assessment to justify the assigned software levels. Even for level A software, it is important to detail what drives the decision and to explain if any additional architectural mitigation is needed. Many projects also explain their plans for supporting stage of involvement (SOI) audits in this section of the PSAC. SOI audits are explained in Chapter 12.

- *Software life cycle*: This section describes the phases of the software development and the integral processes and is typically a summary of the other plans (SDP, SVP, SCMP, and SQAP).

- *Software life cycle data*: This section lists the life cycle data to be developed during the project. Document numbers are normally assigned, although there may be an occasional *TBD* or *XXX*. Frequently, this section includes a table that lists the 22 data items from DO-178C section 11 and includes the document names and numbers. It is also the norm to include an indication if the data will be submitted to the certification authority or just available. Some applicants identify whether the data will be treated as control category #1 (CC1) or control category #2 (CC2) and if the data will be approved or recommended for approval by the designees. As an alternative, the CC1/CC2 information may be in the SCMP. However, if a company-wide SCMP is used, the PSAC may identify the project-specific configuration management details. CC1 and CC2 are discussed in Chapter 10 of this book.

- *Schedule*: This section includes the schedule for the software development and approval. I have not yet seen one of these schedules actually met; therefore, the question often arises: "Why do we need the schedule if no one ever meets it?" The purpose is to help both the project and the certification authority plan their resources. Any changes to the schedule throughout the project should be coordinated with the approving authority (this does not require an update to the PSAC, but if the PSAC is updated for other purposes, the schedule should also be updated). Ordinarily, the schedule in the PSAC is relatively high-level and includes the major software milestones, such as when: plans will be released, requirements will be completed, design will be completed, code will be completed, test cases will be written and reviewed, tests will be executed, and SAS will be written and submitted. Some applicants also include SOI audit readiness dates in their schedule.

- *Additional considerations*: This is one of the most important sections of the PSAC because it communicates any special issues of which the certification authority needs to be aware. In certification it is important to make every effort to *avoid surprises*. Documenting all additional considerations clearly and concisely is a way to minimize surprises and to obtain agreement with the certification authority. DO-178C includes a nonexhaustive list of additional consideration topics (such as, previously developed software, commercial off-the-shelf [COTS] software, and tool qualification). If there is anything about your project that might be considered out of the ordinary, this is the place to disclose it. This might include the use of a partitioned real-time operating system (RTOS), an offshore team, or automation. Additionally, any planned deactivated code or option-selectable software should be explained. Although DO-178C doesn't require it, I recommend that the additional considerations section of the PSAC include a list of all tools that will be used on the project, with a brief description of how the tool will be used and justification for why the tool does or does not need to be qualified. Disclosing such information during planning can help prevent some late discoveries; particularly if a tool should be qualified but has not been identified as one that requires qualification.

As noted in Recommendation #1 earlier, it is useful to include a listing of all the applicable DO-178C (and applicable supplements) objectives in an appendix of the PSAC, along with a brief explanation of how each objective will be satisfied and where in the plans each objective is addressed. The designees and certification authority appreciate this information because it provides evidence that the project thoroughly considered the DO-178C compliance details. As a side note, if you opt to include the objectives mapping in your PSAC, please ensure that it is accurate. Since designees and certification authorities like this information, they typically read it; having it accurate and complete helps to build their confidence.

5.3.2 Software Development Plan

The SDP explains the software development, including requirements, design, code, and integration phases (DO-178C Table A-2). Additionally, the SDP often briefly explains the verification of requirements, design, code, and integration (DO-178C Tables A-3 through A-5). The SDP is written for the developers who will write the requirements, design, and code and perform the integration activities. The SDP should be written so that it guides the developers to successful implementation. This means it needs to be detailed enough to provide them good direction, but not so detailed that it limits their ability to exercise engineering judgment. This is a delicate balance. As noted in Recommendation #10, there may be more detailed procedures or work instructions. If this is the case, the SDP should clearly explain what procedures apply and when they apply (i.e., the SDP points to the procedures rather than including the details in the plan). Occasionally, more flexibility may be needed for the detailed procedures; for example, if the procedures are still being developed after the plans are released. If this is the case, the SDP ought to explain the process for developing and controlling the procedures; however, care should be taken when using this approach, since it can be difficult to ensure that all engineers are following the right version of the procedures.

DO-178C section 11.2 identifies the preferred contents of the SDP. The SDP includes a description of three major items: (1) the standards used for development (occasionally, the standards are even included in the plan), (2) the software life cycle with an explanation of each of the phases and criteria for transitioning between phases, and (3) the development environment (the methods and tools for requirements, design, and code, as well as the intended compiler, linker, loader, and hardware platforms). Each is explained in the following:

1. *Standards*: Each project should identify standards for requirements, design, and code. These standards provide rules and guidelines for the developers to help them write effective requirements, design, and code. The standards also identify constraints to help developers avoid pitfalls that could negatively impact safety or software functionality. The standards should be applicable to the methodology or language used. The SDP typically references the standards, but in some situations, the standards may be included in the SDP (this sometimes occurs for companies that do limited software development or that have small projects). The three development standards are discussed later in this chapter.

2. *Software life cycle*: The SDP identifies the intended life cycle for the software development. This is typically based on a life cycle model.

In addition to identifying the life cycle model by name, it is recommended that the model be explained, since not all life cycle models mean the same to everyone. A graphic of the life cycle and the data generated for each phase is helpful. Some of the life cycle models that have been successfully used on certification projects are waterfall, iterative waterfall, rapid prototyping, spiral, and reverse engineering. I highly recommend avoiding the big bang, tornado, and smoke-and-mirrors life cycle models.*

Unfortunately, some projects identify one life cycle model in their plans but actually follow something else. For example, projects sometimes claim that they use waterfall because they believe that is what DO-178C requires and what the certification authority prefers. However, DO-178C does not require waterfall and to claim waterfall without actually using it causes several challenges. It is important to identify the life cycle model that you actually plan to use, to ensure that it satisfies the DO-178C objectives, and to follow the documented life cycle model. If the documented life cycle model is not what is needed, the plans should be updated accordingly, unless otherwise agreed with the certification authority.

As mentioned earlier, the SDP documents the transition criteria for the software development. This includes the entry criteria and exit criteria for each phase of the development. There are many ways to document transition criteria. A table can be an effective and straightforward way to document such information. The table lists each phase, the activities performed in that phase, the criteria for entering the phase, and the criteria for exiting the phase. Appendix A provides an example transition criteria table for a development effort. It is important to keep in mind that DO-178C doesn't dictate an ordering for the development activities, but the verification efforts will need to be top-down (i.e., verify requirements prior to verifying design prior to verifying code).

3. *Software development environment*: In addition to identifying standards and describing the software life cycle, the SDP identifies the development environment. This includes the tools used to develop requirements, design, source code, and executable object code (e.g., compiler, linker, editor, loader, and hardware platform). If a full list of tools was identified in the PSAC, as suggested, the SDP may simply reference the PSAC. However, the SDP may go into more detail about how the tools are used in the software development. Many times, the PSAC and SDP do not provide the tool part numbers, since that information is included in the Software Life Cycle Environment

---

* *Big bang* is a hypothetical model where everything magically appears at the end of the project. *Tornado* and *smoke-and-mirrors* models are *Leanna-isms*. I use the *tornado model* to describe a project headed for disaster and the *smoke-and-mirrors model* to describe a project that really has no plan.

Configuration Index (SLECI). To avoid redundancy, some projects create the SLECI early in the project and reference it from the SDP. If this is the case, an initial version of the SLECI should be completed with the plans, even though the SLECI will likely need to be updated later as the project matures.

The environment identification provides a means to control it and to ensure that the software can be consistently reproduced. Uncontrolled tools can lead to problems during the implementation, integration, and verification phases. I witnessed one situation where different developers used different compiler versions and settings. When they integrated their modules, things got interesting.

If any development tools require qualification, this should be explained in the SDP. The PSAC may have already provided a summary of qualification, but the SDP should explain how the tool is used in the overall life cycle and how the tool users will know the proper tool operation (such as a reference to Tool Operational Requirements or User's Guide). See Chapter 13 for information on tool qualification.

Chapters 6 through 8 provide more information on software requirements, design, and implementation, respectively.

5.3.3 Software Verification Plan

The primary audience for the SVP is the team members who will perform the verification activities, including testing. The SVP is closely related to the SDP, since the verification effort includes the evaluation of data that were generated during the development phases. As mentioned earlier, the SDP often provides a high-level summary of the requirements, design, code, and integration verification (such as peer reviews). The SVP normally provides additional details on the reviews (including review process details, checklists, required participants, etc.). It is acceptable to include the review details in the SDP and use the SVP to focus on testing and analyses. Regardless of the packaging, it must be clear what plan is covering what activities.

Of all the plans, the SVP tends to vary the most depending on the software level. This is because most of the DO-178C level differences are in the verification objectives. Typically, the SVP explains how the team will satisfy the objectives in DO-178C Tables A-3 through A-7.

The SVP explains the verification team organization and composition, as well as how the required DO-178C independence is satisfied. Although it is not required, most projects have a separate verification team to perform the test development and execution. DO-178C identifies several verification objectives that require independence (they have filled circles [●] in the DO-178C Annex A tables). DO-178C verification independence doesn't require a separate organization but it does require that one or more persons (or maybe a tool) who did not develop the data being verified perform the verification. Independence basically means that another set of eyes and brain (possibly accompanied by a tool) are used to examine the data for correctness, completeness, compliance to standards, etc. Chapter 9 explains more about verification independence.

DO-178C verification includes reviews, analyses, and tests. The SVP explains how reviews, analyses, and tests will be performed. Any checklists used to accompany the verification are also either included in the SVP or referenced from the SVP.

Many of the DO-178C objectives may be satisfied by review. Tables A-3, A-4, and most of A-5 tend to be met using a peer review process (which is discussed further in Chapter 6). Additionally, some of the Table A-7 objectives (such as objectives 1 and 2) are satisfied by review. The SVP explains the review process (including or referencing detailed review procedures), the transition criteria for reviews, and checklists and records used to record the reviews. Either the SVP or the standards normally include (or reference) checklists for reviewing the requirements, design, and code. The checklists are used by engineers to ensure they don't overlook important criteria during the review. Checklists that are brief tend to be most effective; if they are too detailed, they are usually not fully utilized. To create a concise but comprehensive checklist, I recommend separating the checklist items and detailed guidance into separate columns. The checklist column is brief but the guidance column provides detailed information to ensure that the reviewers understand the intent of each checklist item. This approach is particularly useful for large teams, teams with new engineers, or teams using outsourced resources. It helps to set the bar for the reviews and ensure consistency. Checklists typically include items to ensure that required DO-178C objectives are evaluated (including traceability, accuracy, and consistency) and the standards are satisfied, but they are not limited to the DO-178C guidance.

DO-178C Table A-6 is typically satisfied by the development and execution of tests. The SVP explains the test approach; how normal and robustness tests will be developed; what environment will be used to execute the tests; how traceability will occur between requirements, verification cases, verification procedures; how verification results will be maintained; how pass/fail criteria will be identified; and where test results will be documented. In many cases the SVP makes reference to a Software Verification Cases and Procedures document, which details the test plans, specific test cases and procedures, test equipment and setup, etc.

DO-178C Tables A-5 (objectives 6 and 7) and A-7 (objectives 3–8) are usually satisfied (at least partially) by performing analyses. Each planned analysis should be explained in the SVP. Typical analyses include traceability analyses (ensuring complete and accurate bidirectional traceability between system requirements, high-level software requirements, low-level software requirements, and test data), worst-case execution timing analysis, stack usage analysis, link analysis, load analysis, memory map analysis, structural coverage analysis, and requirements coverage analysis. These analyses are explained in Chapter 9. The SVP should identify the intended approach for each analysis and where the procedures and results will be documented.

Since tools are often used for the verification effort, the SVP lists those tools. If the PSAC list is extensive (as previously recommended), it may be possible to make a reference to the PSAC instead of repeating the same list in the SVP. However, the SVP usually provides more detail on how each tool will be used in the software verification and references instructions necessary to properly use the tools. As with the development tools, the SVP may reference the SLECI to identify the tool details (versions and part numbers), rather than including the information in the SVP itself. If this is the case, the SLECI should be completed with the plans and may require revision prior to beginning the formal verification process. The SLECI is explained in Chapter 10.

If an emulator or simulator will be used to verify the software, its use should be explained and justified in the SVP. In some cases, the emulator or simulator may need to be qualified. Chapter 9 discusses the emulation and simulation.

The SVP also identifies the transition criteria for verification activities. Appendix A includes an example of the entry criteria, activities, and exit criteria for the verification activities of a project.

If the software being developed and verified contains partitioning, the SVP should explain how the partitioning integrity will be verified (DO-178C Table A-4 objective 13). Partitioning is discussed in Chapter 21.

DO-178C section 11.3 also mentions that the SVP should discuss assumptions made about the compiler, linker, and loader correctness. If compiler optimization is used, it should be explained in the plans, since it may affect the ability to obtain structural coverage analysis or to perform source-to-object code analysis. Also, the SVP should explain how the accuracy of the linker will be verified. If a loader is used without an integrity check (such as a cyclic redundancy check [CRC]), then the loader functionality will need to be verified. If an integrity check is used, the SVP should explain the approach and justify the adequacy of the check (e.g., mathematically calculate the algorithm accuracy to ensure that the CRC is adequate for the data being protected).

Finally, the SVP should explain how reverification will be performed. If changes are made during the development, will everything be retested or will a regression analysis be performed and only the impacted and changed items retested? The SVP should explain the planned approach, the criteria that will be used, and where the decisions will be documented. Reverification should consider both the changed and impacted software.

If previously developed software (e.g., COTS software or reused software) is used, some reverification may be needed (e.g., if it is installed in a new environment or used in a different way). The SVP should explain this. If no reverification is needed, the SVP should justify why not.

Chapter 9 provides more details on verification.

5.3.4 Software Configuration Management Plan

The SCMP explains how to manage the configuration of the life cycle data throughout the software development and verification effort. Software configuration management (SCM) begins during the planning phase and continues throughout the entire software life cycle—all the way through deployment, maintenance, and retirement of the software.

DO-178C section 11.4 provides a summary of what the SCMP should include. The SCMP explains the SCM procedures, tools, and methods for developmental SCM (used by engineering prior to formal baseline or releases) and formal SCM, as well as the transition criteria for the SCM process.

DO-178C section 7.2 explains the SCM activities, which need to be detailed in the SCMP. A brief summary of what typically goes in the SCMP for each activity is included in the following:

1. *Configuration identification*: The SCMP explains how each configuration item (including individual source code and test files) is uniquely identified. Unique identification typically includes document or data numbering and revisions or versions.

2. *Baselines and traceability*: The SCMP explains the approach for establishing and identifying baselines. If engineering baselines will be established throughout the project, this should be explained. Likewise, the establishment of formal baselines should be explained, including certification and production baselines. The traceability between baselines is also detailed in the SCMP.

3. *Problem reporting*: The problem reporting process is part of the SCM process and should be explained in the SCMP, including an explanation of when the problem reporting process will begin, the required contents of a problem report (PR), and the process for verifying and closing a PR. The problem reporting process is crucial to an effective change management process and should be well defined in the plans. Engineers should also be trained on the PR process. Oftentimes, the SCMP includes a PR form with a brief explanation of how to complete each field. Most PRs include a classification field (to categorize the severity of the PR) and a state field (to identify the state of the PR, such as open, in-work, verified, closed, or deferred). Problem reporting is discussed in Chapters 9 and 10.

   If any other process besides the PR process is used to gather issues or actions, it should also be explained in the SCMP. This may happen when companies have additional change request, deviation, and/or action item tracking processes on top of the problem reporting system.

4. *Change control*: The SCMP explains how changes to life cycle data are controlled to ensure that change drivers are established and approved prior to changing a data item. This is closely related to the problem reporting process.

5. *Change review*: The purpose of the change review process is to make sure that changes are planned, approved, documented, properly implemented, and closed. It is typically monitored by a change review board that approves change implementation and ensures the change has been verified prior to closure. The change review process is closely related to the problem reporting process.

6. *Configuration status accounting*: Throughout the project, it is necessary to know the status of the effort. Configuration status accounting provides this capability. Most SCM tools offer the ability to generate status reports. The SCMP explains what to include in status reports, when status reports will be generated, how status reports will be used, and how the tools (if applicable) support this process. The status and classification of problem reports throughout the development is particularly pertinent to certification. DO-178C section 7.2.6 provides information on what to consider in status accounting reports.

7. *Archival, release, and retrieval*: Many companies have detailed procedures for release, archival, and retrieval. In this case, the company procedures may be referenced in the SCMP, but it should be ensured that such company procedures adequately address the DO-178C guidance in sections 7.2.7 and 11.4.b.7 [1]. Following are the specific items to explain or reference in the SCMP:

    a. *Archival*: The SCMP explains how archiving is achieved. Typically, this includes an off-site archiving process, as well as a description of media type, storage, and refresh rate. The long-term reliability of the media and its readability should be considered.

    b. *Release*: The formal release process for released data is explained in the SCMP. Projects will also maintain data which are not released. The SCMP should indicate how such data will be stored.

    c. *Retrieval*: Additionally, the retrieval processes should be explained or referenced. The retrieval process considers long-term retrieval and media compatibility (e.g., the process may require that certain equipment be archived in order to retrieve data in the future).

8. *Software load control*: The SCMP explains how software is accurately loaded into the target. If an integrity check is used, it should be detailed. If there is no integrity check, the approach for ensuring an accurate, complete, and uncorrupted load should be defined.

9. *Software life cycle environment control*: The software life cycle environment identified in the SDP, SVP, and/or SLECI must be controlled. A controlled environment ensures that all team members are using the approved environment and ensures a repeatable process. The SCMP describes how the environment is controlled.

Normally, this involves a released SLECI and an assessment to ensure that the tools listed in the SLECI are complete, accurate, and utilized. Oftentimes, a configuration audit (or conformity) is required prior to the formal software build and formal test execution steps. SQA may perform or witness the configuration audit.

10. *Software life cycle data control*: The SCMP identifies all software life cycle data to be produced, along with CC1/CC2 categorization. This should also include how the specific project is implementing CC1 and CC2 for their data. DO-178C's CC1/CC2 criteria defines the minimum SCM required, but many projects exceed the minimum, combine data items, or vary some other way from DO-178C's suggestions. If the CC1/CC2 is identified in the PSAC for each configuration item, then the PSAC may be referenced, rather than repeating it in the SCMP. Frequently, the SCMP lists the minimal CC1/CC2 assignment from DO-178C, and the PSAC gives the project-specific assignment. This allows the use of a general, company-wide SCMP. CC1/CC2 is explained in Chapter 10.

If suppliers (including subcontractors or offshore resources) are used, the SCMP also explains the supplier's SCM process. Many times the supplier has a separate SCMP, which is referenced. Or, the supplier may follow the customer's SCMP. If multiple SCMPs are used, they should be reviewed for consistency and compatibility.

The plan for overseeing suppliers' problem reporting processes should also be included in the SCMP (or in another plan referenced from the SCMP). Federal Aviation Administration (FAA) Order 8110.49 section 14-3.a states: "In order to ensure that software problems are consistently reported and resolved, and that software development assurance is accomplished before certification, the applicant should discuss in their Software Configuration Management Plan, or other appropriate planning documents, how they will oversee their supplier's and sub-tier supplier's software problem reporting process" [2]. The Order goes on to provide the FAA's expectations for the SCMP, including an explanation of how supplier problems will be "reported, assessed, resolved, implemented, re-verified (regression testing and analysis), closed, and controlled" [2]. The European Aviation Safety Agency (EASA) identifies similar expectations in their Certification Memorandum CM-SWCEH-002 [3].

Finally, the SCMP identifies SCM-specific data generated, including PRs, Software Configuration Index (SCI), SLECI, and SCM records. All of these SCM data items are discussed in Chapter 10.

After reading dozens of SCMPs, I have identified several common deficiencies, which are listed here for your awareness:

- The problem reporting process is often not explained in enough detail for it to be properly implemented by the development and verification teams. Additionally, it may not be clear when the problem reporting process will start.

- The process for controlling the environment is rarely defined. As a consequence, many projects do not have adequate environment control.

- The approach for archiving is not described, including how the environment (tools used to develop, verify, configure, and manage the software) will be archived.

- The developmental SCM process that is used by engineering on a daily basis is rarely detailed.

- Supplier control is often not adequately explained in the SCMP, and the various SCM processes between companies are rarely reviewed for consistency.

- The plan for establishing developmental baselines is not elaborated.

- SCM tools are frequently not identified or controlled.

Most companies have a company-wide SCMP. When this is the case, the project-specific details still need to be addressed somewhere. This could be in a separate project-specific SCMP that supplements the company-wide SCMP or in the PSAC (or some other logical document). Whatever, the case, it should be clear in the PSAC, as well as the SDP and SVP, so that the team members understand and follow the approved processes.

5.3.5 Software Quality Assurance Plan

The SQAP describes the software quality team's plan for assuring that the software complies with the approved plans and standards, as well as the DO-178C objectives. The SQAP includes the organization of the SQA team within the overall company and emphasizes their independence.

The SQAP also explains the software quality engineer's* role, which typically includes the following:

- Reviewing the plans and standards
- Participating in the peer review process to ensure that the peer review process is properly followed
- Enforcing the transition criteria identified in the plans
- Auditing the environment to ensure developers and verifiers are using the tools identified in the SLECI with the appropriate settings, including compiler, linker, test tools and equipment, etc.
- Assessing compliance to plans
- Witnessing software builds and tests

---

* Most software projects have one SQA engineer assigned. However, large or dispersed projects may have multiple quality engineers.

- Signing/approving key documents
- Closing PRs
- Participating in the change control board
- Performing the software conformity review

SQA is further discussed in Chapter 11. As software quality engineers carry out their responsibilities, they generate SQA records, which explain what they did and any discrepancies discovered. Oftentimes, a common form is used for multiple SQA activities and a blank form is included in the SQAP. The SQAP should explain the approach for SQA record keeping, including what records will be kept, where they will be stored, and how their configuration will be managed.

Many times, the SQAP identifies target percentages for quality's involvement. This can include the percentage of peer review participation and test witnessing. If this is the case, it should be clear how the metrics will be collected.

The SQAP should explain the transition criteria for the SQA process (i.e., when SQA activities begin), as well as any key timing details of when SQA activities will be performed.

Many companies have implemented product quality assurance (PQA) in addition to SQA. The product quality engineer (PQE) oversees the day-to-day activities of the project and focuses on technical quality, as well as process compliance. If PQA is used to satisfy any of the DO-178C Table A-9 objectives, the SQAP should describe the PQA role and how it is coordinated with the SQA role.

If suppliers help with the software development or verification, the SQAP should explain how they will be monitored by SQA. If the supplier has their own SQA team and SQA plans, they should be evaluated for consistency with the higher-level SQAP.

Please note that if the SQAP is a company-wide plan, it should be ensured that it is consistent with the project-specific plans. Oftentimes, there are some project-specific needs not mentioned in the company-wide plan. If this is the case, a separate SQAP, the PSAC, or an SQAP addendum may be used to address the project-specific SQA needs.

5.4 Three Development Standards

In addition to the five plans, DO-178C identifies the need for three standards: requirements standard, design standard, and coding standard. Many companies new to DO-178C confuse these with industry-wide standards (e.g., Institute of Electrical and Electronic Engineers standards). Industry-wide standards may serve as input to the project-specific standards, but each project has specific needs that are not dealt with in the industry-wide standards. If multiple projects use the same methods and languages, company-wide standards may be generated and applied. However, the applicability of the company-wide standards must be evaluated for each project. The standards provide rules and constraints to help developers do their job properly and avoid activities that could have negative safety or functional impacts. Many standards are ineffective for the following reasons:

• The standards are just created to satisfy a check mark for DO-178C compliance and have little actual value to the project.

• The standards are cut and pasted from industry-wide standards but do not meet the project-specific needs.

• The standards are not read until the verification phases (the developers either don't know about the standards or they ignore them).

This section is intended to provide some practical advice for developing effective standards.

Although it's not required, most standards include both mandatory and advisory material. The mandatory items are generally called *rules* and the advisory items are called *guidelines*. In many cases, the applicability of the rules and guidelines varies depending on the software level (e.g., one item may be advisory for level C but mandatory for levels A and B). It should be noted that DO-178C doesn't require standards for level D, although they are not prohibited.

Before jumping into the standards, I want to emphasize their importance. They serve as the instructions to the developers to implement safe and effective practices. If they are written clearly and include rationale and good examples, they can provide an excellent resource to the developers. Once the standards are developed, it is important to provide training to all developers. The training should cover the content of the standards, explain how to use the standards, and emphasize that reading and following the standards is mandatory.

5.4.1 Software Requirements Standards

The software requirements standards define methods, tools, rules, and constraints for the software requirements development [1]. They typically apply to the high-level requirements; however, some projects also apply the requirements standards to the low-level requirements. In general, the requirements standards are a guide for the team writing the requirements. For instance, they explain how to write effective and implementable requirements, use the requirements management tool, perform traceability, handle derived requirements, and create requirements that meet the DO-178C criteria. Requirements standards may also serve as a training tool for engineers in addition to providing success criteria for the requirements review.

Following is a list of items normally included in requirements standards:

- The criteria from DO-178C Table A-3 (in order to proactively address the DO-178C expectations).
- Definitions and examples of high-level requirements, low-level requirements, and derived requirements (for reference purposes).
- Quality attributes of requirements (verifiable, unambiguous, consistent, etc.). Chapters 2 (Section 2.2.3) and 6 (Section 6.7.6.9) explain characteristics of good requirements.
- Traceability approach and instructions.
- Criteria for using the requirements management tool (including an explanation of each attribute and guidance on mandatory versus optional information).
- Criteria to identify requirements (such as numbering scheme or prohibition to reuse a number).
- If tables are used to represent requirements, explanation of how to properly use and identify them (e.g., numbering each row or column).
- If graphics are used to represent or supplement the requirements, description of how to use each type of graphic and any symbols. Additionally, it may be necessary to specify a way to identify and trace each block or symbol. For information on model-based development, see Chapter 14.
- Criteria to distinguish requirements from explanatory material.
- Criteria to document derived requirements (including rationale for the derived requirements to assist the safety personnel in their safety assessment).
- Constraints or limitations on any tools being used.
- Criteria to develop robust requirements.
- Criteria to handle tolerances within the requirements.
- Criteria to use interface control documents and to document requirements that reference them.
- Examples of applied rules and guidelines.

Chapter 6 discusses software requirements. Several of the concepts discussed in Chapter 6 may be suitable to include in the requirements standards.

5.4.2 Software Design Standards

The software design standards define methods, tools, rules, and constraints for developing the software design [1]. In DO-178C, the design includes the low-level requirements and the software architecture.

The design standards are a guide for the team developing the design. The standards explain how to write effective and implementable design, use the design tools, perform traceability, handle derived low-level requirements, and create design data that meets the DO-178C criteria. Design standards may also serve as a training tool for engineers in addition to providing criteria for the design review.

Because design can be presented using different approaches, it is challenging to develop a generic design standard. Many companies have generic design standards, but they are often ineffective for the project-specific needs. Each project should determine their desired methodology and provide instructions to the designers to use the methodology properly. It may be possible to use the company-wide standards as a starting point, but often some kind of tailoring is needed. If the tailoring is minor, it may be feasible to discuss it in the SDP, rather than updating the standard.

Following is a list of items normally included in design standards:

- The criteria from DO-178C Table A-4 (in order to proactively address DO-178C objectives).
- Preferred layout for the design document.
- Criteria for low-level requirements (low-level requirements will have the same quality attributes as high-level requirements; however, low-level requirements are design and will describe *how*, rather than *what*).
- Traceability approach between high-level and low-level requirements.
- Criteria to document derived low-level requirements and their rationale.
- Guidelines for effective architecture, which may include block diagrams, structure charts, state transition diagrams, control and data flow diagrams, flowcharts, call trees, entity-relationship diagrams, etc.
- Naming conventions for modules, which should be consistent with what will be implemented in code.
- Design constraints (e.g., limited level of nested conditions, or prohibition of recursive functions, unconditional branches, reentrant interrupt service routines, and self-modifying instructions).
- Guidelines for robust design.
- Instructions for how to document deactivated code in the design (Chapter 17 provides information on deactivated code).

Chapter 7 provides recommendations for good software design. Several of the concepts discussed in Chapter 7 may be suitable to include in the design standards.

5.4.3 Software Coding Standards

Like the requirements and design standards, the coding standards are used to provide guidelines to the coders. Coding standards explain how to use the specific language properly, constrain some constructs of the language that might not be advisable to use in safety-critical domain, identify naming conventions, explain the use of global data, and develop readable and maintainable code.

Coding standards are relatively commonplace in software development and there are some beneficial industry-wide resources to use when developing coding standards. The Motor Industry Software Reliability Association's C standard (MISRA-C), for example, is an excellent standard to consider as input to C coding standards.

Coding standards are language-specific. If a project uses multiple languages, each language needs to be discussed in the standards. Even assembly language should have guidelines for usage.

Following is a list of items normally included in coding standards:

- The criteria from DO-178C Table A-5 (to proactively address the objectives).
- Approach to document traceability between the code and low-level requirements.
- Module and function or procedure naming conventions.
- Guidelines for using local and global data.
- Guidelines for code readability and maintainability (e.g., use comments and white space, apply abstraction, limit file size, and limit depth of nested conditions).
- Instructions for module structure (including header format and module sections).
- Guidelines for function design (e.g., header format, function layout, unique function identification/name, required nonrecursive and nonreentrant functions, and entry and exit rules).
- Constraints for conditionally compiled code.
- Guidelines for the use of macros.
- Other constraints (e.g., limit or prohibit use of pointers, prohibit reentrant and recursive code).

It is useful to include rationale and examples for each of the guidelines identified in the coding standards. When a coder understands why something is desired or prohibited, it is easier for him or her to apply the guidance.

Chapter 8 provides recommendations for writing safety-critical code. Several of the concepts discussed in Chapter 8 may be suitable to include in the coding standards.

5.5 Tool Qualification Planning

If a tool needs to be qualified, that requires some planning. The planning information may be included in the PSAC; however, for tools with higher tool qualification levels or for tools that will be reused on other projects, additional tool plans may be needed. Tool qualification is discussed in Chapter 13. For now, just be aware that tool qualification requires some special planning.

5.6 Other Plans

In addition to the five plans identified in DO-178C, projects may have some other plans that help with the project management. These may be less formal but are still important to the overall project management. Three such plans are discussed in this section.

5.6.1 Project Management Plan

This plan identifies team members and responsibilities (including names and specific assignments), detailed schedule, status of activities, metrics to gather, etc. It goes into more detail than the SDP and SVP on project organizational details and provides a way to ensure that everything is properly managed.

5.6.2 Requirements Management Plan

This plan is sometimes developed in addition to the SDP and software requirements standards to ensure proper documentation, allocation, and tracing of requirements. It may also help ensure consistent requirements documentation between the systems, hardware, safety, and software teams.

5.6.3 Test Plan

This plan explains the details of the test strategy, including equipment needed, procedures, tools, test case layout, tracing strategy, etc. It may be part of the Software Verification Cases and Procedures document or it may be a separate stand-alone plan. The test plan often includes the test-readiness review checklists and criteria that the team will use to ensure they are ready to begin formal test execution. Test planning is discussed in Chapter 9.

References

1. RTCA DO-178C, *Software Considerations in Airborne Systems and Equipment Certification* (Washington, DC: RTCA, Inc., December 2011).
2. Federal Aviation Administration, *Software Approval Guidelines*, Order 8110.49 (Change 1, September 2011).
3. European Aviation Safety Agency, *Software Aspects of Certification*, Certification Memorandum CM-SWCEH-002 (Issue 1, August 2011).

6

Software Requirements

Acronyms

CASE   computer-aided software engineering
CAST   Certification Authorities Software Team
FAA   Federal Aviation Administration
FAQ   frequently asked question
HLR   high-level requirement
IEEE   Institute of Electrical and Electronic Engineers
LLR   low-level requirement
SWRD   Software Requirements Document
TBD   to be determined

6.1 Introduction

The software requirements are foundational to DO-178C compliance and safety-critical software development. The success or failure of a project depends on the quality of the requirements. As Nancy Leveson writes:

> The vast majority of accidents in which software was involved can be traced to requirements flaws and, more specifically, to incompleteness in the specified and implemented software behavior—that is, incomplete or wrong assumptions about the operation of the controlled system or required operation of the computer and unhandled controlled-system states and environmental conditions. Although coding errors often get the most attention, they have more of an effect on reliability and other qualities than on safety [1].

As the requirements go, so the project goes. The most chaotic projects I've experienced or witnessed started with bad requirements and deteriorated from there. Likewise, the best projects I've seen are ones that spent the effort to get the requirements right. Several fundamentals of effective requirements were elaborated in Chapter 2, when discussing system requirements. Therefore, I encourage you to read or review Section 2.2, if you haven't done so recently. This chapter builds on the Section 2.2 concepts, with emphasis on software requirements rather than system requirements.

Many of the items discussed in this chapter also apply to the system requirements and can augment the material presented in Chapter 2. The line between system requirements and software requirements is often very fuzzy. In general, the software requirements refine the validated system requirements and are used by the software developers to design and implement the software. Also, software requirements identify what the software does, rather than what the system does. When writing the software requirements, errors, deficiencies, and omissions in the system requirements may be identified and should be documented in problem reports and resolved by the systems team.

This chapter examines the importance of good requirements and how to write, verify, and manage requirements. Additionally, the chapter ends with a discussion on prototyping and traceability—two topics closely related to requirements development.

6.2 Defining Requirement

The Institute of Electrical and Electronic Engineers (IEEE) defines a *requirement* as follows [2]:

(1) A condition or capability needed by a user to solve a problem or achieve an objective.
(2) A condition or capability that must be met or possessed by a system or system component to satisfy a contract, standard, specification, or other formally imposed document.
(3) A documented representative of a condition or capability as in 1 or 2.

The DO-178C glossary defines *software requirement, high-level requirements, low-level requirements,* and *derived requirements* as follows [3]:

- *Software requirement*—"A description of what is to be produced by the software given the inputs and constraints. Software requirements include both high-level requirements and low-level requirements."
- *High-level requirements*—"Software requirements developed from analysis of system requirements, safety-related requirements, and system architecture."
- *Low-level requirements*—"Software requirements developed from high-level requirements, derived requirements, and design

constraints from which Source Code can be directly implemented without further information."

- *Derived requirements*—"Requirements produced by the software development processes which (a) are not directly traceable to higher level requirements, and/or (b) specify behavior beyond that specified by the system requirements or the higher level software requirements."

Unlike the IEEE definition, DO-178C defines two levels of software requirements: high-level requirements (HLRs) and low-level requirements (LLRs). This chapter concentrates on HLRs, which will simply be referred to as *requirements* throughout this chapter. DO-178C includes LLRs in the design; therefore, they will be discussed in the next chapter.

In general, requirements are intended to "describe what we're going to have when we're done with a project" [4]. Software requirements normally address the following: functionality, external interfaces, performance, quality attributes (e.g., portability or maintainability), design constraints, safety, and security.

Good requirements do not address design or implementation details, project management details (such as cost, schedule, development methodology), or test details.

6.3 Importance of Good Software Requirements

Let's consider five reasons why requirements are so important to safety-critical software development.

6.3.1 Reason 1: Requirements Are the Foundation for the Software Development

I'm not an architect, but common sense tells me that when building a house, the foundation is extremely important. If the foundation is made of weak or faulty material, is missing sections, or is not level, the house built upon it will have long-term problems. The same is true in software development. If the requirements (the foundation) are weak, it has long-term effects and leads to indescribable problems and difficulties for everyone involved, including the customer.

Reflecting on the projects I've survived over the years, there are some common characteristics that led to the bad requirements. First, the software requirements were developed by inexperienced teams. Second, the teams were pushed to ship something to the customer before they were ready. Third, the system requirements were not validated prior to delivering the software.

These characteristics led to the following common results:

- The customers were not pleased.
- The products suffered from an extremely high number of problem reports.
- The software required at least one complete redesign (in a couple of situations it took two additional iterations).
- The projects were considerably over time and over budget.
- Several leaders were reassigned (right off to other doomed projects) and careers were damaged.

The results of bad requirements lead to what I call the *snowball effect*. The problems and complexity accumulate until the project becomes a huge, unmanageable *snowball*. Figure 6.1 illustrates the problem. When development artifacts are not reviewed and matured before going to the next phase of development, it becomes more difficult and expensive to identify and remove the error(s) later. In some instances it may even become impossible to identify the error, since the root cause is buried so deeply in the data

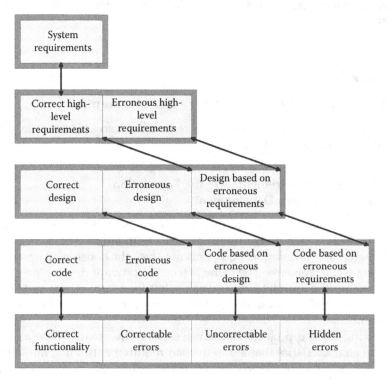

FIGURE 6.1
Effects of bad requirements and inadequate review process.

(the snowball). All development steps are iterative and subject to change as the project progresses; however, building successive development activities on incomplete and erroneous inputs is one of the most common errors and inefficiencies in software engineering.

Requirements will never be perfect the first time, but the goal is to get at least a portion of the requirements as complete and accurate as possible to proceed with design and implementation at an acceptable level of risk. As time progresses, the requirements are updated to add or modify functionality and to make changes based on design and implementation maturity. This iterative approach is the most common way to obtain quality requirements and meet customer needs at the same time. In *Software Requirements* Karl Wiegers writes,

> Iteration is a key to requirements development success. Plan for multiple cycles of exploring requirements, refining high-level requirements into details, and confirming correctness with users. This takes time and it can be frustrating, but it's an intrinsic aspect of dealing with the fuzzy uncertainty of defining a new software product [4].

6.3.2 Reason 2: Good Requirements Save Time and Money

Effective requirements engineering is probably the highest return on investment any project can realize. Multiple studies show that the most expensive errors are those that started in the requirements phase and that the biggest reason for software rework is bad requirements. One study even showed that "requirements errors account for 70 to 85 percent of the rework cost" [5].

A study by the Standish Group gave the following reasons for failures in software projects (several of the reasons are requirements related) [6]:

- Incomplete requirements—13.1%
- Lack of user involvement—12.4%
- Insufficient resources/schedule—10.6%
- Unrealistic expectations—9.9%
- Lack of managerial support—9.3%
- Changing requirements—8.7%
- Poor planning—8.1%
- Software no longer needed—7.4%

Although this study is somewhat dated, the results are consistent with what I see in aviation projects year after year. Good requirements are necessary to obtain successful results. I never cease to be amazed at how many projects do not have time to do a project right the first time, but end up having time and money to do the work two or three times.

6.3.3 Reason 3: Good Requirements Are Essential to Safety

The Federal Aviation Administration (FAA) research report entitled *Requirements Engineering Management Findings Report* states: "Investigators focusing on safety-critical systems have found that requirements errors are most likely to affect the safety of an embedded system than errors introduced during design or implementation" [7]. Without good requirements, it's impossible to satisfy the regulations. The FAA and other regulatory authorities across the world have legally binding regulations requiring that every system on an aircraft show that it meets its intended function under any foreseeable operating condition. This means that the intended functions must be identified and proven. Requirements are the formal method of communicating the safety considerations and the intended functionality.

6.3.4 Reason 4: Good Requirements Are Necessary to Meet the Customer Needs

Without accurate and complete requirements, the customer's expectations will not be met. Requirements are used to communicate between the customer and developer. Poor requirements indicate poor communication which normally leads to a poor product.

One level A project that I reviewed years ago had terrible software requirements and no hope of finishing the level A activities in time to support the aircraft schedule. Because of the software shortcomings, the customer had to redesign part of the aircraft to disable the system in safety-critical operations, add hardware to compensate for what the software was supposed to do, and reduce the reliance on the software to level D. After the initial certification, a new supplier was selected and the original supplier fired. The main reasons for this fiasco were as follows: (1) the software requirements did not comply with the customer's system requirements and (2) when the system requirements were unclear, the software team just improvised rather than asking the customer what they needed. Granted, the system requirements had their issues, but the whole debacle could have been avoided with better communication and requirements development at both companies.

6.3.5 Reason 5: Good Requirements Are Important for Testing

Requirements drive the testing effort. If the requirements are poorly written or incomplete, the following are possible:

- The resulting requirements-based tests may test the wrong thing and/or incompletely test the right thing.
- It may require extensive effort during testing to develop and ferret out the *real* requirements.

- It will be difficult to prove intended functionality.
- It will be challenging or impossible to prove that there is no unintended functionality (i.e., to show that the software does what it is supposed to do and only what it is supposed to do).

Safety hinges upon the ability to show that intended functions are satisfied and that no unintended functionality will impact safety.

6.4 The Software Requirements Engineer

Requirements development is typically performed by one or more requirements engineers (also known as requirements analysts).* Most successful projects have at least two senior requirements engineers who work closely together throughout the project. They work together in developing the requirements, and constantly review each other's work as it progresses. They also perform ongoing *sanity checks* to ensure the requirements are aligned and viable. It's always good to have some junior engineers working with the senior engineers, since the knowledge will be needed for future projects, and organizations need to develop the requirements experts of the future. Organizations should take care when selecting the engineers trusted with requirements development; not everyone is capable of developing and documenting good requirements. The skills needed for an effective requirements engineer are discussed in the following:

Skill 1: Requirements authoring experience. There is no substitute for experience. Someone who has been through multiple projects knows what works and what does not. Of course, it's best if their experience is based on successful projects, but not-so-successful projects can also build experience if the individual is willing to learn from the mistakes.

Skill 2: Teamwork. Since requirements engineers interact with just about everyone on the team, it is important that they be team players who get along well with others. The software requirements engineer will work closely with the systems engineers (and possibly the customers), the designers and coders, the project manager, quality assurance, and testers. Lone Rangers and egomaniacs are difficult to work with and often do not have the best interest of the team in mind.

Skill 3: Listening and observation skills. Requirements engineering often involves discerning subtle clues from the systems engineers or customer. It requires the ability to detect and act on missing elements—it's not just a

---

* A *requirements engineer* will likely have other responsibilities, but this chapter focuses on his or her role as a requirements author; hence, the term *requirements engineer* is used throughout this chapter.

matter of understanding what is there but also determining what is not there and should be there.

Skill 4: Attention to both big picture and details. A requirements engineer must not only be capable of seeing how the software, hardware, and system fit together but also able to visualize and document the details. The requirements engineers needs to be capable of thinking both top-down and bottom-up.

Skill 5: Written communication skills. Clearly, one of the major roles of the requirements engineer is to document the requirements. He or she must be able to write in an organized and clear style. Successful requirements engineers are those who can clearly communicate complex ideas and issues. Additionally, the requirements engineer should be proficient in using graphical techniques to communicate ideas that are difficult to explain in text. Examples of graphical techniques used are tables, flowcharts, data flow diagrams, control flow diagrams, use cases, state diagrams, state charts, and sequence and timing diagrams.

Skill 6: Commitment. The lead requirements engineer(s) should be someone who is committed to seeing the project through to the end. I've seen several projects suffer because the lead engineer took another job, and left with much of the vital knowledge in his head. This is yet another reason why I recommend a team approach to requirements development.

Skill 7: Domain experience. It is advisable to have a requirements engineer who is knowledgeable in the domain. Someone experienced in navigation systems may not know the subtleties required to specify a brake or fuel system. While somewhat subjective, I find that it's best if at least half of the development team has domain experience.

Skill 8: Creative. Brute-forced requirements are typically not the most effective. Good requirements are a mix of art and science. It takes creative thinking (the ability to *think outside the box*) to develop optimal requirements that capture the intended functions and prevent the unintended functions.

Skill 9: Organized. Unless the requirements engineer is organized, his or her output may not effectively communicate what is intended. Also, since the requirements engineers are crucial to the project, they may be frequently distracted with less importance tasks; therefore, they must be able to organize, prioritize, and stay focused.

6.5 Overview of Software Requirements Development

Chapter 5 examined the planning process that precedes the requirements development. There are several requirements aspects to consider during planning, including the requirements methodology and format, use of requirements management tools, development team identification, requirements review process, trace strategy, requirements standards definition, etc. Planning is essential to effective requirements development.

The requirements development effort can be organized into seven activities: (1) gather and analyze input; (2) write requirements; (3) review requirements; (4) baseline, release, and archive requirements; (5) implement requirements; (6) test requirements; and (7) change requirements using the change management process. Figure 6.2 illustrates these activities. Activities 1–3 are discussed in this chapter; activities 4 and 7 are covered in Chapter 10; activity 5 is discussed in Chapters 7 and 8; and activity 6 is addressed in Chapter 9.

Each project goes about the system and software requirements development in a slightly different manner. Table 6.1 summarizes the most common approaches for developing systems and software requirements, along with advantages and disadvantages for each approach. As more systems teams and certification authorities embrace ARP4754A, the state of system requirements should improve. High-quality requirements depend on effective communication and partnership between the stakeholders, including the customer, the systems team (including safety personnel), the hardware team, and the software team.

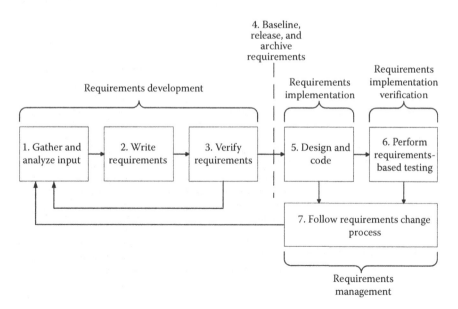

FIGURE 6.2
Requirements development activities.

TABLE 6.1

System and Software Requirements Development Approaches

Approach	Advantages (Pros)	Disadvantages (Cons)
1. *Supplier-driven product*: System and software requirements are developed by the same company.	• Encourages more open communication. • Typically means both systems and software have the same (or very similar) review and release process. • Domain expertise within supplier. • Potential for high reuse and reapplication.	• Customer and systems teams may not specify things in as much detail. • Teams may not be colocated, even though in the same company (e.g., different buildings or different geographical locations). • Oftentimes, the software team is more disciplined in documenting requirements than the systems team (because of the DO-178C requirements).
2. *Customer-driven product*: System requirements are developed by the customer (e.g., aircraft company) and sent to supplier for software implementation.	• Oftentimes, customer requirements are very detailed when they intend to outsource. • Allows customer to select a supplier with appropriate domain expertise, which may not exist in-house.	• There may be a throw-it-over-the-wall mentality. • System requirements may be written at the wrong level, overly prescriptive, and limit the software design options. • Customer may be slow or resistant to updating system requirements, when deficiencies are found. • Supplier may not have the ability to test all requirements.
3. *Combined requirements*: System and software requirements are combined into a single level (by the same company).	• Ensures consistency between system and software requirements. • Potential for less duplication of test effort. • If a simple product, may eliminate artificial requirements layers.	• Makes it challenging to have the appropriate level of granularity and detail in the requirements. • May force the system requirements to be too detailed or may leave the software requirements at an inappropriately high level. • May cause problems with allocation to hardware and software. • Other than a simple product, it is difficult to show the certification authority that all objectives are covered.

6.6 Gathering and Analyzing Input to the Software Requirements

DO-178C assumes that the system requirements given to the software team are fully documented and validated. But, in my experience that is rarely the case. Eventually, the system requirements must be complete, accurate, correct, and consistent; and, the sooner that happens, the better. However, it often falls on the software team to identify requirements deficiencies or ambiguities and to work with the systems team to establish a fully validated set of system requirements.

There is often considerable pressure for the requirements engineer to create the requirements specification without first spending the time analyzing the problem. This results in brute-forced requirements, which are bulky, disorganized, and unclear. Just as an artist takes time to plan a masterpiece, the requirements engineer needs time to gather, analyze, and organize the requirements effort. It's a matter of getting the problem clearly visualized. Once this is done, the actual writing of the specification occurs relatively quickly and requires less rework.

The software requirements engineers typically perform the following gathering and analysis activities in order to develop the software requirements.

6.6.1 Requirements Gathering Activities

Before ever writing a single requirement, the requirements engineers gather data and knowledge in order to comprehend the product they are specifying. Gathering activities include the following:

1. Review and strive to thoroughly understand the system and safety requirements, in whatever state they exist. The software engineers should become intimately familiar with the system requirements. An understanding of the preliminary safety assessment is also necessary to comprehend the safety drivers.

2. Meet with the customers, systems engineers, and domain experts to answer any questions about the system requirements and to fill in missing information.

3. Determine the maturity and completeness of the systems and safety requirements before developing software requirements.

4. Work with systems engineers to make modifications to the system requirements. Before the software team can refine the system requirements into software requirements, the system requirements need to be relatively mature and stable. Some systems teams are quite responsive and work closely with the software team to update the system requirements. However, in many circumstances,

the software team must proactively push the systems engineers for the needed changes.

5. Consider relevant past projects, as well as the problem reports for those projects. Oftentimes, the customer, systems engineers, or the software developers will have some past experience in the domain area. The requirements may not be useable in their previous format, but they can certainly help provide an understanding of what was implemented, what worked, and possibly what did not work.

6. Become fully educated on the requirements standards and certification expectations.

6.6.2 Requirements Analyzing Activities

The analysis process prepares the engineer to write the requirements. Sometimes it is tempting to rush directly into writing the requirements specification. However, without first doing the analysis and considering the problem from multiple views, it may result in significant rework in the future. There will always be some iterations and fine tuning, but the analysis process can help minimize that. Consider the following during requirements analysis:

1. Organize the input gained during the gathering process to be as clear and complete as possible. The requirements engineer must analyze the problem to be solved from multiple views. Oftentimes, use cases* are utilized to consider the problem from the user's perspective.

2. Lay out the framework for the requirements specification task. By identifying what will likely become the table of contents, the requirements can be organized into a logical flow. This framework also serves as a measure of completeness. One common strategy for determining the requirements framework is to list the safety and systems functions to be provided, and then determine what software is needed for each function to operate as intended and to protect against unintended effects.

3. Develop models of the software behavior in order to ensure an understanding of the customer needs. Some of these models will

---

* A use case is a methodology used in analysis to identify, clarify, and organize requirements. Each use case is composed of a set of possible sequences of interactions between the software and its users in a particular environment and related to a certain goal. "A use case can be thought of as a collection of possible scenarios related to a particular goal, indeed, the use case and goal are sometimes considered to be synonymous" [8]. Use cases can be used to organize functional requirements, model user interactions, record scenarios from events to goals, describe the main flow of events, and describe multiple levels of functionality [8].

be refined and integrated into the software requirements and some will just be a vehicle that assists with the requirements development. Some of the models may also be added to the system requirements when deficiencies are noted.

4. In some cases, a prototype may be used to help with the development of the requirements. The requirements analyst may either help develop the prototype or use the prototype to document requirements. A quick prototype can be very beneficial for demonstrating functionality and maturing the requirements details. The risk of a prototype is that it can look impressive on the surface, despite the poor underlying design and code; therefore, project managers or customers insist on using the code from the *working* prototype. When faced with cost and schedule pressures, I've seen several projects abandon their original (and approved) plans and try to use the prototype code for flight test and certification. I have yet to see it work out well; the cost, schedule, and customer relationship all suffer. Section 6.10 provides additional thoughts about prototyping.

6.7 Writing the Software Requirements

The gathering and analysis activities are ongoing. Once sufficient knowledge is gained and the problem sufficiently analyzed, the actual requirements writing commences. Requirements writing involves many parallel and iterative activities. These are presented as *tasks* because they are not serial activities. Six tasks are explained on the next several pages.

6.7.1 Task 1: Determine the Methodology

There are multiple approaches to documenting requirements—all the way from text-only to all-graphical to a combination of text and graphics. Because of the need for traceability and verifiability, many safety-critical software requirements are primarily textual with graphics used to further illustrate the text. However, graphics do play a significant role in the requirements development effort. "Pictures help bridge language and vocabulary barriers …" [4]. At this phase, the graphics focus on the requirements (*what* the software will do) and not design (*how* the software will do it). Many of the graphics may be further elaborated in the design phase, but at the requirements phase they should strive to be implementation-free. Developers may opt to document some design concepts as they write the requirements, but those should be notes for the designer and not part of the software requirements specification.

Just a brief word of caution, be careful to not rely solely on pictures, which are difficult to test. When using graphics to describe the requirements, testability of the requirements should be constantly considered.

Some examples of graphical techniques that enhance the textual descriptions include the following [9]:

- *Context or use case diagrams*—illustrate interfaces with external entities. The details of the interfaces are typically identified in the interface control specification.

- *A high-level data dictionary*—defines data that will flow between processes. The data dictionary will be further refined during the design phase.

- *Entity-relationship or class diagrams*—show logical relationship between entities.

- *State-transition diagrams*—show the transition between states within the software. Each state is usually described in textual format.

- *Sequence diagrams*—show the sequence of events during execution, as well as some timing information.

- *Logic diagrams and/or decision tables*—identify logic decisions of functional elements.

- *Flowcharts or activity diagrams*—identify step-by-step flow and decisions.

- *Graphical user interfaces*—clarify relationships that may be difficult to describe in text.

The model-based development approach strives to increase the graphical representation of the requirements through the use of models. However, to date, even models require some textual description. Model-based development is examined in Chapter 14.

The technique selected should address the connection between textual and graphical representations. The textual requirements normally provide the context and reference to the graphical figures. This helps with traceability and completeness, and facilitates testing. Sometimes the graphics are provided as *reference only* to support the requirements. If this is the case, it should be clearly stated.

Another aspect of methodology to consider before diving into the requirements writing is whether to use computer-aided software engineering (CASE) tools and requirements templates, since these may impact the methodology.

Ideally, the application of the selected methodology is explained and illustrated in the requirements standards. The standards help guide the requirements writers, and ensure that everyone is following the same approach. If there are multiple developers involved in the requirements writing, an example of the methodology and layout should be documented, so that everyone is applying it in the same way. The more examples and details provided, the better.

6.7.2 Task 2: Determine the Software Requirements Document Layout

The end result of the software requirements documentation process is the Software Requirements Document (SWRD). The SWRD

> states the functions and capabilities that a software system must provide and the constraints that it must respect. The SWRS [SWRD] is the basis for all subsequent project planning, design, and coding, as well as the foundation for system testing and user documentation. It should describe as completely as necessary the [software] system's behaviors under various conditions. It should not contain design, construction, testing, or project management details other than known design and implementation constraints [4].*

The SWRD should comprehensively explain the software functionality and limitations; it should not leave room for assumptions. If some functionality or quality does not appear in the SWRD, no one should expect it to magically appear in the end product [4].

Early in the requirements development process, the general layout of the SWRD should be determined. It may be modified later, but it's valuable to have a general outline or framework to start with, especially if there are multiple developers involved. The outline or template helps keep everyone focused on their area's objectives and able to understand what will be covered elsewhere. In order to improve readability and usability of the SWRD, the following suggestions are provided:

- Include a table of contents with subsections.
- Provide an overview of the document layout, including a brief summary of each section and the relationship between sections.
- Define key terms that will be used throughout the requirements and use them consistently. This may include such things as coordinate systems and external feature nomenclature. If the requirements will be implemented or verified by anyone unfamiliar with the language or domain, this is a critical element of the SWRD.
- Provide a complete and accurate list of acronyms.
- Explain the requirements grouping in the document (a graphic, such as a context diagram, might be useful for this).
- Organize the document logically and use section/subsection labels and numbers (typically requirements are organized by features or key functionality).

---

* Added brackets for clarification.

- Identify the environment in which the software will operate.
- Use white space throughout to help readability.
- Use bold, italics, and underlining for emphasis and use them consistently throughout. Be sure to explain the meaning of any conventions, text styles, etc.
- Identify functional requirements, nonfunctional or nonbehavioral requirements, and external interfaces.
- Include any constraints that will apply (e.g., tool constraints, language constraints, compatibility constraints, hardware limitations, or interface conventions).
- Number and label figures and tables, and clearly reference them from the textual requirements.
- Provide cross-references within the specification and to other data as needed.

6.7.3 Task 3: Divide Software Functionality into Subsystems and/or Features

It is important to decompose the software into manageable groups, which make sense and can be easily integrated. In most cases, a context diagram or use case is used to provide the high-level view of the requirements organization.

For larger systems, the software may be divided into subsystems. For smaller systems and subsystems, the software is often further divided into features, with each feature containing specific functionality.

As noted earlier, one way to organize the functions is to work with the safety and systems engineers to define the safety and systems functions to be provided. This input is then used to determine what software is needed for each function to operate as intended, as well as to determine what protections are needed to prevent unintended effects.

There are other ways to organize the requirements. Regardless of the approach selected, reuse and minimized impact of change are normally important characteristics to consider when dividing the functionality.

6.7.4 Task 4: Determine Requirements Priorities

Because projects are often on a tight schedule and are developed using an iterative or spiral life cycle model, it may be necessary to prioritize which requirements must be defined and implemented first. Priorities should be coordinated with the systems team and the customers. After the enabling software (such as boot, execution, and inputs/outputs), software functions that are critical to system functionality or safety or that are highly complex should have the highest priority. Oftentimes, the priorities are based on the urgency defined by the customer and the importance to the overall system functionality. In order to properly prioritize, it is sometimes helpful to divide the subsystems, features, and/or functions into four groups: (1) urgent/important (high priority), (2) not urgent/important (medium priority), (3) urgent/not important (low priority), and (4) not urgent/not important (may not be necessary to implement). For large and complex projects, there are many factors to assess when determining the priority. In general, it is preferable to keep the prioritization process as simple and politically free as possible. The prioritization of subsystems, features, and/or functions should be identified in the project management plan. Keep in mind that priorities may need to be readjusted as the project progresses, depending on the customer feedback and project needs.

6.7.5 A Brief Detour (Not a Task): Slippery Slopes to Avoid

Before discussing the requirements documenting task, let's take a quick detour to discuss a few slippery slopes that many unsuccessful projects seem to attempt. These are mentioned here because projects sometimes start down one or more of these slippery slopes instead of focusing on the requirements. Once a project starts down a slope, it's hard to exit it and refocus on the requirements.

6.7.5.1 *Slippery Slope #1: Going to Design Too Quickly*

One of the most challenging parts of authoring requirements is to avoid designing. By nature, engineers tend to be problem solvers and want to go straight to design. The schedule pressures also force engineers to go to design too quickly—before they've really solidified what it is they're trying to build. However, going to design too quickly can lead to a compromised implementation, since the best implementation is often not apparent until the problem is fully defined. Requirements engineers may consider several potential solutions to highlight the trade-offs. Thinking through options frequently helps mature the requirements and identifies what is really needed in the software. However, the implementation details do not belong in the requirements themselves.

In order to avoid this slippery slope, I suggest that engineers do the following:

1. Clearly define the division between requirements (*what*) and design (*how*) in the standards.

2. Ensure that the requirements process allows for notes or comments. This annotation provides valuable information that will convey the intent without burdening the high-level requirements with design details.

3. Make sure that the process allows for documenting design ideas. Having some potential solutions identified can jump-start the design process.

6.7.5.2 Slippery Slope #2: One Level of Requirements

DO-178C identifies the need for both high-level requirements and low-level requirements. The high-level requirements are documented in the SWRD and the low-level requirements are part of the design. Developers frequently attempt to identify one level of software requirements that they directly code from (i.e., they combine high-level and low-level requirements into a single level of requirements). Good software engineering practices and experience continually warn against this. Requirements and design are separate and distinct processes. Out of the numerous projects I've assessed, I've rarely seen the successful combination of high-level requirements and low-level requirements into one level. Therefore, I recommend that this approach be avoided, except in well-justified cases. I find it more beneficial to document implementation ideas in a draft design document while writing the requirements (i.e., work on the requirements and design in parallel rather than combining them). Some believe modeling will remove the need for two levels of requirements; however, as will be discussed in Chapter 14, this is not the case. I caution anyone who wants to have one level of requirements in order to get to implementation faster. It might look good at first, but tends to fail in the end. The testing and verification phase is one area where issues arise. For the more critical software (levels A and B), it is difficult to get structural coverage when the requirements are not sufficiently detailed (one level of requirements may not have enough details for full structural coverage). Also, for level D projects where the low-level testing is not required, combining the requirements can result in more testing (since merging high-level and low-level requirements tends to result in more detailed requirements than the traditional high-level requirements).

DO-248C's FAQ #81, entitled "What aspects should be considered when there is only one level of requirements (or if high-level requirements and low-level requirements are merged)?" and the Certification Authorities Software Team (CAST)* paper CAST-15 entitled "Merging High-Level and Low-Level Requirements" warn against merging software requirements into a single level [10,11]. There are a few projects where only one level of software requirements is needed (e.g., a low-level functionality like parts of an operating system or a math library function), but they are in the minority.

6.7.5.3 Slippery Slope #3: Going Straight to Code

Over the last few years there has been an increased tendency for projects to start coding right away. They might have some concepts and a few requirements documented, but when pressured to get functional software in the

---

* CAST is a team of international certification authorities who strive to harmonize their positions on airborne software and aircraft electronic hardware in CAST papers.

field, they just implement it the best they can. Later, the developers are pressured to use the kludged or *ad hoc* code for production. The code usually doesn't align with the requirements that are eventually developed, it has none of the design decisions documented, and it often does not consider off-nominal conditions. Furthermore, the prototyped code is normally brute-forced and is not the best and safest solution. I have seen months and even years added to the project schedule, as the project tries to reverse engineer the design decisions and missing requirements from the prototype code. Oftentimes, during the reverse engineering effort, the project discovers that the code wasn't complete or robust, let alone safe.

Let's now exit the slippery slopes and consider what is involved in documenting the high-level software requirements (hopefully these will help teams avoid the slopes).

6.7.6 Task 5: Document the Requirements

One of the challenges of writing requirements is determining the level of detail. Sometimes the system requirements are very detailed, so it forces a lower level of detail in the software requirements than preferred. At other times, the system requirements are far too vague and require additional work and decomposition by the software requirements authors. The level of detail is a judgment call; however, the requirements should sufficiently explain what the software will do but not get into the implementation details. When writing the software high-level requirements, it is important to remember that the design layer (which includes the software low-level requirements) is still to occur. The software high-level requirements should provide adequate detail for the design effort, but not get into the design.

6.7.6.1 Document Functional Requirements

The majority of the requirements in the SWRD are functional requirements (also known as, behavioral requirements). Functional requirements

> define precisely what inputs are expected by the software, what outputs will be generated by the software, and the details of relationships that exist between those inputs and outputs. In short, behavioral requirements describe all aspects of interfaces between the software and its environment (that is, hardware, humans, and other software) [12].

Basically, the functional requirements define what the software does. As previously discussed, they are generally organized by subsystem or feature and are documented using a combination of natural language text and graphics.

The following concepts should be considered when documenting the functional requirements:

- Organize the requirements into logical groupings.
- Aim for understandability by customers and users who are not software experts.
- Document requirements that are clear to designers (use comments or notes to expand potentially challenging areas).
- Document requirements with focus on external software behavior, not the internal behavior (save that for design).
- Document requirements that can be tested.
- Use an approach that can be modified (including requirements numbering and organization).
- Identify the source of the requirements (see Sections 6.7.6.6 and 6.11 for more on traceability).
- Use text and graphics consistently.
- Identify each requirement (see Section 6.7.6.4 on unique identification).
- Minimize redundancy. The probability of discrepancies increases each time the requirements are restated.
- Follow the requirements standards and agreed upon techniques and template. If a standard, technique, or template does not meet the specific need, determine if an update or waiver to plans, standards, or procedures is needed.
- If a requirements management tool is used, follow the agreed upon format and complete all appropriate fields proactively (Section 6.9 discusses this more).
- Implement characteristics of good requirements (see Section 6.7.6.9).
- Coordinate with teammates to get early feedback and ensure consistency across the team.
- Identify safety requirements. Most companies find it beneficial to identify requirements that directly contribute to safety. These are requirements with a direct tie to the safety assessment and that support ARP4754A compliance.
- Document derived requirements and include the reason for their existence (i.e., rationale or justification). (Derived requirements are discussed later.)
- Include and identify robustness requirements. Robustness is "the degree to which a system continues to function properly when confronted with invalid inputs, defects in connected software or hardware components, or unexpected operating conditions" [4].

For each requirement, consider whether there are any potential abnormal conditions (e.g., invalid inputs or invalid states) and ensure that there is a defined behavior for each of the conditions.

6.7.6.2 Document Nonfunctional Requirements

Nonfunctional (nonbehavioral) requirements are those that "define the overall qualities or attributes to be exhibited by the resulting software" [12]. Even though they do not describe functionality, it is important to document these requirements, since they are expected by the customer and they drive design decisions. These requirements are important because they explain how well the product will work. They include characteristics such as speed of operation, ease of use, failure rates and responses, and abnormal conditions handling [4]. Essentially, the nonfunctional requirements include constraints that the designers must understand.

When nonfunctional requirements vary by feature or function, they should be specified with the feature or function. When the nonfunctional requirements apply across all features or functions, then they are usually included in a separate section. Nonfunctional requirements should be identified as such, either by a separate section of the requirements or with some kind of attribute, since they will usually not trace down to code and will impact the test strategy. Nonfunctional requirements still need to be verified, but they are often verified by analysis or inspection rather than test, since they may not exhibit testable functionality.

Following are some of the common requirements types that are identified as nonfunctional requirements:

1. *Performance requirements* are probably the most common class of nonfunctional requirements. They include information to help designers, for example, response times, computational accuracy, timing expectations, memory requirements, and throughput.
2. *Safety requirements* that might not be part of functionality are documented as nonfunctional requirements. Examples include the following:
   a. *Data protection*—preventing loss of or corruption of data.
   b. *Safety regulations*—specifying specific regulatory guidance or rules that must be satisfied.
   c. *Availability*—defining the time the software will be available and fully operational.
   d. *Reliability*—identifying when some aspect of the software is used to support system reliability.
   e. *Safety margins*—defining margins or tolerances needed to support safety, for example, timing or memory margin requirements.

    f. *Partitioning*—ensuring that partitioning integrity is maintained.

    g. *Degradation of service*—explaining how software will *degrade gracefully* or act in the presence of a failure.

    h. *Robustness*—identifying how the software will respond in the presence of abnormal conditions.

    i. *Integrity*—protecting data from corruption or improper execution.

    j. *Latency*—protecting against latent failures.

3. *Security requirements* may be needed to support safety, ensure system reliability, or protect proprietary information.

4. *Efficiency requirements* are a measure of how well the system utilizes processor capacity, memory, or communication [4]. They are closely related to performance requirements but may identify other important characteristics of the software.

5. *Usability requirements* define what characteristics are needed to make the software user-friendly; this includes human factors considerations.

6. *Maintainability requirements* describe the need to easily modify or correct the software. This includes maintainability during the initial development, during integration, and after the software has gone into production.

7. *Portability requirements* address the need to easily move the software to other environments or target computers.

8. *Reusability requirements* define the need to use the software for other applications or systems.

9. *Testability requirements* describe what capabilities need to be built into the software for test, including systems or software development testing, integration testing, customer testing, aircraft testing, and production testing.

10. *Interoperability requirements* document how well the software can exchange data with other components. Specific interoperability standards may apply.

11. *Flexibility requirements* describe the need to easily add new functionality to the software, both during the initial development and over the life of the product.

6.7.6.3 Document Interfaces

Interfaces include user interfaces (e.g., in display systems), hardware interfaces (as in communication protocol for a specific device), software interfaces (such as an application programmer interface or library interface), and communication interfaces (e.g., when using a databus or network).

Requirements for interfaces with hardware, software, and databases need to be documented. Oftentimes, the SWRD references an interface control document. In some cases, explicit SWRD requirements, independent standards, or a data dictionary describe the data and control interfaces. The interfaces should be documented in a manner to support the data and control coupling analysis, which is discussed in Chapter 9.

Any interface documents referenced from the requirements need to be under configuration control, since they affect the requirements, testing, system operations, and software maintenance.

6.7.6.4 Uniquely Identify Each Requirement

Each requirement should have a unique tag (also known as a number, label, or identifier). Most organizations use *shall* to identify requirements. Each *shall* identifies one requirement and has a tag. Using this approach helps the requirements engineer distinguish between what is actually required and what is commentary or support information.

Some tools automatically assign requirements tags, while others allow manual assignment of the tag. The identification approach should be documented in the standards and closely followed. Once a tag has been used, it should not be reassigned, even if the requirement is deleted. Additionally, it's important to ensure that each tag only has one requirement. That is, don't lump multiple requirements together, since this leads to ambiguity and makes it difficult to confirm test completeness.

6.7.6.5 Document Rationale

It is a good practice to include rationale with the requirements, since it can improve the quality of a requirement, reduce the time required to understand a requirement, improve accuracy, reduce time during maintenance, and be useful to educate engineers on the software functionality. The process of writing the rationale not only improves the reader's comprehension of the requirements but also helps the authors write better requirements. The FAA *Requirements Engineering Management Handbook* states:

> Coming up with the rationale for a bad requirement or assumption can be difficult. Forcing the specifier to think about why the requirement is necessary or why the assumption is being made will often improve the quality of the requirement... Requirements document what a system will do. Design documents how the system will do it. Rationale documents why a requirement exists or why it is written the way it is. Rationale should be included whenever there is something in the requirement that may not be obvious to the reader or that might help the reader to understand why the requirement exists [13].

The handbook goes on to provide recommendations for writing the rationale, including the following [13]:

- Provide rationale throughout the requirements development to explain why the requirement is needed and why specific values are included.

- Avoid specifying requirements in the rationale. If the information in the rationale is essential to the required system behavior, it should be part of the requirements and not the rationale.

- Provide rationale when the reason for a requirement's existence is not obvious.

- Include rationale for environmental assumptions upon which the system depends.

- Provide rationale for values and ranges in each requirement.

- Keep each rationale short and relevant to the requirement being explained.

- Capture rationale as soon as possible to avoid losing the train of thought.

6.7.6.6 Trace Requirements to Their Source

Each requirement should trace to one or more parent requirements (the higher level requirements from which the requirement was decomposed). Requirements engineers must ensure that each system requirement allocated to software is fully implemented by the software requirements that trace to it (i.e., its children).

Traceability should be documented as the requirements are written. It is virtually impossible to go back and correct the tracing later. Many requirements management tools provide the capability to include trace data, but the tools do not automatically know the trace relationships. Developers must be disciplined at tracing the requirements as they are developed.

In addition to tracing up, the requirements should be written in such a way that they are traceable down to the low-level requirements and test cases. Section 6.11 provides additional information on traceability.

6.7.6.7 Identify Uncertainties and Assumptions

It is common to have unknown information during requirements definition. Those can be identified with a *TBD* (to be determined) or some other clear notation. It's a good practice to include a note or footnote to identify who is responsible for addressing the *TBD* and when it will be completed. All *TBD*s should be addressed before the requirements are formally reviewed and before implementation. Likewise, any *assumptions* should be documented, so that they can be confirmed and verified by the appropriate teams (e.g., systems, hardware, verification, or safety).

6.7.6.8 Start a Data Dictionary

Some projects try to avoid having a data dictionary because it can be tedious to maintain. However, a data dictionary is extremely valuable for data-intensive systems. Most of the data dictionary is completed during design. However, it is beneficial to start documenting shared data (including data meaning, type, length, format, etc.) during requirements definition. The data dictionary helps with integration and overall consistency. It also helps prevent errors caused by inconsistent understanding of data [4]. Depending on the project details, the data dictionary and interface control document may be integrated.

6.7.6.9 Implement Characteristics of Good Requirements

Chapter 2 identified the characteristics of good system requirements. Software requirements should have those same characteristics, which include atomic, complete, concise, consistent, correct, implementation-free, necessary, traceable, unambiguous, verifiable, and viable. Additional suggestions for writing high-quality software requirements are provided here:

- Use concise and complete sentences with correct grammar and spelling.
- Use one *shall* for each requirement.
- Use an active voice.
- Emphasize important items using graphics, bolding, sequencing, white space, or some other method.
- Use terms consistently as identified in the SWRD glossary or definitions section.
- Avoid ambiguous terms. Examples of ambiguous terms include the following: *as a goal, to the extent practical, modular, achievable, sufficient, timely, user-friendly*, etc. If such terms are used, they need to be quantified.
- Write requirements at the appropriate level of granularity. Usually the appropriate level is one that can be tested by one or just a few tests.
- Keep the requirements at a consistent level of granularity or detail.
- Minimize or avoid the use of words that indicate multiple requirements, such as *unless* or *except*.
- Avoid using *and/or* or using the slash (/) to separate two words, since this can be ambiguous.

- Use pronouns cautiously (e.g., *it* or *they*). It is typically better to repeat the noun.
- Avoid *i.e.* (which means *that is*) and *e.g.* (which means *for example*) since many people get the meanings confused.
- Include rationale and background for requirements in a notes or comment field (see Section 6.7.6.5). There is nothing better than a good comment to get inside the author's head.
- Avoid negative requirements, since they are difficult to verify.
- Ensure that the requirements fully define the functionality by looking for omissions (i.e., things that are not specified that should be).
- Build robustness into the requirements by thinking through how the software will respond to abnormal inputs.
- Avoid words that sound alike or similar.
- Use adverbs ending in *-ly* cautiously (e.g., reasonably, quickly, significantly, and occasionally), since they may be ambiguous.

Leveson emphasizes the importance of complete requirements, when she writes:

> The most important property of the requirements specification with respect to safety is completeness or lack of ambiguity. The desired software behavior must have been specified in sufficient detail to distinguish it from any undesired program that might be designed. If a requirements document contains insufficient information for the designers to distinguish between observably distinct behavioral patterns that represent desired and undesired (or safe and unsafe) behavior, then the specific is ambiguous or incomplete [1].

6.7.7 Task 6: Provide Feedback on the System Requirements

Writing software requirements involves scrutiny of the system requirements. The software team often finds erroneous, missing, or conflicting system requirements. Any issues found with the system requirements should be documented in a problem report, communicated to the systems team, and followed up to confirm that action is taken. In many programs, the software team assumes that the systems team fixed the system requirements based on verbal or e-mail feedback; however, in the final throes of certification, it is discovered that the system requirements were not updated. To avoid this issue, the software team should proactively ensure that the system requirements are updated by writing problem reports against the system requirements and following up on each problem report. Failure to follow through on issues could lead to an inconsistency between system requirements and software functionality that may stall the certification process, since the requirements disconnect is considered a DO-178C noncompliance.

6.8 Verifying (Reviewing) Requirements

Once the requirements are mature and stable, they are verified. This normally occurs by performing one or more peer reviews. The purpose of the review process is to catch errors before they are implemented; therefore, it is one of the most important and valuable activities of safety-critical software development. When done properly, reviews can prevent errors, save significant time, and reduce cost. The peer review is performed by a team of one or more reviewers.

To optimize the review process, I recommend two stages of peer reviews: informal and formal. The informal stage is first and helps to mature requirements as quickly as possible. In fact, as previously mentioned, I support the concept of team development, where at least two developers jointly develop the requirements and continuously consult each other and check each other's work. The goal is to perform frequent informal reviews early on, in order to minimize issues discovered during the formal peer review.

During the formal requirements review(s), reviewers use a checklist (which is typically included in the software verification plan or requirements standards). Based on DO-178C, the following items (as a minimum) are usually included in the checklist and assessed during the requirements review [3]:*

- Entry criteria identified in the plans for the review have been satisfied.† In most cases, this requires release of the system requirements, release of the software requirements standards, release of the software development and verification plans, and configuration control of the software requirements.

- High-level software requirements comply with the system requirements. This ensures that the high-level requirements fully implement the system requirements allocated to software.

- High-level software requirements trace to the system requirements. This is a bidirectional trace: all system-level requirements allocated to software should have high-level requirements implementing the system requirements, and all high-level requirements (except for derived requirements) should trace to system-level requirements. Section 6.11 provides additional thoughts on traceability.

- High-level software requirements are accurate, unambiguous, consistent, and complete. This includes ensuring that inputs and outputs are clearly defined and in quantitative form (including units of measure, range, scaling, accuracy, and frequency of arrival), both

---

* These are based on DO-178C, sections 6.3.1 and 11.9, unless otherwise noted.
† Per DO-178C section 8.1.c.

normal and abnormal conditions are addressed, any diagrams are accurate and clearly labeled, etc.

- High-level software requirements conform to the requirements standards. As recommended earlier, the requirements standards should include the attributes of good requirements mentioned earlier. Chapter 5 discussed the requirements standards.
- High-level software requirements are verifiable, for example, requirements that involve measurement include tolerances, only one requirement per identifier/tag, quantifiable terms, and no negative requirements.
- High-level software requirements are uniquely identified. As noted earlier, each requirement should have a *shall* and a unique identifier/tag.
- High-level requirements are compatible with the target computer. The purpose is to ensure that high-level requirements are consistent with the target computer's hardware/software features—especially with respect to response times and input/output hardware. Oftentimes, this is more applicable during design reviews than during requirements reviews.
- Proposed algorithms, especially in the area of discontinuities, have been examined to ensure their accuracy and behavior.
- Derived requirements are appropriate, properly justified, and have been provided to the safety team.
- Functional and operational requirements are documented for each mode of operation.
- High-level requirements include performance criteria, for example, precision and accuracy.
- High-level requirements include timing requirements and constraints.
- High-level requirements include memory size constraints.
- High-level requirements include hardware and software interfaces, such as protocol, formats, and frequency of inputs and outputs.
- High-level requirements include failure detection and safety monitoring requirements.
- High-level requirements include partitioning requirements to specify how the software components interact with each other and the software levels of each partition.

During the formal reviews, the requirements may be verified as a whole or as functional groups. If reviews are divided by functionality, there still needs to be a review that examines the consolidated groups for consistency and cohesion.

For a review to be effective, the right reviewers should be assembled. Reviewers should include qualified technical personnel, including software developers, systems engineers, test engineers, safety personnel, software quality assurance engineer, and certification liaison personnel.* It is imperative that every reviewer read and thoroughly understand the checklist items before carrying out the review. If the reviewers are unfamiliar with the checklist, training should be provided with guidance and examples for each of the checklist items.

The comments from the formal review should be documented, categorized (for instance, significant issue, minor issue, editorial comment, duplicate comment, no change), and dispositioned. The commenter should agree with the action taken before the review is closed. The requirements checklist should also be successfully completed prior to closure of the review. More recommendations for the peer review process are included in the following.

6.8.1 Peer Review Recommended Practices

Although not required, most projects use a formal peer review process in order to verify their plans, requirements, design, code, verification cases and procedures, verification reports, configuration index, accomplishment summary, and other key life cycle data. A team with the right members can often find errors that an individual might miss. The same peer review process can be used across the multiple life cycle data items (it isn't limited to requirements). In other words, the review process can be standardized, so that only the checklists, data to be reviewed, and reviewers change for the actual reviews. This section identifies some of the recommended practices to integrate into the peer review process:

- Assign a moderator or technical lead to schedule the review, provide the data, make assignments, gather and consolidate the review comments, moderate the review meeting, ensure that the review checklist is completed, make certain that all review comments are addressed prior to closure of the review, etc.
- Ensure that the data to be reviewed is under configuration management. It may be informal or developmental configuration management, but it should be controlled and the version identified in the peer review records.
- Identify the data to be reviewed and the associated versions in the peer review record.
- Identify the date of the peer review, reviewers invited, reviewers who provided comments, and amount of time each reviewer spent

---

* Normally quality and certification personnel participate at their discretion. Additionally, safety may review the requirements without participating in the review meeting. The mandatory reviewers should be clearly specified in the plans.

on the review. The reviewer information is important to provide evidence of independence and due diligence.

- Involve the customer, as needed or as required by contract or procedures.

- Use a checklist for the review and ensure that all reviewers have been trained on the checklist and the appropriate standards for the data under review. The checklists are typically included or referenced in the approved plans or standards.

- Provide the review package (including data to be evaluated with line or section numbers, checklist, review form, and any other data needed for the review) to the reviewers, and identify required and optional reviewers.

- Allocate responsibilities for each reviewer (e.g., one person to review traceability, one person to review compliance to standards, and so on). Ensure that required reviewers are covering all aspects of the checklist, reviewers perform their tasks, and reviewers are qualified for the task assigned. If a required reviewer is gone or cannot carry out his or her task, someone else equally qualified may need to perform the review, or the review may need to be rescheduled.

- Provide instructions to the reviewers (e.g., identify comment due dates, meeting dates, roles, focus areas, open issues, file locations, reference documents).

- Give reviewers adequate notice and time to perform the review. If a required reviewer needs more time, reschedule the review.

- Ensure that the proper level of independence is achieved. The DO-178C Annex A tables identify by software level when independence is required. Chapter 10 discusses verification independence.

- Use qualified reviewers. Key technical reviewers include those who will use the data (e.g., tester and designer) and one or more independent developers (when independence is required). As noted earlier, for requirements reviews, it's recommended that systems and safety personnel be involved. The review will only be as good as the people performing it, so it pays off over the life of a project to use the best and most qualified people for technical roles. Junior engineers can learn by performing support roles.

- Invite software quality assurance and certification liaison personnel, as well as any other support personnel needed.

- Keep the team size to a reasonable number. This is subjective and tends to vary depending on the software level and the significance of the data being reviewed.

- Provide a method for reviewers to document comments (a spreadsheet or comment tool is typical). The following data are normally entered by the reviewer: reviewer name, document identification and version, section or line number of document, comment number, comment, and comment classification (significant, minor, editorial, etc.).

- Schedule a meeting to discuss nontrivial comments or questions. Some companies prefer to limit the number of meetings and only use them for controversial topics. If meetings are not held, ensure there is a way to obtain agreement from all reviewers on the necessary actions. In my experience, a brief meeting to discuss technical issues is far more effective and efficient than ongoing e-mail threads.

- Assuming there is a meeting, give the author of the data time to review and propose a response for each comment prior to the meeting. The meeting time should focus on the nontrivial items that require face-to-face interaction.

- If a team is spread out geographically, use electronic networking features (e.g., WebEx, NetMeeting, or Live Meeting) and teleconferencing to involve the appropriate people.

- Limit the amount of time discussing issues. Some companies impose a 2-minute rule; any discussions that require more than 2 minutes are tabled for future discussion. If an item cannot be resolved in the time allocated, set up a follow-on meeting with the right stakeholders. The moderator helps to keep the discussions on track and on schedule.

- Identify a process to discuss controversial issues, such as an escalation path, a lead arbitrator, or a product control board.

- Complete the review checklist. Several potential approaches may be used: the checklist(s) may be completed by each team member for their part, by the team during the peer review meeting, or by a qualified reviewer. It is typically not possible to successfully complete the checklist until all of the review comments are addressed.

- Ensure that all issues are addressed and closed, before the review is closed. If an issue needs to be addressed but cannot be addressed prior to closing the review, a problem report should be generated. The problem report number should be included in the peer review records to ensure that the issue is eventually addressed or properly dispositioned.

- Break large documents into smaller packages and review high-risk areas first. Once all of the individual packages are reviewed, an experienced engineer or team should perform an integration review to make sure all of the packages are consistent and accurate. That is, look at the data together.

- Have an organized approach to store and retrieve review records and checklists, since they are certification evidence.

Here are some common issues that arise during the actual implementation of a peer review process, which can be avoided by proper management of the peer reviews:

- Considering the activity as a *check mark*, rather than a tool to work out technical issues early, add value to the end product, and save time and money over the life of the project.
- Not giving reviewers time to thoroughly review the data.
- Having an overly large review team.
- Not using qualified and well-trained reviewers.
- Not closing comments before proceeding to the next phase.
- Not completing the required checklist completely or promptly.

6.9 Managing Requirements

6.9.1 Basics of Requirements Management

A vital part of software development is requirements management. No matter how thorough the planning and the diligence in writing the requirements, change will happen. An organized requirements management process is essential to managing the inevitable change. Requirements management includes "all activities to maintain the integrity, accuracy, and currency of the requirements agreement as the project progresses" [4].

In order to manage the requirements, the following should be done:

- *Develop requirements to be modifiable.* As previously mentioned, modifiability is a characteristic of good requirements. Modifiable requirements are well organized, at the appropriate level of granularity, implementation-free, clearly identified, and traceable.
- *Baseline the requirements.* Both the functional and nonfunctional requirements should be baselined. The baseline typically occurs after the requirements have been through the peer review. For large projects, it is useful to have version control of the individual requirements as well as sections of or the entire SWRD.
- *Manage all changes to the baseline.* This typically occurs through the problem reporting process and change control board. Changes to requirements are identified, approved by the change control board, implemented, and rereviewed.
- *Update requirements using approved process.* The updates to the requirements should use the same requirements process defined in the plans. That is, follow the standards, implement quality attributes,

perform reviews, etc. Some companies diligently follow the process the first time around but get lax during the updates. Because of this tendency, external auditors and quality assurance engineers tend to look closely at the thoroughness of modifications.

- *Rereview the requirements.* Once changes are implemented, the changes and requirements affected by the changes should be re-reviewed. If multiple requirements are changed, a team review may be appropriate. If the number of requirements changed or impacted is small and straightforward, the review may be performed by an individual. The appropriate level of independence is still needed.

- *Track status.* The status of each requirement should be tracked. The typical states of the requirements changes are: proposed, approved, implemented, verified, deleted, or rejected [9]. Oftentimes, the status is managed through the problem reporting process in order to avoid having two status systems. The problem reporting process typically includes the following states: open (requirements change has been proposed), in-work (change has been approved), implemented (change has been made), verified (change has been reviewed), cancelled (change not approved), and closed (change fully implemented, reviewed, and under configuration management).

Change management is further discussed in Chapter 10.

6.9.2 Requirements Management Tools

Most companies use a commercially available requirements management tool to document and help manage their requirements; however, some do have their own homegrown tool. Customers may mandate a specific tool in order to promote a consistent requirements management approach at all hierarchical levels. Whatever requirements management tool is selected, it should have the capability to do the following, as a minimum:

- Easily add requirements attributes or fields
- Export to a readable document format
- Accommodate graphics (e.g., tables, flowcharts, user interface graphics)
- Baseline requirements
- Add or delete requirements
- Handle multiple users in multiple geographic locations
- Document comments or rationale for requirements
- Trace up and down
- Generate trace reports

- Protect from unauthorized change (e.g., password)
- Be backed up
- Reorder requirements without renumbering them
- Manage multiple levels of requirements (such as system, high-level software, low-level software)
- Add levels of requirements if needed
- Support multiple development programs

Typical requirements fields or attributes that are included in the requirements management tool are as follows:

- *Requirements identification*—a unique identifier/tag for the requirement.
- *Requirements applicability*—if multiple projects are involved, some requirements may or may not apply.
- *Requirements description*—states the requirement.
- *Requirements comment*—explains important things about the requirement such as rationale and related requirements. For derived requirements, this includes rationale for why the derived requirement is needed.
- *Status*—to identify the status of each requirement (such as approved by change control board, in-work, implemented, verified, deleted, or rejected).
- *Change authority*—identifies the problem report, change request number, etc. used to authorize the requirement implementation or change.
- *Trace data*—documents trace up to parent requirements, down to child requirements, and out to test cases and/or procedures.
- *Special fields*—identify safety requirements, derived requirements, robustness requirements, approval status, etc.

A document or spreadsheet can be used for the requirements documentation. However, the more complex the project and larger the development team, the more benefit there is to using a requirements management tool.

If a requirements management tool is used, the following are important to have:

- Technical support and training. The developers should be trained on how to use the tool properly.
- Project-specific instructions to explain how to use the tool properly in the given environment (such information is often included in the standards or a process manual).
- Frequently asked questions and examples to help address common issues that will be encountered.

Requirements management tools can be powerful. They can help manage requirements and versions, support larger teams, facilitate traceability, track status of requirements, support use of requirements on multiple projects, and much more. However, even with a requirements management tool, good requirements change management is needed. The tool will not "compensate for lack of process, discipline, experience, or understanding" [4].

6.10 Requirements Prototyping

Alan Davis explains: "Prototyping is the technique of constructing a partial implementation of a system so that customers, users, or developers can learn more about a problem or solution to that problem" [12]. Prototypes can help to get customer feedback on key functionality, explore design options to determine feasibility, mature the requirements, identify ambiguous or incomplete requirements, minimize requirements misunderstandings, and improve the requirements robustness.

However, when I hear the word *prototype*, I tend to cringe (at least a little). Many projects not only use the prototype to help with requirements development but also try to salvage the prototype code. This is rarely successful. Prototype code is often developed quickly in order to prove a concept. It typically is not robustly designed and does not comply with the development standards. Prototyping can be a great way to mature the requirements, get customer feedback, and determine what does and doesn't work. The problem arises when the customer or management wants to use the prototype code for integration and certification. This can lead to reverse engineering the design and requirements and considerable rework of the code. It ends up taking longer and having more code issues than discarding the code and starting over.

Despite this, there are exceptions and prototyping can be successfully used—especially when prototype code is planned and part of an organized process, and not just a last-minute idea to recover lost time. There are two common approaches to successful prototyping [4,12]:

1. *Throwaway prototype*: This is a quick prototype developed without firm requirements and design in order to determine functional feasibility, obtain customer feedback, and identify missing requirements. The throwaway prototype is used to mature the requirements and then discard the code. In order to avoid the temptation to keep the prototype, (1) only implement part of the functionality, (2) establish a firm agreement up front that the code will be discarded, and (3) use a different environment. Without these protections, some customers

or project managers will be tempted to try to use the prototype for certification. Throwaway prototypes are not built to be robust or efficient, nor are they designed to be maintainable. Keeping a throwaway prototype is a big mistake.

2. *Evolutionary prototype*: This approach is entirely different. It is developed with the intent to use the code and supporting data later. The evolutionary prototype is usually a partial implementation of a key functionality. Once the functionality is proven, it is cleaned up and additional functionality is added. The evolutionary prototype is intended to be used in the final product; therefore, it uses a rigorous process, considers the quality attributes of requirements and design, evaluates the multiple design options, implements robustness, follows requirements and design standards, and uses code comments and coding standards. The prototype forms the foundation of the final product and is also closely related to the spiral life cycle model.

Both prototype approaches may be used, but should be explained in the plans, agreed with the certification authority, and implemented as agreed.

6.11 Traceability

DO-178C requires bidirectional traceability between system and software high-level requirements, software high-level requirements and software low-level requirements, software low-level requirements and code, requirements and test cases, test cases and test procedures, and test procedures and test results [3]. This section examines (1) the importance and benefits of traceability, (2) top-down and bottom-up traceability, (3) what DO-178C says about traceability, and (4) trace challenges to avoid.

6.11.1 Importance and Benefits of Traceability

Traceability between requirements, design, code, and test data is essential for compliance to DO-178C objectives. There are numerous benefits of good traceability.

Benefit 1: Traceability is needed to pass a certification authority audit. Traceability is vital to a software project and DO-178C compliance. Without good traceability, the entire claim of development assurance falls apart because the certification authority is not *assured*. Many of the objectives and concepts of DO-178C and good software engineering build upon the concept of traceability.

Benefit 2: Traceability provides confidence that the regulations are satisfied. Traceability is important because, when it is done properly, it ensures that *all* of the requirements are implemented and verified, and that *only* the requirements are implemented. This directly supports regulatory compliance, since the regulations require evidence that intended functionality is implemented.

Benefit 3: Traceability is essential to change impact analysis and maintenance. Since changes to requirements, design, and code are essential to software development, engineers must consider how to make the software and its supporting life cycle data changeable. When a change occurs, traceability helps to identify what data are impacted, need to be updated, and require reverification.

Benefit 4: Traceability helps with project management. Up-to-date trace data help project managers know what has been implemented and verified and what remains to be done.

Benefit 5: Traceability helps determine completion. A good bidirectional trace scheme enables engineers to know when they have completed each data item. It also identifies data items that have not yet been implemented or that were implemented without a driver. When a phase (e.g., design or test case development) is completed, the trace data show that the software life cycle data are complete and consistent with the previous phase's data and are ready to be used as input to the next phase.

6.11.2 Bidirectional Traceability

In order to achieve the all-requirements-and-only-requirements implementation goal, two kinds of traceability are needed: forward traceability (top-down) and backward traceability (bottom-up). Figure 6.3 illustrates the bidirectional traceability concepts required by DO-178C.

Bidirectional traceability doesn't just happen, it must be considered throughout the development process. The best way to enforce it is to implement and check the bidirectional traceability at each phase of the development effort, as noted here:

- During the *review of the software requirements*, verify the top-down and bottom-up tracing between the system requirements and high-level software requirements.

- During the *review of the design description*, verify the bidirectional tracing between the high-level software requirements and the low-level software requirements.

- During the *code reviews*, verify the bidirectional tracing between the low-level software requirements and the source code.

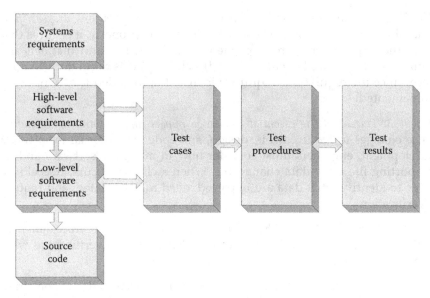

FIGURE 6.3
Bidirectional traceability between life cycle data.

- During the *review of the test cases and procedures*, verify the bidirectional tracing between the test cases and requirements (both high-level and low-level) and between the test cases and test procedures.
- During the *review of the test results*, verify the bidirectional tracing between the test results and the test procedures.

Both directions should be considered during the reviews. Just because one direction is complete doesn't necessarily mean the trace is bidirectional. For example, all system requirements allocated to software may trace down to high-level software requirements; however, there may be some high-level software requirements that do not trace up to a system requirement. Derived requirements should be the only requirements that don't have parents.

The tracing activity shouldn't just look for the completeness of the tracing but should also evaluate the technical accuracy of the tracing. For example, when evaluating the trace between a system requirement and its children (high-level software requirements), consider these questions:

- Do these high-level software requirements completely implement the system requirement?
- Is there any part of the system requirement not reflected in the high-level software requirements?
- Is the relationship between these requirements accurate and complete?

- Are there any missing traces?

- If the high-level software requirements trace to multiple system requirements, is the relationship of the requirements group accurate and complete?

- Is the granularity for each level of the requirements appropriate? For example, is the ratio of system to high-level software requirements appropriate? There isn't a magic number, but a significant number of requirements with a 1:1 or 1:>10 might indicate a granularity issue.

- If there are many-to-many traces, are they appropriate? An overabundance of many-to-many traces (i.e., children tracing to many parents and parents tracing to many children) may indicate a problem (this is discussed in Section 6.11.4).

6.11.3 DO-178C and Traceability

DO-178B identified traceability as a verification activity but was somewhat vague about how that trace information should be documented. Most applicants include trace information as part of their verification report; however, some include it in the developed data itself (i.e., in the requirements, design, code, and test cases and procedures). DO-178C is still flexible regarding where the trace information is documented; however, it does require an artifact called *trace data* during the development of requirements, design, code, and tests.

DO-178C section 5.5 identifies the trace activities that occur during software development. It explicitly requires bidirectional tracing between (1) system requirements allocated to software and the high-level software requirements, (2) high-level software requirements and low-level software requirements, and (3) low-level software requirements and source code [3]. DO-178C Table A-2 identifies *trace data* as the evidence of the trace activity and an output of the development process.

Similarly, DO-178C section 6.5 explains the trace activities during the verification process and requires bidirectional tracing between (1) software requirements and test cases, (2) test cases and test procedures, and (3) test procedures and test results. DO-178C Table A-6 identifies *trace data* as an output of the testing process.

DO-178B did not include the term *bidirectional traceability*, although it alluded to it and essentially required it. DO-178B section 6 discussed forward traceability (sections 6.3.1.f, 6.3.2.f, 6.3.4.e, and 6.4.4.1), whereas the objectives in DO-178B Tables A-3 (objective 6), A-4 (objective 6), A-5 (objective 5), and A-7 (objectives 3 and 4) alluded to backward traceability. Thus, both have been required by the certification authorities. However, DO-178C is more explicit in this area. DO-178C specifically identifies the need for *bidirectional traceability*, as well as the development of *trace data*.

6.11.4 Traceability Challenges

Like everything else in software engineering, implementing good traceability has its challenges. Some of the common challenges are as follows:

Challenge 1: Tracing proactively. Most software engineers love to solve problems and create design or code. However, few of them enjoy the paperwork and record keeping that goes with the job. In order to have accurate and complete traceability, it is essential that traceability be documented as the development takes place. In other words, as the software high-level requirements are written, the tracing to and from the system requirements should be noted; as the design is documented, the tracing to and from high-level requirements should be written down; as the code is being developed, the tracing to and from the low-level requirements should be documented; etc.

If tracing doesn't occur proactively by the author of the data, it is difficult for someone not as familiar with the data to do it later because he or she may not know the thought-process, context, or decisions of the original developers, and may be forced to make guesses or assumptions (which may be wrong). Additionally, if the tracing is done after the fact, there tend to be holes in the bidirectional traceability. To say it another way, some requirements may only be partially implemented and some functions may be implemented that aren't in the requirements. Instead of properly fixing the requirements, the after-the-fact, cleanup engineers may partially trace (to make it look complete), call many requirements derived (because they don't have parents), or trace to general requirements (ending up with a one-to-too-many or many-to-many dilemma). This seems to especially be the case when inexperienced engineers are used to perform the after-the-fact tracing.

Challenge 2: Keeping the trace data current. Some projects do very well at creating the initial trace data; however, they fail to update it when changes are made. Traceability should be evaluated and appropriately updated anytime the data are modified. Current and accurate trace data are critical to requirements management decisions and the change impact assessments (which are discussed in Chapter 10).

Challenge 3: Doing bidirectional tracing. It can be tempting to think everything is complete, when the traceability in one direction is complete. However, as previously noted, just because the forward tracing is complete doesn't mean the backward trace is, and vice versa. The trace data must be considered from both top-down (forward) and bottom-up (backward) perspectives.

Challenge 4: Many-to-many traces. This occurs when parent requirements trace to multiple children and children requirements trace to multiple parents. While there are definitely situations where many-to-many tracing is accurate and appropriate, an overabundance of many-to-many traces tends to be the symptom of a problem. Oftentimes, the problem is that the requirements are not well organized or that the tracing was done after the fact. While there are no certification guidelines against many-to-many traces, they can be very confusing to the developers and certifying authority, and they are often very difficult to justify and maintain. Many-to-many tracing should be minimized as much as possible.

Challenge 5: Deciding what is derived and what is not. Derived requirements can be a challenge. More than once I've evaluated a project where the *derived* flag was set because there were missing higher level requirements and the project didn't want to add requirements. Derived requirements should be used cautiously. They are not plugs for missing higher level requirements, but represent design details that aren't significant at the higher level. Derived requirements should not be added to compensate for missing functionality at the higher level. One technique to help avoid inaccurate classification of derived requirements is to justify why each requirement is needed and why it is classified as derived.* If the requirement can't be explained or justified, it may not be needed or a higher level requirement may be missing. Keep in mind that all derived requirements need to be evaluated by the safety team, and all code must be traceable to requirements—there is not a category called *derived code.*

Challenge 6: Weak links. Oftentimes there will be some requirements that are debatable as to whether they are traced or derived. They may be related to a higher level requirement but not the direct result of the higher level requirement. If one decides to include the trace, it is helpful to include rationale about the relationship between the higher and lower level requirements. For lack of a better description, I call these debatable traces *weak links.* By providing a brief note about why the *weak links* are included, it helps all of the users of the requirements better understand the connection and to evaluate the impact of future changes. The notes also help the certification and software maintenance efforts by explaining what may not be obvious relationships. The explanation does not need to be verbose; a short sentence or two is usually quite adequate. Even if the requirement is classified as derived, it's still recommended to mention the relationship, since it may help change impact analysis and change management later (i.e., it supports modifiability).

Challenge 7: Implicit traceability. Some projects use a concept of *implicit traceability.* For example, the tracing may be assumed by the naming convention or the document layout. Such approaches have been accepted by the certification authorities. The new challenge is DO-178C's requirement for *trace data.* Implicit tracing is built in and doesn't result in a separate trace artifact. Therefore, implicit tracing should be handled cautiously. Here are a few suggestions:

- Include the trace rules in the standards and plans, so that developers know the expectations.
- Document the reviews of the tracing well to ensure that the implicit tracing is accurate and complete.

---

* DO-178C section 5.1.2.h identifies this as an activity.

- Identify the approach in the software plans and get the certification authority's buy-in.
- Be consistent. Do not mix implicit and explicit tracing in a section (unless well documented).

Challenge 8: Robustness testing traces. Sometimes developers or testers argue that robustness testing does not need to be traceable to the requirements, since they are trying to *break* the software rather than prove intended functionality. They contend that by limiting the testers to just the requirements, they might miss some weakness in the software. As will be discussed in Chapter 9, the *break-it* mentality is important for effective software testing. Testers should not limit themselves to the requirements when they are thinking through their testing effort. Testers should, however, identify when the requirements are incomplete, rather than create robustness tests that don't trace to requirements. Frequently, the testers identify missing scenarios that should be reflected in the requirements. It's highly recommended that the testers participate in requirements and design reviews to proactively identify potential requirements weaknesses. Likewise, developers may want to be involved in test reviews to quickly fill any requirements gaps and to ensure that testers understand the requirements.

References

1. N. Leveson, *Safeware: System Safety and Computers* (Reading, MA: Addison-Wesley, 1995).
2. IEEE, *IEEE Standard Glossary of Software Engineering Terminology*, IEEE Std-610-1990 (Los Alamitos, CA: IEEE Computer Society Press, 1990).
3. RTCA DO-178C, *Software Considerations in Airborne Systems and Equipment Certification* (Washington, DC: RTCA, Inc., December 2011).
4. K. E. Wiegers, *Software Requirements*, 2nd edn. (Redmond, WA: Microsoft Press, 2003).
5. D. Leffingwell, Calculating the return on investment from more effective requirements, *American Programmer* 10(4), 13–16, 1997.
6. Standish Group Study (1995) referenced in I. F. Alexander and R. Stevens, *Writing Better Requirements* (Harlow, U.K.: Addison-Wesley, 2002).
7. D. L. Lempia and S. P. Miller, *Requirements Engineering Management Findings Report*, DOT/FAA/AR-08/34 (Washington, DC: Office of Aviation Research, June 2009).
8. SearchSoftwareQuality.com, Software quality resources. http://searchsoftwarequality.techtarget.com/definition/use-case (accessed on 5/1/2012).
9. K. E. Wiegers, *More about Software Requirements* (Redmond, WA: Microsoft Press, 2006).

10. RTCA DO-248C, *Supporting Information for DO-178C and DO-278A* (Washington, DC: RTCA, Inc., December 2011).
11. Certification Authorities Software Team (CAST), Merging high-level and low-level requirements, Position Paper CAST-15 (February 2003).
12. A. M. Davis, *Software Requirements* (Upper Saddle River, NJ: Prentice-Hall, 1993).
13. D. L. Lempia and S. P. Miller, *Requirements Engineering Management Handbook*, DOT/FAA/AR-08/32 (Washington, DC: Office of Aviation Research, June 2009).

Recommended Readings

1. K. E. Wiegers, *Software Requirements*, 2nd edn. (Redmond, WA: Microsoft Press, 2003). While not written for safety-critical software, this book is an excellent resource for requirements authors.
2. D. L. Lempia and S. P. Miller, *Requirements Engineering Management Handbook*, DOT/FAA/AR-08/32 (Washington, DC: Office of Aviation Research, June 2009). This handbook was sponsored by the FAA and provides valuable guidelines for those writing safety-critical software requirements.

7

Software Design

Acronyms

CSPEC	control specification
HLR	high-level requirement
LLR	low-level requirement
PSPEC	process specification
UML	Unified Modeling Language

7.1 Overview of Software Design

DO-178C takes a detour from other software development literature in the area of design. DO-178B, and now DO-178C, explains that the software design contains the software architecture and low-level requirements (LLRs). Software architecture is a commonly understood part of design; however, the term LLR has caused considerable confusion. During the DO-178C committee deliberations, there was an attempt to adjust the terminology to align with other domains and common software engineering methods, but consensus was not achieved on the adjustment. Therefore, the term LLR remains and is discussed in this chapter.

The design serves as the blueprint for the software implementation phase. It describes both *how* the software will be put together (architecture) and *how* it will perform the desired functionality (LLRs). To understand the DO-178C guidance on design, it is important to understand the two elements of the software design: the architecture and the LLRs.

7.1.1 Software Architecture

DO-178C defines *software architecture* as: "The structure of the software selected to implement the software requirements" [1]. Roger Pressman provides a more comprehensive definition:

> In its simplest form, architecture is the structure or organization of program components (modules), the manner in which these components interact, and the structure of data that are used by the components. In a broader sense, however, components can be generalized to represent major system elements and their interactions [2].

Architecture is a crucial part of the design process. Some things to keep in mind while documenting the architecture are noted here:

First, DO-178C compliance requires that the architecture be compatible with the requirements. Therefore, some means to ensure the compatibility is needed. Oftentimes traceability or a mapping between the requirements and architecture is used.

Second, the architecture should be documented in a clear and consistent format. It's important to consider the coder who will use the design to implement the code, as well as the developers who will maintain the software and its design in the future. The architecture must be clearly defined in order to be accurately implemented and maintained.

Third, the architecture should be documented in such a way that it can be updated as needed and possibly implemented in iterations. This may be to support an iterative or evolutionary development effort, configuration options, or the safety approach.

Fourth, different architectural styles exist. For most styles the architecture includes components and connectors. However, the type of these components and connectors depends on the architectural approach used. Most real-time airborne software uses a functional structure. In this case, components represent functions and connectors show the interfaces between the functions (either in the form of data or control).

7.1.2 Software Low-Level Requirements

DO-178C defines LLRs as: "Software requirements developed from high-level requirements, derived requirements, and design constraints from which Source Code can be directly implemented without further information" [1]. The LLRs are a decomposition of the high-level requirements (HLRs) to a level from which code can be directly written. DO-178C essentially puts the detailed engineering thought process in the design phase and not in the coding phase. If the design is documented properly, the coding effort should be relatively straightforward. This philosophy is not without controversy. The need for two levels of software requirements (high-level and low-level) has been heavily debated. Following are 10 concepts to keep in mind when planning for and writing LLRs:

Concept 1: LLRs are design details. The word *requirements* can be misleading, since LLRs are part of design. It is helpful to think of them as implementation steps for the coder to follow. Sometimes LLRs are even represented as pseudocode or models. Some projects require the use of *shall* for their LLRs and some do not. I have seen both approaches used successfully.

Concept 2: LLRs must be uniquely identified. Since the LLRs must be traced up to HLRs and down to code, the LLRs need to be uniquely identified (this is why some organizations prefer to include *shall* in the requirements).

Concept 3: LLRs should have the quality attributes described in Chapters 2 and 6. However, the LLRs focus on *how*, not on *what*. The LLRs are the implementation details and should get into the details of how the software will be implemented to carry out the functionality documented in the HLRs.

Concept 4: LLRs must be verifiable. One of the reasons the word *requirements* was retained in DO-178C is that for levels A, B, and C, the LLRs need to be tested. This is important to consider when writing the requirements.

Concept 5: Occasionally, LLRs may not be needed at all. In some projects the HLRs are detailed enough that one can directly code from them. This is the exception and not the rule. And, as noted in Chapter 6, this approach is not recommended. HLRs should focus on functionality, whereas LLRs focus on implementation. However, if the HLRs are indeed detailed enough (which means they have likely mixed *what* and *how*), DO-178C section 5.0 does allow a single level of software requirements, if that single level can satisfy both the high-level and low-level requirements objectives (i.e., DO-178C Table A-2 objectives 1, 2, 4 and 5; Table A-3; and Table A-4). Additionally, the contents of the single level of requirements combined with the software architecture should address the guidance identified in DO-178C section 11.9 (software requirements) and 11.10 (software design). Be warned, however, that this is not always accepted by the certification authorities and could result in project restart if not properly coordinated. If this approach seems feasible for your project, be sure to explain it in your plans, justify why it will work, and detail how the DO-178C objectives will be satisfied.

Concept 6: Occasionally, LLRs may not be needed in some areas. In a few projects, the HLRs will have adequate detail to code from in most areas but might need additional refinement in some areas. That is, some HLRs need to be decomposed into LLRs and some do not. If this approach is used, it should be clear which requirements are HLRs and which are LLRs. It should also be clear when the HLRs will not be further decomposed, so the programmer knows what requirements form the foundation for the coding effort. This approach can be quite tricky to carry out in reality, so use it with extreme caution.

Concept 7: More critical software tends to need more detailed LLRs. In my experience, the higher the software criticality (levels A and B), the more detailed the LLRs need to be to completely describe the requirements and obtain the needed structural coverage. The more critical the software is, the more rigorous the criteria for structural coverage. Structural coverage is discussed in Chapter 9; however, it is something to keep in mind during the design phase.

Concept 8: Derived LLRs must be handled with care. There may be some derived LLRs that are identified during the design phase. DO-178C defines derived requirements as: "Requirements produced by the software development processes which (a) are not directly traceable to higher level requirements, and/or (b) specify behavior beyond that specified by the system requirements or the higher level software requirements" [1]. Derived LLRs are not intended to compensate for holes in the HLRs but instead represent implementation details that were not yet known during the requirements phase. The derived LLRs do not trace up to HLRs but will trace down to code. A derived LLR should be documented, identified as derived, justified as to why it is needed, and evaluated by the safety assessment team to ensure it does not violate any of the system or safety assumptions. To support the safety team, the justification or rationale should be written so that someone unfamiliar with the details of the design can understand the requirement and evaluate its impact on safety and overall system functionality.

Concept 9: LLRs are usually textual but may be represented as models. The LLRs are often expressed in textual format with tables and graphics to communicate the details as needed. As will be discussed in Chapter 14, the LLRs may be captured as models. If this is the case, the model would be classified as a *design model* and the guidance of DO-331 would apply.

Concept 10: LLRs may be represented as pseudocode. Sometimes, LLRs are represented as pseudocode or are supplemented with pseudocode. DO-248C FAQ #82, entitled "If pseudocode is used as part of the low-level requirements, what issues need to be addressed?" provides certification concerns with using pseudocode as the LLRs [3]. When LLRs are represented as pseudocode, the following concerns exist [3]:

- This approach may result in a large granularity jump between the HLRs and the LLRs which could make it difficult to detect unintended and missing functionality.
- This approach could result in insufficient architectural detail, which impacts verification, including data coupling and control coupling analysis.
- Unique identification of LLRs may be difficult.

- Bidirectional tracing to and from HLRs may be challenging.
- Performing structural coverage using low-level testing is generally inadequate, since the code and pseudocode are so similar; such a testing process does not effectively detect errors, identify missing functionality, or find unintended functionality.

7.1.3 Design Packaging

The design packaging varies from project to project. Some projects integrate the architecture and LLRs, while others put them in entirely separate documents. Some projects include them in the same document but locate them in different sections. DO-178C does not dictate the packaging preference. It often depends on the methodology used. Regardless of the packaging decision, the relationship between the architecture and requirements must be clear. The design (including the LLRs) will be used by those implementing the software; therefore, the coders should be kept in mind as the design is documented.

7.2 Approaches to Design

There are a variety of techniques employed by designers to model the software architecture and behavior. The two design approaches used in aviation software are structure-based and object-oriented. Some projects combine the concepts from both approaches.

7.2.1 Structure-Based Design (Traditional)

The structure-based design is common for real-time embedded software and uses some or all of the following representations:

- *Data context diagram*—the top-level diagram which describes the functional behavior of the software and shows the data input and output from the software.
- *Data flow diagram*—a graphical representation of the processes performed by the software, showing the flow of data between processes. It is a decomposition of the data context diagram. The data flow diagram is typically represented in multiple levels, each level going into more detail. Data flow diagrams include processes, data flows, and data stores.

- *Process specification (PSPEC)*—accompanies the data flow diagram and shows how the output for the processes are generated from the given inputs [4].

- *Control context diagram*—the top-level diagram which shows the control of the system by establishing the control interfaces between the system and its environment.

- *Control flow diagram*—the same diagram as the data flow diagram, except the flow of control through the system is identified rather than the flow of data.

- *Control specification (CSPEC)*—accompanies the control flow diagram and shows how the output for the processes are generated from the given inputs [4].

- *Decision table (also called a truth table)*—shows the combinations of decisions made based on given input.

- *State transition diagram*—illustrates the behavior of the system by showing its states and the events that cause the system to change states. In some designs this might be represented as a state transition table instead of a diagram.

- *Response time specification*—illustrates external response times that need to be specified. It may include event-driven, continuous, or periodic response times. It identifies the input event, output event, and the response time for each external input signal [4].

- *Flowchart*—graphically represents sequence of software actions and decisions.

- *Structure chart*—illustrates the partitioning of a system into modules, showing their hierarchy, organization, and communication [5].

- *Call tree (also called call graph)*—illustrates the calling relationships between software modules, functions, or procedures.

- *Data dictionary*—defines the data and control information that flow through the system. Typically includes the following information for each data item: name, description, rate, range, resolution, units, where/how used, etc.

- *Textual details*—describes implementation details (e.g., the LLRs).

- *Tasking diagram*—shows the characteristics of tasks (e.g., sporadic or periodic), task procedures, input/output of each task, and any interactions with the operating system (such as semaphores, messages, and queues).

7.2.2 Object-Oriented Design

Object-oriented design techniques may use these representations [2]:

- *Use case*—accompanies the requirements to graphically and textually explain how a user interacts with the system under specific circumstances. It identifies actors (the people or devices that use the system) and describes how the actor interacts with the system.

- *Activity diagram*—supplements the use case by graphically representing the flow of interaction within a scenario. It shows the flow of control between actions that the system performs. An activity diagram is similar to a flowchart, except the activity diagram also shows concurrent flows.

- *Swimlane diagram*—a variation of the activity diagram; it shows the flow of activities described by the use case and simultaneously indicates which actor is responsible for the action described by an activity. It basically shows the activities of each actor in a parallel fashion.

- *State diagram*—like the state transition diagram described earlier, the object-oriented state diagram shows the states of the system, actions performed depending on those states, and the events that lead to a state change.

- *State chart*—an extension of the state diagram with added hierarchy and concurrency information.

- *Class diagram*—a Unified Modeling Language (UML) approach which models classes (including their attributes, operations, and relationships and associations with other classes) by providing a static or structural view of a system.

- *Sequence diagram*—shows the communications between objects during execution of a task, including the temporal order in which messages are sent between the objects to accomplish that task.

- *Object-relationship model*—a graphical representation of the connections between classes.

- *Class-responsibility-collaborator model*—provides a way to identify and organize the classes that are relevant to the requirements. Each class is represented as a box, sometimes referred to as an *index card*; each box includes class name, class responsibilities, and collaborators. The responsibilities are the attributes and operations relevant for the class. Collaborators are those classes that are required to provide information to another class to complete its responsibility. A collaboration is either a request for information or a request for some action.

Chapter 15 provides more information on object-oriented technology.

7.3 Characteristics of Good Design

DO-178C offers flexibility for documenting the design. Rather than go into detail on design techniques, which are available in many other books, let's consider the characteristics that a good software design possesses.

Characteristic 1: Abstraction. A good design implements the concept of abstraction at multiple levels. Abstraction is the process of defining a program (or data) with a representation similar to its meaning (semantics), while hiding the implementation details. Abstraction strives to reduce and factor out details so that the designer can focus on a few concepts at a time. When abstraction is applied at each hierarchical level of the development, it allows each level to only deal with the details that are pertinent to that level. Both procedural and data abstraction are desirable.

Characteristic 2: Modularity. A modular design is one where the software is logically partitioned into elements, modules, or subsystems (often referred to as *components*). (A *component* may be a single code module or a group of related code modules.) The overall system is divided by separating the features or functions. Each *component* focuses on a specific feature or function. By separating the features and functionality into smaller, manageable components, it makes the overall problem less difficult to solve. Pressman writes:

> You modularize a design (and the resulting program) so that development can be more easily planned; software increments can be defined and delivered; changes can be more easily accommodated; testing and debugging can be conducted more efficiently; and long-term maintenance can be conducted without serious side effects [2].

When a design is properly modularized, it is fairly simple to understand the purpose of each component, verify the correctness of each component, understand the interaction between components, and assess the overall impact of each component on the software structure and operation [6].

Characteristic 3: Strong cohesion. To make a system truly modular, the designer strives for functional independence with each component. This is carried out by cohesion and coupling. Good designs strive for strong cohesion and loose coupling. "Cohesion may be viewed as the glue that keeps the component together" [7]. Even though DO-178C doesn't require an evaluation of a component's cohesiveness, it should be considered during design because it will affect the overall quality of the design. Cohesion is a measure of the component's strength and acts like a chain holding the component's activities together [5]. Yourdon and Constantine define seven layers of cohesion, with cohesion becoming weaker as you go down the list [8]:

- *Functional cohesion*: All elements contribute to a single function; each element contributes to the execution of only one task.
- *Sequential cohesion*: The component consists of a sequence of elements where the output of one element serves as input to the next element.
- *Communicational cohesion*: The elements of a component use the same input or output data but order is not important.
- *Procedural cohesion*: The elements are involved in different and possibly unrelated activities that must be executed in a given order.
- *Temporal cohesion*: The elements are functionally independent but their activities are related in time (i.e., they are carried out at the same time).
- *Logical cohesion*: Elements include tasks that are logically related. A logically cohesive component contains a number of activities of the same general kind; the user picks what is needed.
- *Coincidental cohesion*: Elements are grouped into components in a haphazard way. There is no meaningful relationship between the elements.

Characteristic 4: Loose coupling. Coupling is the degree of interdependence between two components. A good design minimizes coupling by eliminating unnecessary relationships, reducing the number of necessary relationships, and easing the tightness of necessary relationships [5]. Loose coupling helps to minimize the ripple effect when a component is modified, since the component is easier to comprehend and adapt. Therefore, loose coupling is desired for effective and modular design.

DO-178C defines two types of coupling [1]:

- *Data coupling*: "The dependence of a software component on data not exclusively under the control of that software component."
- *Control coupling*: "The manner or degree by which one software component influences the execution of another software component."

However, software engineering literature identifies six types of coupling, which are listed in the following starting with the tightest coupling and going to the loosest [5,7]:

- *Content coupling*: One component directly affects the working of another component, since one component refers to the inside of the other component.
- *Common coupling*: Two components refer to the same global data area. That is, the components share resources.

- *External coupling*: Components communicate through an external medium, such as a file or database.

- *Control coupling (not same as DO-178C definition)*: One component directs the execution of another component by passing the necessary control information.

- *Stamp coupling*: Two components refer to the same data structure. This is sometimes called *data structure coupling*.

- *Data coupling (not same as DO-178C definition)*: Two components communicate by passing elementary parameters (such as a homogeneous table or a single field).

Unfortunately, DO-178C overloads the terms *data coupling* and *control coupling*. The DO-178C use of *data coupling* is comparable to the mainstream software engineering concepts of data, stamp, common, and external coupling. The DO-178C use of *control coupling* is covered by software engineering concepts of control and content coupling [9]. Data and control coupling analyses are discussed in Chapter 9 since these analyses are part of the verification phase.

A well-designed software product strives for loose coupling and strong cohesion. Implementing these characteristics helps to simplify the communication between programmers, make it easier to prove correctness of components, reduce propagation of impact across components when a component is changed, make components more comprehensible, and reduce errors [7].

Characteristic 5: Information hiding. Information hiding is closely related to the concepts of abstraction, cohesion, and coupling. It contributes to modularity and reusability. The concept of information hiding suggests that "modules [components] should be specified and designed so that information (e.g., algorithms and data) contained within a module [component] is inaccessible to other modules [components] that have no need for such information" [2].*

Characteristic 6: Reduced complexity. Good designs strive to reduce complexity by breaking the larger system into smaller, well-defined subsystems or functions. While this is related to the other characteristics of abstraction, modularity, and information hiding, it requires a conscientious effort by the designer. Designs should be documented in a straightforward and understandable manner. If the complexity is too great, the designer should look for alternate approaches to divide the responsibilities of the function into several smaller functions. Overly complex designs result in errors and significant impacts when change is needed.

Characteristic 7: Repeatable methodology. Good design is the result of a repeatable method driven by requirements. A repeatable method uses well-defined notations and techniques that effectively communicate what is intended. The methodology should be identified in the design standards.

---

* Brackets added for consistency.

Characteristic 8: Maintainability. The overall maintenance of the project should be considered when documenting the design. Designing for maintainability includes designing the software to be reused (wholly or partially), loaded, and modified with minimal impact.

Characteristic 9: Robustness. The overall robustness of the design should be considered during the design phase and clearly documented in the design description. Examples of robust design considerations include off-nominal functionality, interrupt functionality and handling of unintended interrupts, error and exception handling, failure responses, detection and removal of unintended functionality, power loss and recovery (e.g., cold and warm start), resets, latency, throughput, bandwidth, response times, resource limitations, partitioning (if required), deactivation of unused code (if deactivated code is used), and tolerances.

Characteristic 10: Documented design decisions. The rationale for decisions made during the design process should be documented. This allows proper verification and supports maintenance.

Characteristic 11: Documented safety features. Any design feature used to support safety should be clearly documented. Examples include watchdog timers, cross channel comparisons, reasonability checks, built-in test processing, and integrity checks (such as cyclic redundancy check or checksum).

Characteristic 12: Documented security features. With the increasing sophistication of hackers, the design should include protection from vulnerabilities and detection mechanisms for attacks.

Characteristic 13: Reviewed. Throughout the design phase, informal technical reviews should be performed. During these informal reviews, the reviewers should seriously evaluate the quality and appropriateness of the design. Some designs (or at least portions of them) may need to be discarded in order to arrive at the best solution. Iterative reviews with technically savvy engineers help to identify the best and optimal design sooner, hence reducing issues found later during the formal design review and during testing.

Characteristic 14: Testability. The software should be designed to be testable. Testability is how easily the software can be tested. Testable software is both visible and controllable. Pressman notes that testable software has the following characteristics [2]:

- *Operability*—the software does what it is supposed to as per the requirements and design.
- *Observability*—the software inputs, internal variables, and outputs can be observed during execution to determine if tests pass or fail.

- *Controllability*—the software outputs can be altered using the given inputs.
- *Decomposability*—the software is constructed as components that can be decomposed and tested independently, if needed.
- *Simplicity*—the software has functional simplicity, structural simplicity, and code simplicity. For example, simple algorithms are more testable than complex algorithms.
- *Stability*—the software is not changing or is only changing slightly.
- *Understandability*—good documentation helps testers to understand the software and therefore to test it more thoroughly.

Some features that may help make the software testable are the following [10]:

- *Error or fault logging.* Such functionality in the software may provide the testers a way to better understand the software behavior.
- *Diagnostics.* Diagnostic software (such as code integrity checks or memory checks) can help identify problems in the system.
- *Test points.* These provide hooks into the software that can be useful for testing.
- *Access to interfaces.* This can help with interface and integration testing.

Coordination between the test engineers and the designers can help make the software more testable. In particular, test engineers should be involved during the design reviews and even sooner if possible.

Characteristic 15: Avoids undesired features. Typically the following are prohibited in safety-critical designs:

- *Recursive function execution*, that is, functions that can call themselves, either directly or indirectly. Without extreme caution and specific design actions, the use of recursive procedures may result in unpredictable and potentially large use of stack space.
- *Self-modifying code* is code that alters its own instructions while it is executing, normally to reduce the instruction path length, to improve performance, or to reduce repetitively similar code.
- *Dynamic memory allocation*, unless the allocation is done only once in a deterministic fashion during system initialization. DO-332 describes concerns and provides guidance on dynamic memory allocation.

7.4 Design Verification

As with the software requirements, the software design needs to be verified. This is typically carried out through a peer review. The recommendations for peer reviews in Chapter 6 also apply to design reviews. During the design review, both the LLRs and the architecture are evaluated. The DO-178C Table A-4 objectives for the design verification are listed and explained in the following [1]:

- *DO-178C Table A-4 objective 1*: "Low-level requirements comply with high-level requirements." This ensures that the LLRs completely and accurately implement the HLRs. That is, all functionality identified in the HLRs has been identified in the LLRs.

- *DO-178C Table A-4 objective 2*: "Low-level requirements are accurate and consistent." This ensures that the LLRs are error free and consistent with themselves, as well as with the HLRs.

- *DO-178C Table A-4 objective 3*: "Low-level requirements are compatible with target computer." This verifies any target dependencies of the LLRs.

- *DO-178C Table A-4 objective 4*: "Low-level requirements are verifiable." This typically focuses on the testability of the LLRs. For levels A–C the LLRs will need to be tested. Therefore, verifiability needs to be considered during the initial development of the LLRs. (Characteristic #14 in Section 7.3 provides additional discussion on testability.)

- *DO-178C Table A-4 objective 5*: "Low-level requirements conform to standards." Chapter 5 discusses the development of the design standards. During the review, the LLRs are evaluated for their conformance to the standards.*

- *DO-178C Table A-4 objective 6*: "Low-level requirements are traceable to high-level requirements." This is closely related to Table A-4 objective 1. The bidirectional traceability between HLRs and LLRs supports the compliance of the LLRs to the HLRs. During the review, the accuracy of the traces is also verified. Any traces that are unclear should be evaluated and either modified or explained in the rationale. Traceability concepts are discussed in Chapter 6.

- *DO-178C Table A-4 objective 7*: "Algorithms are accurate." Any mathematical algorithms should be reviewed by someone with the appropriate background to confirm the accuracy of the algorithm. If an algorithm is being reused from a previous system that was thoroughly reviewed and the algorithm is unchanged, the review evidence from

---

* It should be noted that some projects apply the requirements standards to the LLRs rather than the design standards.

the previous development may be used. The reuse of such verification evidence should be noted in the plans.

- *DO-178C Table A-4 objective 8*: "Software architecture is compatible with high-level requirements." Oftentimes, there is a tracing or mapping between the requirements and architecture to help confirm their compatibility.

- *DO-178C Table A-4 objective 9*: "Software architecture is consistent." This verification objective ensures that the components of the software architecture are consistent and correct.

- *DO-178C Table A-4 objective 10*: "Software architecture is compatible with target computer." This confirms that the architecture is appropriate for the specific target on which the software will be implemented.

- *DO-178C Table A-4 objective 11*: "Software architecture is verifiable." As previously noted, testability should be considered when developing the architecture. (Characteristic #14 in Section 7.3 provides additional discussion on testability.)

- *DO-178C Table A-4 objective 12*: "Software architecture conforms to standards." The design standards should be followed when developing the architecture. For levels A, B, and C, both the architecture and the LLRs are assessed for compliance with the design standards.

- *DO-178C Table A-4 objective 13*: "Software partitioning integrity is confirmed." If partitioning is used, it must be addressed in the design and verified. Chapter 21 examines the partitioning topic.

A checklist is usually used during the design review to assist engineers in their thorough evaluation of the design. The earlier recommendations (Section 7.3) and the DO-178C Table A-4 objectives provide a good starting point for the design standards and review checklist.

References

1. RTCA DO-178C, *Software Considerations in Airborne Systems and Equipment Certification* (Washington, DC: RTCA, Inc., December 2011).
2. R.S. Pressman, *Software Engineering: A Practitioner's Approach*, 7th edn. (New York: McGraw-Hill, 2010).
3. RTCA DO-248C, *Supporting Information for DO-178C and DO-278A* (Washington, DC: RTCA, Inc., December 2011).
4. K. Shumate and M. Keller, *Software Specification and Design: A Disciplined Approach for Real-Time Systems* (New York: John Wiley & Sons, 1992).

5. M. Page-Jones, *The Practical Guide to Structured Systems Design* (New York: Yourdon Press, 1980).
6. J. Cooling, *Software Engineering for Real-Time Systems* (Harlow, U.K.: Addison-Wesley, 2003).
7. H.V. Vliet, *Software Engineering: Principles and Practice*, 3rd edn. (Chichester, U.K.: Wiley, 2008).
8. E. Yourdon and L. Constantine, *Structured Design: Fundamentals of a Discipline of Computer Program and Systems Design* (New York: Yourdon Press, 1975).
9. S. Paasch, Software coupling, presentation at *Federal Aviation Administration Designated Engineering Representative Conference* (Long Beach, CA: September 1998).
10. C. Kaner, J. Bach, and B. Pettichord, *Lessons Learned in Software Testing* (New York: John Wiley & Sons, 2002).

8

Software Implementation: Coding and Integration

Acronyms

ACG	autocode generator
ANSI	American National Standards Institute
ARM	Ada Reference Manual
CAST	Certification Authorities Software Team
CPU	central processing unit
CRC	cyclic redundancy check
DoD	Department of Defense
FAQ	frequently asked question
I/O	input/output
IEC	International Electrical Technical Commission
ISO	International Standards Organization
LRM	Language Reference Manual
MISRA	Motor Industry Software Reliability Association
PPP	Pseudocode Programming Process
PSAC	Plan for Software Aspects of Certification
RTOS	real-time operating system
SCI	Software Configuration Index
SLECI	Software Life Cycle Environment Configuration Index
WORA	write once, run anywhere

8.1 Introduction

This chapter examines two aspects of software implementation: (1) coding and (2) integration. Additionally, the verification of the coding and integration processes is discussed.

Coding is the process of developing the source code from the design description. The terms *construction* or *programming* are often preferred because they carry the connotation that coding is not just a mechanical step, but is one that requires forethought and skill—like constructing a building or bridge. However, DO-178C uses the term *coding* throughout. DO-178C follows the philosophy that most of the construction activity is part of design, rather than coding. DO-178C does not, however, prohibit the designer from also writing the code. Regardless of who determines the code construction details, the importance of the source code development should not be undermined. Requirements and design are critical, but the compiled and linked code is what actually flies.

Integration is the process of building the executable object code (using a compiler and linker) and loading it into the target computer (using a loader). Even though the integration process is sometimes considered trivial, it is a vitally important process in the development of safety-critical software.

8.2 Coding

Since the coding process generates the source code that will be converted to the executable image that actually ends up in the safety-critical system, it is an exceptionally important step in the software development process. This section covers the DO-178C guidance for coding, common languages used to develop safety-critical software, and recommendations for programming in the safety-critical domain. The C, Ada, and assembly languages are briefly explored, because they are the most commonly used languages in embedded safety-critical systems. Some other languages are briefly mentioned but are not detailed. This section ends by discussing a couple of code-related special topics: libraries and autocode generators (ACGs).

Please note that the subject of parameter or configuration data is closely related to coding; however, it is discussed in Chapter 22, rather than in this chapter.

8.2.1 Overview of DO-178C Coding Guidance

DO-178C provides guidance on the planning, development, and verification of source code.

First, during the planning phase the developer identifies the specific programming language, coding standards, and compiler (per DO-178C section 4.4.2). Chapter 5 discusses the overall planning process and the coding standards. This chapter provides some coding recommendations that should be considered in the company-specific coding standards during the planning phase.

Second, DO-178C provides guidance on the code development (DO-178C section 5.3). DO-178C explains that the primary objective of the coding process is developing source code that is "traceable, verifiable, consistent, and correctly implements low-level requirements" [1]. To accomplish this objective, the source code must implement the design description and only the design description (including the low-level requirements and the architecture), trace to the low-level requirements, and conform to the identified coding standards. The outputs of the coding phase are source code and trace data. The build instructions (including compile and link data) and load instructions are also developed during the coding phase but are treated as part of integration, which is discussed in Section 8.4.

Third, the source code is verified to ensure its accuracy, consistency, compliance to design, compliance to standards, etc. DO-178C Table A-5 summarizes the objectives for verifying the source code, which are discussed in Section 8.3.

8.2.2 Languages Used in Safety-Critical Software

At this time, there are three primary languages used in airborne safety-critical systems: C, Ada, and assembly. There are some legacy products out there with other languages (including FORTRAN and Pascal). C++ has been used in some projects but is typically severely limited so that it is really like C. Java and C# have been used in several tool development efforts and work quite well for that; however, to date they are still not ready to use when implementing safety-critical systems. Other languages have been used or proposed. Because C, Ada, and assembly are the most predominant, they are briefly described in the following sections. Table 8.1 summarizes other languages as well. It was tempting to provide a more detailed discussion of the languages; however, there are other resources that describe them well.

8.2.2.1 Assembly Language

Assembly language is a low-level programming language used for computers, microprocessors, and microcontrollers. It uses a symbolic representation of the machine code to program a given central processing unit (CPU). The language is typically defined by the CPU manufacturer. Therefore, unlike high-level languages, assembly is not portable across platforms; however, it is often portable within a processor family. A utility program known as an *assembler* is used to translate the assembly statements into the machine code for the target computer. The assembler performs a one-to-one mapping from the assembly instructions and data to the machine instructions and data.

There are two types of assemblers: one pass and two pass. A one-pass assembler goes through the source code once and assumes that all symbols will be defined before any instruction that references them. The one-pass assembler has speed advantages. Two-pass assemblers create a table with all symbols and their values in the first pass, and then they use the table in a second pass to generate code. The assembler must at least be able to determine the length of each instruction on the first pass so that the addresses of symbols can be calculated. The advantage of the two-pass assembler is that symbols can be defined anywhere in the program source code. This allows

TABLE 8.1

Languages Occasionally Used in Aviation Software

Language	Description
C++	C++ was introduced in 1979 by Bjarne Stroustrup at Bell Labs as an enhancement to C. C++ was originally named C with Classes and was later renamed to C++ in 1983. It has both high-level and low-level language features. C++ is a statically typed, free-form, multiparadigm, general-purpose, high-level programming language. It is used for applications, device drivers, embedded software, high-performance server and client applications, and hardware design. C++ adds the following object-oriented enhancements to C: classes, virtual functions, multiple inheritance, operator overloading, exception handling, and templates. Most C++ compilers will also compile C. C++ (and a subset Embedded C++) has been used in several safety-critical systems, but many of the object-oriented features were not used.
C#	C# was introduced in 2001 by a team led by Anders Hejlsberg at Microsoft. C# was initially designed to be a general-purpose, high-level programming language that was simple, modern, and object-oriented. Characteristics of C# include strongly typed, imperative, functional, declarative, generic, object-oriented, and component-oriented. C# has been used for software tools used in aviation but is not yet considered mature enough for safety-critical software.
FORTRAN	FORTRAN was introduced in 1957 by John Backus at IBM. FORTRAN is a general-purpose, high-level programming language that is designed for numeric computation and scientific computing. High-performance computing is one of its desired features. Other characteristics include procedural and imperative programming, with later versions adding array programming, modular programming, object-oriented programming, and generic programming. It was used in avionics in the past and still exists on a few legacy systems.
Java	Java was introduced in 1995 by James Gosling at Sun Microsystems. Java was based on C and C++. It is a general-purpose, high-level programming language designed with very few implementation dependencies. Other characteristics include class-based, concurrent, and object-oriented. Java uses the philosophy "write once, run anywhere" (WORA), meaning code running on one platform does not need recompilation to run on a different platform. Java applications are typically compiled into bytecode, which is stored in a class file. This bytecode can then be executed on any Java virtual machine regardless of the computer architecture. A real-time Java has been developed and a safety-critical subset is in work. Java has been used for some aviation software tools but is still not considered mature enough for safety-critical software.
Pascal	Pascal was introduced in 1970 by Niklaus Wirth and was based on the ALGOL programming language. Pascal is an imperative and procedural high-level programming language. It was designed to promote good programming practices using structured programming and data structuring. Additional characteristics of Pascal include enumerations, subranges, records, dynamically allocated variables with associated pointers, and sets, which allow programmer-defined complex data structures such as lists, trees, and graphs. Pascal strongly types all objects, allows nested procedure definitions to any depth, and allows most definitions and declarations inside functions and procedures. It was used in avionics in the past and still exists on a few legacy systems.

Source: Wikipedia, Programming languages, http://en.wikipedia.org/wiki/List_of_programming_languages, accessed on April 2012.

the programs to be defined in more logical and meaningful ways and makes two-pass assembler programs easier to read and maintain [2].

In general, assembly should be avoided when possible. It is difficult to maintain, has extremely weak data typing, has limited or no flow control mechanisms, is difficult to read, and is generally not portable. However, there are occasions when it is needed, including interrupt handling, hardware testing and error detection, interface to processor and peripheral devices, and performance support (e.g., execution speed in critical areas) [3].

8.2.2.2 Ada

Ada was first introduced in 1983 as what is known as Ada-83. It was influenced by the ALGOL and Pascal languages and was named after Ada Lovelace, who is believed to be the first computer programmer. Ada was originally developed at the request of the U.S. Department of Defense (DoD) and was mandated by the DoD for several years. Prior to Ada's arrival, there were literally hundreds of languages used on DoD projects. The DoD desired to standardize a language to support embedded, real-time, and mission-critical applications. Ada includes the following features: strong typing, packages (to provide modularity), run-time checking, tasks (to allow parallel processing), exception handling, and generics. Ada 95 and Ada 2005 add the object-oriented programming capability. The Ada language has been standardized through the International Standards Organization (ISO), American National Standards Institute (ANSI), and International Electrical Technical Commission (IEC) standardization efforts. Unlike most ISO/IEC standards, the Ada language definition* is publicly available for free to be used by programmers and compiler manufactures.

Ada is generally favored by those most committed to safety, because it has such a strong feature set available. It "supports run-time checks to protect against access to unallocated memory, buffer overflow errors, off-by-one errors, array access errors, and other detectable bugs" [4]. Ada also supports many compile-time checks that are not detectable until run-time in other languages or that would require explicit checks to be added to the source code.

In 1997 the DoD effectively removed its mandate to use Ada. By that time, the overall number of languages used had declined and those languages remaining were mature enough to produce quality products (i.e., a few mature languages were available rather than hundreds of immature ones).

8.2.2.3 C

C is one of the most popular programming languages in history. It was originally developed between 1969 and 1973 by Dennis Ritchie at Bell Telephone Laboratories for use with the UNIX operating system. By 1973, the language

---

* It is known as the *Ada Reference Manual* (ARM) or *Language Reference Manual* (LRM).

was powerful enough that most of the UNIX operating system kernel was rewritten in C. This made UNIX one of the first operating systems implemented in a nonassembly language [5]. C provides low-level access to memory and has constructs that map well to machine instructions. Therefore, it requires minimal run-time support and is useful for applications that were previously coded in assembly.

Some are hesitant to call C a *high-level* language; instead they prefer to call it a *mid-level* language. It does have high-level language features such as structured data, structured control flow, machine independence, and operators [6]. However, it also has low-level constructs (e.g., bit manipulation). C uses *functions* to contain all executable code. C has weak typing, may hide variables in nested blocks, can access computer memory using pointers, has a relatively small set of reserved keywords, uses library routines for complex functionality (e.g., input/output [I/O], math functions, and string manipulation), and uses several compound operators (such as, + =, − =, * =, ++, etc.). Text for a C program is free-format and uses the semicolon to terminate a statement.

C is a powerful language. With that powerful capability comes the need for extreme caution when using it. The Motor Industry Software Reliability Association's C standard (MISRA-C) [7] provides excellent guidelines to safely implement C and is frequently used as input to company-specific coding standards.

8.2.3 Choosing a Language and Compiler

When selecting a language and compiler to be used for a safety-critical project or projects, several aspects should be considered.

Consideration 1: Capabilities of the language and compiler. The language and compiler must be capable of doing their job. I once worked with a company that developed a compiler to implement a safe subset of Ada. They even went through the effort and expense to qualify it as a level A development tool. However, the Ada subset was not extensive enough to support projects in the real world; therefore, it was not utilized by the industry. Some of the basic language capabilities that are important for most projects are as follows:

- Readability of the code—consider such things as case sensitivity and mixing, understandability of reserved words and math symbols, flexible format rules, and clear naming conventions.
- Ability to detect errors at compile-time, such as typos and common coding errors.
- Ability to detect errors at run-time, including memory exhaustion checks, exception handling constructs, and math error handling (e.g., number overflow, array bound violations, and divide-by-zero) [3].
- Portability across platforms (discussed more in Consideration #9).

- Ability to support modularity, including encapsulation and information hiding.
- Strong data typing.
- Well-defined control structures.
- Support to interface with a real-time operating system (RTOS), if using an RTOS on current or future project.
- Ability to support real-time system needs, such as multitasking and exception handling.
- Ability to interface with other languages (such as assembly).
- Ability to compile and debug separately.
- Ability to interact with hardware, if not using an RTOS.

Consideration 2: Software criticality. The higher the criticality, the more controlled the language must be. A level D project *might* be able to certify with Java or C#, but level A projects require a higher degree of determinism and language maturity. Some compiler manufactures provide a real-time and/or safety-critical subset of their general-purpose language.

Consideration 3: Personnel's experience. Programmers tend to think in the language they are most familiar with. If an engineer is an Ada expert, it will be difficult to switch to C. Likewise, programming in assembly requires a special skill set. Programmers are capable of programming in multiple languages, but it takes time to become proficient and to fully appreciate the capabilities and pitfalls of each language.

Consideration 4: Language's safety support. The language and compiler must be able to meet the applicable DO-178C objectives, as well as the required safety requirements. Level A projects are required to verify the compiler output to prove that it does not generate unintended code. This typically leads organizations to select mature, stable, and well-established compilers.

Consideration 5: Language's tool support. It is important to have tools to support the development effort. Here are some of the points to consider:

- The compiler should be able to detect errors and support safety needs.
- A dependable linker is needed.
- A good debugger is important. The debugger must be compatible with the selected target computer.
- The testability of the language should be considered. Test and analysis tools are often language specific and sometimes even compiler specific.

Consideration 6: Interface compatibility with other languages. Most projects use at least one high-level language, as well as assembly. One project that I worked on used Ada, C, and assembly for the airborne software, as well as C++, Java, and C# for the tools. The language and compiler selected must be capable of interfacing with assembly and code from other utilized languages. Typically, assembly files are linked in with the compiled code of the high-level language. Jim Cooling states it well: "One highly desirable feature is the ability to reference high-level designators (e.g. variables) from assembly code (and vice versa). For professional work this should be mandatory. The degree of cross-checking performed by the linker on the high-level and assembler routines (e.g. version numbers) is important" [3].

Consideration 7: Compiler track record. Normally, a compiler that is compatible with the international standards (e.g., ANSI or ISO) is desired. Even if only a subset of the compiler is utilized, it is important to ensure that the features used properly implement the language. Most safety-critical software developers select a mature compiler with a proven track record. For level A projects, it is necessary to show that the compiler-generated code is consistent with the source code. Not all compilers can comply with these criteria.

Consideration 8: Compatibility with selected target. Most compilers are target specific. The selected compiler and environment must produce code compatible with the utilized processor and peripheral devices. Typically, the following capabilities to access processor and devices are needed: memory access (for control of data, code, heap, and stack operations), peripheral device interface and control, interrupt handling, and support of any special machine operations [3].

Consideration 9: Portability to other targets. Most companies consider the portability of the code when selecting a language and compiler. A majority of aviation projects build upon existing software or systems, rather than developing brand new code. As time progresses, processors become more capable and older ones become obsolete. Therefore, it is important to select a language and compiler that will be somewhat portable to other targets. This is not always predictable, but it should at least be considered. For example, assembly is typically not portable, so every time the processor changes, the code needs to be modified. If the same processor family is used, the change may be minimal, but it still must be considered. Ada and C tend to be more portable but may still have some target dependencies. Java was developed to be highly portable, but, as discussed before, it is not currently ready for use in the safety-critical domain.

8.2.4 General Recommendations for Programming

This section is not intended to be a comprehensive programming guide. Instead, it provides a high-level overview of safety-critical programming practices based on the thousands of lines of code in multiple languages and across multiple companies that I have examined. These recommendations are applicable to any language and may be considered in company coding standards.

Recommendation 1: Use good design techniques. Design is the blueprint for the code. Therefore, a good design is important for the generation of good software. The programmer should not be expected to make up for the shortcomings of the requirements and design. Chapter 7 provides characteristics of good design (e.g., loose coupling, strong cohesion, abstraction, and modularity).

Recommendation 2: Encourage good programming practices. Good programmers take pride in being able to do what many consider impossible; they pull off small miracles on a regular basis. However, programmers can be an odd lot. One engineer that I worked with spent his evenings reading algorithm books. I am a geek and have a house full of books, but I do not find algorithm books *that* enjoyable. The following recommendations are offered to promote good programming practices:

1. *Encourage teamwork.* Teamwork helps to filter out the bad practices and unreadable code. There are several ways to implement teamwork, including pair programming (where coding is performed by a pair of programmers), informal reviews on a daily or weekly basis, or a mentor–trainee arrangement.

2. *Hold code walkthroughs.* Formal reviews are essentially required for higher software levels. However, there are great benefits to having less formal reviews by small teams of programmers. In general, it is good to have every line of code reviewed by at least two other programmers. There are several benefits to this review process. First, it provides some healthy competition among peers—no one wants to look foolish in front of their peers. Second, reviews help to standardize the coding practices. Third, reviews help to increase the continuous improvement; when a programmer sees how someone else solved a problem, he or she can implement that into his or her own bag of tricks. Fourth, reviews can also increase reusability. There may be some routines that can be used among functions. Finally, reviews provide some continuity if someone leaves the team.

3. *Provide good code examples.* This can function as a training manual for the team. Some companies keep a *best code listing.* The listing is updated with good examples. This provides a training tool and encourages programmers to develop good code that might make it into the listing.

4. *Require adherence to the coding standard.* DO-178C requires compliance with the coding standards for levels A to C. Too often, the standards are ignored until it is time for the formal code review, when it is difficult to make changes. Programmers should be trained on the standards and required to follow them. This means having reasonable standards that can be applied.

5. *Make code available to the entire team.* When people know that others will be looking at their work, they are more prone to keep it cleaned up.

6. *Reward good code.* Recognize those who generate good code. The code quality determination is often based on the feedback of peers, the number of defects found during review, the speed of the code development, and the overall stability and maturity of the code throughout the project. The reward should be something they want. At the same time, rewards should be treated with caution, since people tend to optimize what they are being measured on. For example, basing performance on lines of code generated may actually encourage inefficient programming.

7. *Encourage the team to take responsibility for their work.* Hold each team member accountable for his or her work. Be willing to admit mistakes and avoid the blame game. Excuses are harmful to the team. Instead of excuses, propose solutions [8].

8. *Provide opportunities for professional development.* Support training and anything that will promote the professional development of your programmers.

Recommendation 3: Avoid software deterioration. Software deterioration occurs when disorder increases in the software. The progression from clean code to convoluted and incorrect code starts slowly. First, one just decides to wait to add the comments *later*; then he or she puts in some trial code to see how it works and forgets to remove it; etc. Before long, an unreadable and non-repairable set of codes appears. Code deterioration occurs when details are neglected. It is remedied by being proactive. When something looks awry, deal with it; do not ignore it. If there is no time to deal with it, keep an organized listing of issues that need to be dealt with and address them before the code review.

Recommendation 4: Keep maintainability in mind throughout. Steve McConnell writes: "Keep the person who has to modify your code in mind. Programming is communicating with another programmer first and communicating with the computer second" [6]. The majority of a coder's time is spent reviewing and modifying code, either on the first project or on a maintenance effort. Therefore, code should be written for humans, not just for machines. Many of the suggestions provided in this section will help with the maintainability. Readability and well-ordered program structures should be a priority. Decoupled code also supports maintainability.

Recommendation 5: Think the code through ahead of time. For each chapter of this book, I have spent hours researching the subject and organizing my thoughts into an outline. The actual writing occurs rather quickly once the research is done and the organization determined. The same is true when writing code.

In my experience, the most problematic code evolves when the programmers are forced to create code fast. The requirements and design are either immature or nonexistent, and the programmers just crank out code. It is like building a new house without a blueprint and without quality materials; it is going to crumble—you just do not know when.

A solid design document provides a good starting point for the programmers. However, very few will be able to directly go from design to code without some kind of intermediate step. This step could be formal or informal. It may become part of the design or may become the comments in the code itself.

McConnell recommends the Pseudocode Programming Process (PPP) [6]. The PPP uses pseudocode to design and check a routine prior to coding, reviewing, and testing the routine. The PPP uses English-like statements to explain the specific operations of the routine. The statements are language independent, so the pseudocode can be programmed in any language, and the pseudocode is at a higher level of abstraction than the code itself. The pseudocode communicates intent rather than implementation in the target language. The pseudocode may become part of the detailed design and is often easier to review for correctness than the code itself. An interesting benefit of PPP is that the pseudocode can become the outline for the source code itself and can be maintained as comments in the code. Also, pseudocode is easier to update than the code itself. For detailed steps on the PPP, see chapter 9 of *Code Complete* by Steve McConnell. PPP is just one of many possible approaches for creating routines or classes, but it can be quite effective. As noted in Chapter 7, using pseudocode as LLRs may present some certification challenges, as discussed in DO-248C frequently asked question (FAQ) #82. See Section 7.1.2 for a summary of the concerns.

Recommendation 6: Make code readable. Code readability is extremely important. It affects the ability to understand, review, and maintain the code. It helps reviewers to understand the code and ensure that it satisfies the requirements. It is also critical for code maintenance long after the programmer has transitioned to another project.

One time I was asked to review the code for a communication system. I read the requirements and design for a particular functionality and crawled through the code. The C code was very clever and concise. There were no comments, and white space was scarce. It was extremely difficult to determine what the code did and how it related to the requirements. So, I finally asked to have the programmer explain it. It took him a while to recall his intentions, but he was able to eventually walk me through the code. After I understood his thought process, I said something like: "the code is really hard to read." To which he responded without hesitation: "It was hard for me to write; I figured it should be hard for others to read." Not exactly the wisest words to say to an auditor from the Federal Aviation Administration, but at least the guy was honest. And, I think he summarized very well the mentality of many programmers.

Some programmers consider it *job security* to write cryptic code. However, it is really a liability to the company. It cannot be emphasized enough: code should be written for humans as well as the computer. McConnell writes: "The smaller part of the job of programming is writing a program so that the computer can read it; the larger part is writing it so that other humans can read it" [6].

Writing code to be readable has an impact on its comprehensibility, review-ability, error rate, debug ability, modifiability, quality, and cost [6]. All of these are important factors for real projects.

There are two aspects of coding that significantly impact the readability: (1) layout and (2) comments. Following is a summary of layout and comment recommendations to consider as a bare minimum. The "Recommended Reading" section of this chapter provides some other resources to consider.

1. Code Layout Recommendations
   a. *Show the logical structure of the code.* As a general rule, indent statements under the statement to which they are logically subordinate. Studies show that indention helps with comprehension. Two-space to four-space indentions are typically preferred, since any more than that decreases the comprehension [6].

   b. *Use whitespace generously.* Whitespace includes blank lines between blocks of code or sections of related statements that are grouped together, as well as indentions to show logical structure of the code. For example, when coding a routine, it is helpful to use blank lines between the routine header, data declarations, and body.

   c. *Consider modifiability in layout practices.* When determining the style and layout practices, use something that is easy to modify. Some programmers like to use a certain number of asterisks to make a header look pretty; however, it can take unnecessary time and resources to modify. Programmers tend to ignore things that waste their time; therefore, layout practices should be practical.

   d. *Keep closely related elements together.* For example, if a statement breaks lines, break it in a readable place and indent the second line, so it is easy to see they are together. Likewise, keep related statements together.

   e. *Use only one statement per line.* Many languages allow multiple statements on a single line; however, it becomes difficult to read. One statement per line promotes readability, complexity assessment, and error detection.

   f. *Limit statement length.* In general, statements should not be over 80 characters. Lines more than 80 characters are hard to read. This is not a hard rule but a recommended general practice to increase readability.

g. *Make data declarations clear.* To clearly declare data the following are suggested: use one data declaration per line, declare variables close to where they are first used, and order declarations by type [6].

h. *Use parentheses.* Parentheses help to clarify expressions involving more than two terms.

2. Code Comment Recommendations

Code commenting is often lacking, inaccurate, or ineffective. Here are some recommendations to effectively comment code.

a. *Comment the "why."* In general, comments should explain *why* something is done—not *how* it is done. Comments ought to summarize the purpose and goal(s) (i.e., the intent) of the code. The code itself will show how it is done. Use the comments to prepare the reader for what follows. Typically, one or two sentences for each block of code are about right.

b. *Do not comment the obvious.* Good code should be self-documenting, that is, readable without explanation. "For many well-written programs, the code is its own best documentation" [9]. Use comments to document things that are not obvious in the code, such as the purpose and goal(s). Likewise, comments that just echo back the code are not useful.

c. *Comment routines.* Use comments to explain the purpose of routines, the inputs and outputs, and any important assumptions. For routines, it is advisable to keep the comments close to the code they describe. If the comments are all included at the top of the routine, the code may still be difficult to read and the comments may not get updated with the code. It is preferable to briefly explain the routine at the top and then include specific comments throughout the routine body. If a routine modifies global data, this is important to explain.

d. *Use comments for global data.* Global data should be used cautiously; however, when they are used, they should be commented in order to ensure that they are used properly. Comment global data when they are declared, including the purpose of the data and why they are needed. Some developers even choose to use a naming convention for global data (e.g., start variables with g_). If a naming convention is not used, comments will be very important to ensure proper use of the global data [6].

e. *Do not use comments to compensate for bad code.* Bad code should be avoided—not explained.

f. *Comments should be consistent with the code.* If the code is updated, the comments are also updated.

    g. *Document any assumptions.* The programmer should document any assumptions being made and clearly identify them as assumptions.

    h. *Document anything that could cause surprises.* Unclear code should be evaluated to decide if it needs a rewrite. If the code is appropriate, a comment should be included to explain the rationale. As an example, performance issues sometimes drive some *clever* code. However, this should be the exception and not the rule. Such code should be explained with a comment.

    i. *Align comments with the corresponding code.* Each comment should align with the code it explains.

    j. *Avoid endline comments.* In general, it is best to put comments on a different line, rather than adding them at the end of the code line. The possible exceptions are data declarations and end-of-block notes for long blocks of code [6].

    k. *Precede each comment with a blank line.* This helps with the overall readability.

    l. *Write comments to be maintainable.* Comments should be written in a style that is easy to maintain. Sometimes in an effort to make things look *nice*, it can also make it difficult to maintain. Coders should not have to spend their precious time counting dashes and aligning the stars (asterisks).

    m. *Comment proactively.* Commenting should be an integral part of the coding process. If used properly, it can even become the outline for the code that the programmer then uses to organize the code. Keep in mind that if the code is difficult to comment, it might not be good code to start out with and may need to be modified.

    n. *Do not overcomment.* Just as too few comments can be bad, so can too many. I rarely see overcommented code, but when I do, it tends to be the result of unnecessary redundancy. Comments should not just repeat the code, but should explain why the code is needed. Studies at IBM showed that one comment per every 10 statements is the clarity peak for commenting. More or less comments reduced understandability [10]. Obviously, this is just a general guideline, but it is worth noting. Be careful not to concentrate too much on the number of comments, rather evaluate if the comments describe why the code is there.

Recommendation 7: Manage and minimize complexity. Managing and minimizing complexity is one of the most important technical topics in software development. Addressing complexity starts at the requirements and design level, but code complexity should also be carefully monitored and controlled.

When the code grows too complex, it become unstable and unmanageable, and productivity moves in the negative direction.

On one particularly challenging project, I actually had to read the convoluted code to understand what the software did. The requirements were useless, the design was nonexistent, and the code was hacked together. It took years to clean up the code, generate the requirements and design, and prove that it really did what it was supposed to do. Complexity was only one of the many challenges on this particular project.

To avoid such a scenario, I offer the following suggestions to help manage and minimize complexity:

- Use the concept of modularity to break the problem into smaller, manageable pieces.
- Reduce coupling between routines.
- Use overloading of operators and variable names cautiously or prohibit altogether, when possible.
- Use a single point of entry and exit for routines.
- Keep routines short and cohesive.
- Use logical and understandable naming conventions.
- Reduce the number of nested decisions, as well as inheritance tree depth.
- Avoid clever routines that are hard to understand. Instead, strive for simple and easy-to-understand.
- Convert a nested *if* to a set of *if–then–else* statements or a *case* statement, if possible.
- Use a complexity measurement tool or technique to identify overly complex code. Then rework code as needed.
- Have people unfamiliar with the code review it to ensure it is understandable. It is easy to get so engrained in the data that you lose sight of the bigger picture. Independent reviews by technically competent people can provide a valuable sanity check.
- Use modularity and information hiding to separate the low-level details from the module use. This essentially allows one to divide the problem into smaller modules (or routines, components, or classes) and hide complexity so it does not need to be dealt with each time. Each module should have a well-defined interface and a body. The body implements the module, whereas the interface is what the user sees.
- Minimize use of interrupt-driven and multitasking processing where possible.
- Limit the size of files. There is no magic number, but in general, files over 200–250 lines of code become difficult to read and maintain.

Recommendation 8: Practice defensive programming. DO-248C, FAQ #32 states: "Defensive programming practices are techniques that may be used to prevent code from executing unintended or unpredictable operations by constraining the use of structures, constructs, and practices that have the potential to introduce failures in operation and errors into the code" [11]. DO-248C goes on to recommend avoidance of input data errors, nondeterminism, complexity, interface errors, and logical errors during the programming process. Defensive programming increases the overall robustness of the code in order to avoid undesirable results and to protect against unpredictable events.

Recommendation 9: Ensure the software is deterministic. Safety-critical software must be deterministic; therefore, coding practices that could lead to nondeterminism must be avoided or carefully controlled (e.g., self-modifying code, dynamic memory allocation/deallocation, dynamic binding, extensive use of pointers, multiple inheritance, or polymorphism). Well-defined languages, proven compilers, limited optimization, and limited complexity also help with determinism.

Recommendation 10: Proactively address common errors. It can be beneficial to keep a log of common errors to help with training and raise awareness for the entire team, along with guidelines for how to avoid them. As an example, interface errors are common. They can be addressed by minimizing complexity of interfaces, consistently using units and precision, minimizing use of global variables, and using assertions to identify mismatched interface assumptions. As another example, common logic and computation errors might be avoided by examining accuracy and conversion issues (such as fixed-point scaling), watching for loop count errors, and using proper precision for floating-point numbers.

In his book *The Art of Software Testing*, Myers lists 67 common coding errors and breaks them into the following categories: data reference errors, data declaration errors, computation errors, comparison errors, control flow errors, interface errors, input/output errors, and other errors [12]. Similarly, in the book *Testing Computer Software*, Cem Kaner et al. identify and categorize 480 common software defects [13]. These sources provide a starting point for a common-errors list; however, they are just a starting point. The common errors will vary depending on language used, experience, system type, etc.

Recommendation 11: Use assertions during development. Assertions may be used to check for conditions that should never occur, whereas error-handling code is used for conditions, which could occur. Following are some guidelines for assertions [6,8]:

- Assertions should not include executable code, since assertions are normally turned off at compile-time.
- If it seems it can never happen, use an assertion to ensure that it will not.

- Assertions are useful for verifying pre- and post-conditions.
- Assertions should not be used to replace real error handling.

Recommendation 12: Implement error handling. Error handling is similar to assertions, except error handling is used for conditions, which could occur. Error handling checks for bad input data; assertions check for bad code (bugs) [6]. Data will not always come in proper format or with acceptable values; therefore, one must protect against invalid input. For example, check values from external sources for range tolerance and/or corruption, look for buffer and integer overflows, and check values of routine input parameters [11]. There are a number of possible responses to an error, including return a neutral value, substitute the next piece of valid data, return the same answer as the previous time, substitute the closest legal value, log a warning message to a file, return an error code, call an error-processing routine, display an error message, handle the error locally, or shut down the system [6]. Obviously, the response to the error will depend on the criticality of the software and the overall architecture. It is important to handle errors consistently throughout the program. Additionally, ensure that the higher level code actually handles the errors that are reported by the lower level code [6]. For example, applications (higher level code) should handle errors reported by the real-time operating system (lower level code).

Recommendation 13: Implement exception handling. Exceptions are errors or fault conditions that make further execution of a program meaningless [3]. When exceptions are thrown, an exception handling routine should be called. Exception handling is one of the most effective methods for dealing with run-time problems [3]. However, languages vary on how they implement exceptions and some do not include the exception construct at all. If exceptions are not part of the language, the programmer will need to implement checks for error conditions in the code. Consider the following tips for exceptions [6,8]:

- Programs should only throw exceptions for exceptional conditions (i.e., ones that cannot be addressed by other coding practices).
- Exceptions should notify other parts of the program about errors that require action.
- Error conditions should be handled locally when possible, rather than passing them on.
- Exceptions should be thrown at the right abstraction level.
- The exception message should identify the information that led to the exception.
- Programmers should know the exceptions that the libraries and routines throw.
- The project should have a standard approach for using exceptions.

Recommendation 14: Employ common coding practices for routines, variable usage, conditionals, loops, and control. Coding standards should identify the recommended practices for routines (such as functions or procedures), variable usage (e.g., naming conventions and use of global data), conditionals, control loops, and control issues (e.g., recursion, use of *goto*, nesting depth, and complexity).*

Recommendation 15: Avoid known troublemakers. Developing an exhaustive list of everything that can go wrong in the coding process is impossible. However, there are a few known troublemakers out there and practices to avoid them, as noted in the following:

1. *Minimize the use of pointers.* Pointers are one of the most error-prone areas of programming. I cannot even begin to describe the hours I have spent tracking down issues with pointers—always staring some insane deadline in the face. Some companies choose to avoid pointers altogether. If you can find a reasonable way to work around pointers, it is definitely recommended. However, that may not always be an option. If pointers are used, their usage should be minimized, and they should be used cautiously.†

2. *Limit use of inheritance and polymorphism.* These are related to object-oriented technology, which is discussed in Chapter 15.

3. *Be careful when using dynamic memory allocation.* Most certification projects prohibit dynamic memory allocation. DO-332 (the object-oriented supplement) provides some recommendations for how to handle it safely, if it is used.

4. *Minimize coupling and maximize cohesion.* As noted in Chapter 7, it is desirable to minimize the data and control coupling between code components and to develop highly cohesive components. The design sets the stage for this concept. However, the programmers will be the ones who actually implement it. Therefore, it is worth emphasizing again. Andrew Hunt and David Thomas recommend: "Write shy code—modules that don't reveal anything unnecessary to other modules and that don't rely on other modules' implementation" [8]. This is the concept of decoupled code. In order to minimize coupling, limit module interaction; when it is necessary for modules to interact, ensure that it is clear why the interaction is needed and how it takes place.

5. *Minimize use of global data.* Global data are available to all design sections and, therefore, may be modified by any individual as work

---

* Although not specific for safety-critical software, Steve McConnell's book, *Code Complete* [6], provides detailed recommendations on each of these topics.
† Steve McConnell's *Code Complete* [6] provides some suggestions for avoiding errors in pointer usage.

progresses. As already noted, global data should be minimized in order to support loose coupling and to develop more deterministic code. Some of the potential issues with global data are as follows: the data may be inadvertently changed, code reuse is hindered, initialization ordering of the global data can be uncertain, and the code becomes less modular. Global data should only be used when absolutely necessary. When it is used, it helps to distinguish it somehow (perhaps by a naming convention). Also, it is useful to implement some kind of lock or protection to control access to global data. Additionally, an accurate and current data dictionary with the global data names, descriptions, types, units, readers, and writers is important. An accurate documentation of the global data will be essential when analyzing the data coupling (which is discussed in Chapter 9).

6. *Use recursion cautiously or not at all.* Some coding standards completely prohibit recursion, particularly for more critical software levels. If recursion is used, possibly for lower levels of criticality, explicit safeguards should be included in the design to prevent stack overrun due to unlimited recursion. DO-178C section 6.3.3.d promotes the prevention of "unbounded recursive algorithms" [1]. DO-248C FAQ #39 explains: "An unbounded recursive algorithm is an algorithm that directly invokes itself (self-recursion) or indirectly invokes itself (mutual recursion) and does not have a mechanism to limit the number of times it can do this before completing" [11]. The FAQ goes on to explain that recursive algorithms need an upper bound on the number of recursive calls and that it should be shown that there is adequate stack space to accommodate the upper bound [11].

7. *Reentrant functions should be used cautiously.* Similar to recursion, many developers prohibit the use of reentrant code. If it is allowed, as is often the case in multithreaded code, it must be directly traceable to the requirements and should not assign values to global variables.

8. *Avoid self-modifying code.* Self-modifying code is a program that modifies its instruction stream at run-time. Self-modifying code is error prone and difficult to read, maintain, and test. Therefore, it should be avoided.

9. *Avoid use of goto statement.* Most safety-critical coding standards prohibit the use of *goto* because it is difficult to read, can be difficult to prove proper code functionality, and can create *spaghetti code.* That being said, if it is used, it should be used sparingly and very cautiously.*

---

* In his book, *Code Complete* [6], Steve McConnell provides some suggestions on how to use *goto* conscientiously.

10. *Justify any situations where you opt to not follow these recommendations.* These are merely recommendations and there may be some situations that warrant violating one or more of these recommendations. However, it should also be noted that these recommendations are based on several years of practice and coordinating with international certification authorities. When opting not to follow these recommendations, be sure to technically justify it and to coordinate with the certification authority.

Recommendation 16: Provide feedback when issues are noted in requirements or design. During the coding phase, programmers are encouraged to provide feedback on any issues identified with the requirements or design. Design and coding phases are closely related and sometimes overlap. In some projects the designer is also the programmer. When the designers and programmers are separate, it is useful to include the programmers in the requirements and design reviews. This allows the programmers the opportunity to understand the requirements and design and to provide early feedback.

Once coding begins, there should be an organized way for programmers to identify issues with requirements and design and to ensure that appropriate action is taken. The problem reporting process is normally used to note issues with the requirements or design; however, there needs to be a proactive response to the identified issues. Otherwise, there is a risk that requirements, design, and code may become inconsistent.

Recommendation 17: Proactively debug the code. As will be discussed in Chapter 9, debug and developmental testing (e.g., unit testing and static code analysis) should occur during the coding phase. Do not wait until formal testing to find the code errors.

8.2.5 Special Code-Related Topics

8.2.5.1 Coding Standards

As noted in Chapter 5, coding standards are developed during the planning process to define how the selected programming language will be used on the project.* It is important to create complete coding standards and train the team how to use the standards. Typically, a company spends considerable time on their coding standards because the standards will be used on multiple projects throughout the company. The information provided in this chapter provides some concepts and issues to address in the coding standards.

I recommend including the rationale for each rule or recommendation in the standards, along with examples. Programmers are more apt to apply the standards if they understand *why* the suggestions are there.

---

* DO-178C section 11.8 explains the expected contents of the standards.

Too frequently, code is written without proper attention to the standards. Then, during code reviews, significant problems are found and the code must be reworked. It is more efficient to ensure that the coders understand and apply the standards from the start.

8.2.5.2 Compiler-Supplied Libraries

Most compiler manufacturers supply libraries with their compiler to be used by the programmers to implement functionality in the code (e.g., math functions). When called, such library functions become part of the airborne software and need to satisfy the DO-178C objectives, just like other airborne software. The Certification Authorities Software Team (CAST)* position on this topic is documented in the CAST-21 paper, entitled "Compiler-Supplied Libraries." The position basically requires the library code to meet the DO-178C objectives (i.e., it requires requirements, design, and tests for the library functions) [14]. Typically, manufacturers either develop their own libraries, including the supporting artifacts, or they reverse engineer the compiler-supplied library code to develop the requirements, design, and test cases. Functions that do not have the supporting artifacts (requirements, design, source code, tests, etc.) should either be removed from the library or purposely deactivated (removal is preferred, so the functions are not unintentionally activated in the next use of the library). Many companies develop an entire library, so it can be used on multiple projects. To make the libraries reusable, it is advisable to separate the library requirements, design, and test data from the other airborne software. Depending on the project, the libraries may need to be retested on the subsequent project (due to compiler setting differences, processor differences, etc.). For level C or D applications, it might be feasible to use testing and service history to demonstrate the library functionality. It is advisable to explain the approach for libraries in the Plan for Software Aspects of Certification (PSAC) in order to secure the certification authority's agreement.

8.2.5.3 Autocode Generators

This chapter has mostly focused on handwritten source code. If an ACG is used, many of the concerns in this chapter should be considered when developing the ACG. Additionally, Chapter 13 provides some additional insight into the tool qualification process that may be required if the code generated by an ACG is not reviewed.

---

* CAST is a team of international certification authorities who strive to harmonize their positions on airborne software and aircraft electronic hardware in CAST papers.

8.3 Verifying the Source Code

DO-178C Table A-5 objectives 1–6 and section 6.3.4 address the source code verification. Most of these objectives are satisfied with a code peer review (using the same basic review process discussed in Chapter 6 but with the focus on code). Each DO-178C Table A-5 objective, along with a brief summary of what is expected, is included in the following [1]:*

- *DO-178C Table A-5 objective 1*: "Source Code complies with low-level requirements." Satisfying this objective involves a comparison of the source code and low-level requirements to ensure that the code accurately implements the requirements and only the requirements. This objective is closely related to Table A-5 objective 5, since traceability helps with the compliance determination.

- *DO-178C Table A-5 objective 2*: "Source Code complies with software architecture." The purpose of this objective is to ensure that the source code is consistent with the architecture. This objective ensures that the data and control flows in the architecture and code are consistent. As will be discussed in Chapter 9, this consistency is important to support data and control coupling analyses.

- *DO-178C Table A-5 objective 3*: "Source Code is verifiable." This objective focuses on the testability of the code itself. The code needs to be written to support testing. Chapter 7 identifies characteristics of testable software.

- *DO-178C Table A-5 objective 4*: "Source Code conforms to standards." The purpose of this objective is to ensure that the code conforms to the coding standards identified in the plans. This chapter and Chapter 5 discuss the coding standards. Normally, a peer review and/or a static analysis tool are used to ensure that the code satisfies the standards. If a tool is used, it may need to be qualified.

- *DO-178C Table A-5 objective 5*: "Source Code is traceable to low-level requirements." This objective confirms the completeness and accuracy of the traceability between source code and low-level requirements. The traceability should be bidirectional. All requirements should be implemented and there should be no code that does not trace to one or more requirements. Generally, low-level requirements trace to source code functions or procedures. (See Chapter 6 for more information on bidirectional traceability.)

- *DO-178C Table A-5 objective 6*: "Source Code is accurate and consistent." This objective is a challenging one. Compliance with it involves a review of the code to look for accuracy and consistency.

---

* DO-178C Table A-5 objective 7 is explained in Chapter 9.

However, the objective reference also states that the following are verified: "stack usage, memory usage, fixed point arithmetic overflow and resolution, floating-point arithmetic, resource contention and limitations, worst-case execution timing, exception handling, use of uninitialized variables, cache management, unused variables, and data corruption due to task or interrupt conflicts" [1]. This verification activity involves more than a code review. Chapter 9 explains some of the additional verification activities that are needed to evaluate stack usage, worst-case execution timing, memory usage, etc.

8.4 Development Integration

DO-178C identifies two aspects of integration. The first is the integration during the development effort; that is, the compiling, linking, and loading processes. The second aspect is the integration during the testing effort, which includes software/software integration and software/hardware integration. Integration typically starts by integrating software modules within a functional area on a host, then integrating multiple functional areas on the host, and then integrating the software on the target hardware. The effectiveness of the integration is proven through the testing effort. This section considers the integration during the development effort (see Figure 8.1). Chapter 9 considers integration during the testing phase.

8.4.1 Build Process

Figure 8.1 provides a high-level view of the integration process, which includes the code compilation, the linking, and the loading onto the target computer. The process of using the source code to develop executable

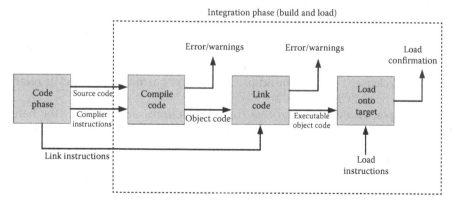

FIGURE 8.1
The code and integration phase.

object code is called the *build process*. The output of the coding phase includes the source code, as well as the compile and link instructions. The compile and link instructions are documented in *build instructions*. The build instructions must be well documented with repeatable steps, because they document the process used to build the executable image that will be used for safety-critical operation. DO-178C section 11.16.g suggests that the build instructions be included in the Software Configuration Index (SCI).

The build instructions often include multiple scripts (such as makefiles). Because these scripts have a crucial role in the development of the executable image(s), they should be under configuration control and reviewed for accuracy, just like the source code. Unfortunately, the review of compile and link data is frequently overlooked during the planning phases. Some organizations take great care with the source code but neglect the scripts that enable the build process altogether. When explaining *source code* DO-178C section 11.11 states: "This data consists of code written in source language(s). The Source Code is used with the compiling, linking, and loading data in the integration process to develop the integrated system or equipment" [1]. Therefore, the compile and link data should be carefully controlled along with the source code. This requires verification and configuration management of the data. The software level determines the extent of the verification and configuration management.

The build process relies on a well-controlled development environment. The development environment lists all tools (with versions), hardware, and settings (including compiler or linker settings) of the build environment. DO-178C suggests that this information be documented in a Software Life Cycle Environment Configuration Index (SLECI). The SLECI is discussed in Chapter 10.

Prior to building the software for release, most companies require a clean build. In this situation, the build machine is *cleaned* by removing all of its software. The build machine is then loaded with the approved software using the *clean build procedure*. This clean build ensures that the approved environment is used and that the build environment can be regenerated (which is important for maintenance). After the build machine is properly configured, the software build instructions are followed to generate the software for release. The clean build procedures are normally included or referenced in the SLECI or the SCI.

One aspect of the build process that is sometimes overlooked is the handling of compiler and linker warnings and errors. The build instructions should require that warnings and errors be examined after compiling and linking. The build instructions should also identify any acceptable warnings or the process for analyzing warnings to determine if they are acceptable. Errors are generally not acceptable.

In my experience, the clean build procedures and software build instructions are often not documented well; hence, they are not repeatable.

The build process frequently relies on the engineer who performs the build on an almost daily basis. To address this common deficiency, it is beneficial to have someone who did not write the procedures and who does not normally perform the build to execute the procedures to confirm repeatability. I recommend doing this as part to build procedures review.

8.4.2 Load Process

The load process controls the loading of the executable image(s) onto the target. There are typically load procedures for the lab, the factory, and the aircraft (if the software is field-loadable). The load procedures should be documented and controlled. DO-178C section 11.16.k explains that the procedures and methods used to load the software into the target hardware should be documented in the SCI.*

The load instructions should identify how a complete load is verified, how incomplete loads are identified, how corrupted loads are addressed, and what to do if an error occurs during the loading process.

For aircraft systems, many manufacturers use the ARINC 615A [15] protocol and a high-integrity cyclic redundancy check to ensure that the software is properly loaded onto the target.

8.5 Verifying the Development Integration

The verification of the development integration process typically includes the following activities to ensure that the integration process is complete and correct:

- Review the compile data, link data, and load data (e.g., scripts used to automate the build and load).
- Review the build and load instructions, including an independent execution of the instructions, to ensure completeness and repeatability.
- Analyze link data, load data, and memory map to ensure hardware addresses are correct, there are no memory overlaps, and there are no missing software components. This addresses DO-178C Table A-5 objective 7, which states: "Output of software integration process is complete and correct" [1]. These analyses are further discussed in Chapter 9.

---

* This was not included in DO-178B.

References

1. RTCA DO-178C, *Software Considerations in Airborne Systems and Equipment Certification* (Washington, DC: RTCA, Inc., December 2011).
2. Wikipedia, Assembly language, http://en.wikipedia.org/wiki/Assembly_language (accessed on January 2011).
3. J. Cooling, *Software Engineering for Real-Time Systems* (Harlow, U.K.: Addison-Wesley, 2003).
4. Wikipedia, Ada programming language, http://en.wikipedia.org/wiki/Ada_(programming_language) (accessed on January 2011).
5. Wikipedia, C programming language, http://en.wikipedia.org/wiki/C_(programming_language) (accessed on January 2011).
6. S. McConnell, *Code Complete*, 2nd edn. (Redmond, WA: Microsoft Press, 2004).
7. The Motor Industry Software Reliability Association (MISRA), *Guidelines for the Use of the C Language in Critical Systems*, MISRA-C:2004 (Warwickshire, U.K.: MISRA, October 2004).
8. A. Hunt and D. Thomas, *The Pragmatic Programmer* (Reading, MA: Addison-Wesley, 2000).
9. C. M. Krishna and K. G. Shin, *Real-Time Systems* (New York: McGraw-Hill, 1997).
10. C. Jones, *Software Assessments, Benchmarks, and Best Practices* (Reading, MA: Addison-Wesley, 2000).
11. RTCA DO-248C, *Supporting Information for DO-178C and DO-278A* (Washington, DC: RTCA, Inc., December 2011).
12. G. J. Myers, *The Art of Software Testing* (New York: John Wiley & Sons, 1979).
13. C. Kaner, J. Falk, and H. Q. Nguyen, *Testing Computer Software*, 2nd edn. (New York: John Wiley & Sons, 1999).
14. Certification Authorities Software Team (CAST), Compiler-supplied libraries, Position Paper CAST-21 (January 2004, Rev. 2).
15. Aeronautical Radio, Inc., Software data loader using ethernet interface, ARINC REPORT 615A-3 (Annapolis, MD: Airlines Electronic Engineering Committee, June 2007).

Recommended Reading

1. S. McConnell, *Code Complete*, 2nd edn. (Redmond, WA: Microsoft Press, 2004).
2. A. Hunt and D. Thomas, *The Pragmatic Programmer* (Reading, MA: Addison-Wesley, 2000).
3. B. W. Kernighan and R. Pike, *The Practice of Programming* (Reading, MA: Addison-Wesley, 1999).
4. The Motor Industry Software Reliability Association (MISRA), *Guidelines for the Use of the C Language in Critical Systems*, MISRA-C:2004 (Warwickshire, U.K.: MISRA, October 2004).

5. ISO/IEC PDTR 15942, *Guide for the Use of the Ada Programming Language in High Integrity Systems*, http://anubis.dkuug.dk/JTC1/SC22/WG9/documents.htm (July 1999).

6. L. Hattan, *Safer C: Developing Software for High-Integrity and Safety-Critical Systems* (Maidenhead, U.K.: McGraw-Hill, 1995).

7. S. Maguire, *Writing Solid Code* (Redmond, WA: Microsoft Press, 1993).

9

Software Verification

Acronyms

BVA	boundary value analysis
CAST	Certification Authorities Software Team
CC1	control category #1
CC2	control category #2
CCB	change control board
DC/CC	data coupling and control coupling
DP	discussion paper
FAA	Federal Aviation Administration
IMA	integrated modular avionics
ISR	interrupt service routine
MC/DC	modified condition/decision coverage
OOT	object-oriented technology
PR	problem report
PSAC	Plan for Software Aspects of Certification
RAM	random access memory
SAS	Software Accomplishment Summary
SCMP	Software Configuration Management Plan
SDP	Software Development Plan
SLECI	Software Life Cycle Environment Configuration Index
SQA	software quality assurance
SQAP	Software Quality Assurance Plan
SVCP	Software Verification Cases and Procedures
SVP	Software Verification Plan
SVR	Software Verification Report
TRR	test readiness review
WCET	worst-case execution timing

9.1 Introduction

Verification is an integral process that applies throughout the entire software life cycle. It starts in the planning phase and goes all the way through production release and even into maintenance.

DO-178C glossary defines verification as: "The evaluation of the outputs of a process to ensure correctness and consistency with respect to the inputs and standards provided to that process" [1]. The verification guidance in DO-178C includes a combination of reviews, analyses, and tests. Reviews and analyses assess the accuracy, completeness, and verifiability of the outputs of each of the life cycle phases (including planning, requirements, design, code/integration, test development, and test execution). In general, a review provides a qualitative assessment of correctness, whereas an analysis provides repeatable evidence of correctness [1]. Testing is "the process of exercising a system or system component to verify that it satisfies specified requirements and to detect errors" [1]. All three approaches are used extensively when verifying safety-critical software.

When performing verification, the terms *error*, *fault*, and *failure* are commonly used. The DO-178C glossary defines each term as follows [1]:

- "Error—With respect to software, a mistake in requirements, design, or code."
- "Fault—A manifestation of an error in software. A fault, if it occurs, may cause a failure."
- "Failure—The inability of a system or system component to perform a required function within specified limits. A failure may be produced when a fault is encountered."

The primary purpose of verification is to identify errors, so that they can be corrected before they become faults or failures. In order to identify errors as early as possible, verification must start early in the software life cycle.

9.2 Importance of Verification

Ronald Reagan is credited with, or at least popularizing, the saying: "Trust but verify." George Romanski puts it this way: "Many people can write software, but not many people would trust their lives with it until it was verified." Verification is an important process in any software life cycle but is especially so for safety-critical software. In DO-178C, over half of the objectives are classified as verification objectives. The more critical the software, the more verification activities are required, and the more confidence we have that errors have been identified and removed.

According to the many program managers that I talk to and work with, for safety-critical software, over half of the software project budget is dedicated to verification. Unfortunately, in too many projects, the verification activity is seen as a necessary evil; therefore, it is understaffed and is approached as just a check mark activity. Despite the fact that volumes of evidence show that early error detection saves time and money, many projects continue to just *get through* the reviews to satisfy the certification authority and push testing off to the end. Too often junior engineers are thrown at the reviews, analyses, and tests to *just get it done*. I do not have anything against junior engineers; in fact, I used to be one. However, developing good verification skills takes time and proper training. Verification should include experienced engineers. It is fine to use less experienced engineers as well; but, those who have been through several projects are typically the ones who find the errors that will wreak havoc on the project if left undetected.

A colleague told me about how he once walked into the lab and observed a junior engineer performing a test of a minor function. While the small part of the system being tested was working properly, the engineer failed to notice the cockpit warnings that were being displayed from a major fault in the system. This was a case of an inexperienced engineer getting a *check mark* for his assigned task but missing the basic system understanding. It is like having a mechanic check your car and say the tires are properly inflated, while at the same time smoke is pouring from the engine. The challenge is to know the entire system and to not have target fixation (tunnel vision).

From a safety perspective, verification is absolutely essential. It is used to satisfy the regulations by confirming that the software performs its intended function—and only its intended function. Basically, without good verification, development assurance has no merit, because it is verification that builds confidence in the product.

9.3 Independence and Verification

Before going into some of the details of verification, let us briefly examine the topic of independence. If you have ever tried to edit your own work, you know how hard it is to find those sneaky typos. You know how it is supposed to read, so you just overlook them. The same situation is also true when verifying software data. Therefore, as the criticality of the software increases, the required independence between activities increases.

DO-178C defines independence as follows:

> Separation of responsibilities which ensures the accomplishment of objective evaluation. (1) For software verification process activities, independence is achieved when the verification activity is performed by a person(s) other than the developer of the item being verified, and a tool(s) may be used to achieve equivalence to the human verification activity. (2) For the software quality assurance process, independence also includes the authority to ensure corrective action [1].

This chapter is concerned with the first half of the definition, which addresses *verification independence*. As can be seen from the definition, verification independence does not require a separate organization—only a separate person(s) or tool(s). Chapter 11 will explore the second half of the definition, which addresses *software quality assurance independence*.

DO-178C Tables A-3 to A-7 identify the verification objectives that require independence (shown with a filled circle [●]). For level A, 25 verification objectives require independence; for level B, only 13 verification objectives need to be satisfied with independence; and for levels C and D no verification objectives require independence. Independence for DO-178C Tables A-3 through A-5 is typically satisfied by someone reviewing the data that did not write the data. However, DO-178C Table A-5 objectives 6 and 7 tend to require some analysis and/or test as well. The two independence objectives in DO-178C Table A-6 are normally satisfied by having someone who did not write the code write the tests. For DO-178C Table A-7, all level A and three level B objectives require independence. Table A-7 independence is normally satisfied by a combination of reviews (objectives 1–4) and analyses (objectives 5–9).*

DO-248C discussion paper (DP) #19 explains the typical interpretation of verification independence. This DP was based on a Certification Authorities Software Team (CAST)† paper (CAST-26); therefore, it provides insight into how certification authorities normally interpret verification independence [2]. It should be noted that the CAST paper promoted more independence between development activities than DO-248C DP #19 identifies. DO-248C clarifies that development independence is only needed between the developer of the source code and the test specifications; other development activities do not require an independent person (or tool) [3]. DO-178C also clarifies this in section 6.2.e, which states: "For independence, the person who created a set of low-level requirements-based test cases should not be the same person who developed the associated Source Code from those low-level requirements" [1].

---

* The details of the verification objectives are discussed later; for now, the focus is on which objectives require independence.

† CAST is a team of international certification authorities who strive to harmonize their positions on airborne software and aircraft electronic hardware in CAST papers.

As mentioned elsewhere, the verification is only as good as the person or tool doing the verifying. Therefore, it is important to use skilled personnel or effective tools. In some cases the tools may need to be qualified (see Chapter 13).

There are a few DO-178C objectives that do not require independence; however, it is still a good practice to have it—particularly for levels A and B software. Experience shows that independent verification is one of the most effective ways to find errors early. Many mature companies use independence for their requirements and design reviews at all levels, even if it is not required, because it is effective at finding errors and saves time and money in the long run. Also, as will be discussed shortly, verifiers tend to have a different mentality than developers, so they are often able to find more errors than a developer reviewing his or her own work.

9.4 Reviews

Let us now examine the three verification approaches in DO-178C. This section considers reviews; the next two sections will examine analysis and test, respectively.

As noted earlier, a review is a qualitative assessment of an artifact for correctness and compliance to the required objectives [1]. Chapters 6 through 8 discussed reviews of requirements, design, and code. Most companies use a peer review process to review their artifacts. The peer review involves a team of reviewers—each with a specific purpose and focus. Chapter 6 provided suggestions for conducting an effective peer review. Typically, a review references the applicable standard(s) and includes a checklist to guide the reviewers. The reviewers document their comments and then each comment is appropriately addressed or dispositioned and verified. If the review leads to significant updates, it may be necessary to completely re-review the data.

9.4.1 Software Planning Review

Chapter 5 discussed the development of five software plans and three standards. For levels A, B, and C DO-178C Table A-1 objectives 6 and 7 require that plans comply with DO-178C and are coordinated [1]. These objectives are normally satisfied with a review of the plans and standards to ensure consistency and compliance with DO-178C guidance. The preferred practice is to review each document separately; followed by a review of all the plans and standards together once they are all written. Without the review of the documents together, inconsistencies and gaps may go undetected. I often emphasize: "Do not let the certification authority or authorized designee be the first to read your plans and standards together."

9.4.2 Software Requirements, Design, and Code Reviews

Chapters 6 through 8 discussed the characteristics of good requirements, design, and code. Additionally, these chapters identified the objectives (from DO-178C Tables A-3 through A-5) for verifying each of the artifacts. The review of each life cycle data item should happen as early as possible in the life cycle to identify and correct errors proactively. Requirements, design, and code are often informally reviewed several times before the formal peer review. Additionally, coders normally perform their own debug tests prior to the code peer review. The informal reviews and debug activities sift out the big and most obvious errors earlier, reducing rework later.

9.4.3 Test Data Reviews

Please note that peer reviews are also used to verify test cases and procedures, analysis procedures and results, and test results. These reviews are discussed later in this chapter.

9.4.4 Review of Other Data Items

Reviews are also used to ensure accuracy and correctness of other important data items (e.g., Software Life Cycle Environment Configuration Index, Software Configuration Index, and Software Accomplishment Summary).

9.5 Analyses

An analysis is a verification activity that provides repeatable evidence of correctness [1]. There are several types of analyses performed during the safety-critical software life cycle. Two main categories of analysis are needed for compliance with DO-178C: (1) code and integration analyses and (2) coverage analyses. Other analyses are also frequently conducted, depending on the selected verification approach. The typical code/integration analyses are explained in this section. Coverage analysis is discussed later in this chapter (see Section 9.7).

Engineers (myself included) sometimes use the terms *analysis* or *analyze* loosely. However, *analysis* for DO-178C compliance has a specific meaning—it is to be *repeatable*, therefore, it needs to be well documented. When reviewing data as an FAA designee, I frequently find substantial issues with analyses. Oftentimes, the so-called *analysis* is not written down; therefore, it is not repeatable. Likewise, when analyses *are* written, they are often missing criteria to determine success. *Smoke-and-mirrors, hand-waving,* or *black-magic* analyses are not acceptable. An analysis should have procedures and results. The procedures include the following [4]:

- Purpose, criteria, and related requirements
- Detailed instructions for conducting the analysis
- Analysis acceptability and completion criteria

The analysis results include the following [4]:

- Identification of the analysis procedure
- Identification of data item analyzed
- Identification of who performed the analysis
- Analysis results and supporting data
- Corrective actions generated as a result of analysis
- Analysis conclusion with the substantiating data

Analyses should be performed with the appropriate level of independence, as identified by the DO-178C Annex A tables. If an analysis is used in lieu of a test, the same level of independence required for the testing objective is needed.

Analyses should be started as soon as feasible in order to identify issues as early as possible. Normally, once a code baseline is established, the analysis activities can begin. The final analyses will not be performed until the software is finalized, but preliminary analyses may uncover some issues. I recently consulted on a project that discovered their worst-case execution time (WCET) was about 110%–130% of what was required; essentially, they had negative margins, which meant that functions might not run. Unfortunately, the problem was found a few weeks before the targeted certification date and caused lengthy delays. When it comes to analysis, the ideal and reality often collide (usually because of staff shortage and schedule compression); however, the longer a team waits to start analyses, the more risk they incur.

The typical integration analyses performed for DO-178C compliance are briefly discussed.* The results of all integration analyses are normally summarized in a software verification report, which is explained in Section 9.6.7. Some analyses are also summarized in the Software Accomplishment Summary (SAS) as software characteristics. The analyses summarized in the SAS are noted later. The SAS itself is discussed in Chapter 12.

---

* DO-178C sections 6.3.4.f and 6.3.5 allude to several of these as part of code and integration verification.

9.5.1 Worst-Case Execution Time Analysis

Knowing a program's timing characteristics is essential for the successful design and execution of real-time systems. A critical timing measure is the WCET of a program. WCET is the longest possible time that it may take to complete the execution of a set of tasks on a given processor in the target environment. WCET analysis is performed to verify that the worst-case time stays within its allocation. Although the approach taken depends on the software and system architecture, WCET is typically both analyzed and measured.

When analyzing WCET, each branch and loop in the code is generally analyzed to determine the worst-case execution path(s) through the code. The time for each branch and loop in the worst-case execution path are then summed together to give the WCET.* This time is verified against time allocated in the requirements. The analysis needs to be confirmed with actual timing measurements. There are several factors that complicate WCET analysis, including interrupts, algorithms with multiple decision steps, data or instruction cache usage, scheduling methods, and use of a real-time operating system [3]. These factors should be considered during the development of the software in order to ensure the WCET will fall within the required margins. Use of cache memory (such as L1, L2, and L3 caches) or pipelining further complicates the WCET analysis and requires additional analysis to ensure that the cache and pipeline impact on timing is understood.†

Oftentimes, tools are used to help identify the worst-case path. When this is the case, the accuracy of the tool needs to be determined. There are several approaches to confirming the tool accuracy, including (1) manually verifying the tool output, (2) qualifying the tool, or (3) running an independent tool in parallel and comparing results. Be warned that tools have similar challenges as manual analyses, when cache or pipelining is used.

The WCET analysis approach and results are summarized in the SAS as part of the *software characteristics* (as will be discussed in Chapter 12).

9.5.2 Memory Margin Analysis

A memory margin analysis is performed to ensure that there are adequate margins for production operation and future growth. All used memory is analyzed, including nonvolatile memory (NVM), random access memory (RAM), heap, stack, and any dynamically allocated memory (if used).

---

* Care must be used when gathering WCET data. It is quite possible to have a WCET greater than 100% yet have the system perform safely. For example, the worst-case branches in a procedure may be mutually exclusive. The problem gets worse when the mutual exclusion occurs across components, that is, the worst-case time in component A cannot occur at the same time as the worst-case time in component B. Yet if both components are in the same execution frame, the WCET data may sum them both.

† CAST-20, entitled *Addressing cache in airborne systems and equipment* [5] provides additional details on the concerns regarding WCET when cache and/or pipelining is used.

The memory margin analysis approach and results are summarized in the SAS as part of the *software characteristics*.

As an example, stack usage is typically performed by analyzing the source code and determining the deepest level function call tree during routine processing and during interrupt processing. These functions are then used to determine the greatest amount of stack memory from the combined call trees. If this is performed at the source level, then further analysis is required to determine how much additional data are used in each function for saved registers, formal parameters, local variables, return address data, and any compiler-required intermediate results. Stack analysis performed on the executable image will factor this in. The analyzed stack usage amount is then compared with the amount of stack available to determine that there is adequate margin. For partitioned or multitasked software, this is repeated for all partitions and tasks as each will have its own stack. As with the timing analysis, the stack usage analysis is typically confirmed using actual measurements, unless it is performed by a qualified stack analysis tool.

9.5.3 Link and Memory Map Analysis

Link analysis verifies that the modules for the software build are mapped correctly into the processor memory segments defined by the corresponding linker command files. The link analysis typically involves an inspection of the memory map file to verify the following:

- The origin, maximum length, and attributes of each segment allocated by the linker correspond to the segment definition specified in the linker command file.
- None of the allocated segments overlap.
- The actual total allocated length of each segment is less than or equal to the maximum length specified in the linker command file.
- The various sections defined by each source code module are mapped into the proper segments by the linker.
- Only the expected object modules from the source code appear.
- Only the expected procedures, tables, and variables from the linked object modules appear.
- The linker assigns correct values to the output linker symbols.

In an integrated modular avionics (IMA) system, some of this checking may be performed against the configuration data as well.*

---

* IMA systems are briefly discussed in Chapters 20 and 21. Likewise, configuration data is examined in Chapter 22.

9.5.4 Load Analysis

The load analysis is sometimes performed in conjunction with the link analysis. It verifies the following:

- All software components are built and loaded into the correct location.
- Invalid software is not loaded onto the target.
- Incorrect or corrupted software will not be executed.
- Loading data are correct.

9.5.5 Interrupt Analysis

For real-time systems, an interrupt analysis is often performed to verify that (1) all interrupts enabled by the software are correctly handled by the corresponding interrupt service routines (ISRs) defined in the software and (2) there are no unused ISRs. The interrupt analysis includes examining the source code to determine the set of enabled interrupts and verifying that an ISR exists for each enabled interrupt. Each ISR is generally analyzed to confirm the following:

- The ISR is located in memory correctly.
- At the beginning of the ISR the system context is saved.
- The operations performed within the ISR are appropriate for the corresponding physical interrupt, for example, no blocking operations are permitted in the interrupt context.
- All time-critical operations occurring within the ISR complete before any other interrupt can occur.
- Interrupts are disabled prior to passing data to control loops.
- The time that interrupts are disabled for passing data is minimized.
- The system context is restored at the end of the ISR.

9.5.6 Math Analysis

While not always necessary, some teams perform a *math analysis* to ensure that mathematical operations do not have a negative impact on the software operation. The math analysis is sometimes performed as part of the code review, and therefore it is not always documented as a separate analysis (if this is the case, it should be noted in the code review checklist). The analysis typically involves reviewing each arithmetic or logical operation in the code to do the following:

- Identify the variables that comprise each arithmetic/logical operation.
- Analyze each variable's declaration (including its scaling) to verify that the variables used in the arithmetic/logic operation are properly declared and correctly scaled (with appropriate resolution).
- Verify that overflow from the arithmetic operation cannot occur.

In some projects, static code analyzers are used to perform or supplement this analysis. Additionally, if math libraries are used, they will need to be verified for correctness as well (see Section 8.2.5.2). Of particular interest is the behavior of the floating point operations at the boundaries. A typical modern processor may have floating point states such as plus and minus zero, plus and minus infinity, and denormalized numbers known as *NaN*'s (not a number).* The behavior of the floating point algorithms which could get such values should be specified in robustness requirements and be verified.

9.5.7 Errors and Warnings Analysis

As noted in Chapter 8, during the build procedure, the compiler and linker may generator errors and warnings. Errors should be resolved; however, some warning may be acceptable. Any unresolved warnings not justified in the build procedures need to be analyzed to confirm that they do not impact the software's intended functionality.

9.5.8 Partitioning Analysis

If a system includes partitioning, a partitioning analysis is often performed to confirm the robustness of the partitioning. This analysis is particularly pertinent to IMA systems and partitioned operating systems. (See Chapter 21 for discussion of partitioning analysis.)

9.6 Software Testing

High-level and low-level requirements-based testing are major activities that are required for compliance with DO-178C.† DO-178C Table A-6 summarizes the objectives for testing. The objectives include the development and execution of test cases and procedures to verify the following [1]:

- *DO-178C Table A-6 objective 1*: "Executable Object Code complies with high-level requirements."
- *DO-178C Table A-6 objective 2*: "Executable Object Code is robust with high-level requirements."
- *DO-178C Table A-6 objective 3*: "Executable Object Code complies with low-level requirements."

---

* A NaN can be a quiet NaN or a signaling NaN.
† It should be noted that some technologies, such as formal methods, may alleviate the need for some of the testing activities. Formal methods are discussed in Chapter 16.

- *DO-178C Table A-6 objective 4*: "Executable Object Code is robust with low-level requirements."
- *DO-178C Table A-6 objective 5*: "Executable Object Code is compatible with target computer."

Testing is only part of the overall verification effort, but it is an important part and requires a tremendous amount of effort, particularly for higher software levels. Therefore, four sections are dedicated to the subject:

- *Section 9.6* provides a lengthy discussion of the testing topics, including (1) the purpose of software testing, (2) an overview of DO-178C's test guidance, (3) a survey of testing strategies, (4) test planning, (5) test development, (6) test execution, (7) test reporting, (8) test traceability, (9) regression testing, (10) testability, and (11) test automation.
- *Section 9.7* explains verification of the test activities, including (1) review of test cases and procedures, (2) review of test results, (3) requirements coverage analysis, and (4) structural coverage analysis. This is referred to as *verification of verification* in DO-178C Table A-7.
- *Section 9.8* provides suggestions for problem reporting, since the testing effort is intended to identify problems.
- *Section 9.9* ends the chapter by providing some recommendations for the overall test program.

9.6.1 Purpose of Software Testing

The purpose of software testing is to uncover errors that were made during the development phases. Although some see this as a negative viewpoint, experience shows that *success-based testing* is ineffective. If someone is out to prove the software works correctly, it likely will. However, that does not help the overall quality of the product. As I often say, " 'Yes, Sir!' men and women are generally not good testers."

Many people consider a *successful* test one that does not find any errors. However, from a testing perspective, the opposite is true: a successful test is one that finds errors. Testing is normally the only part of the project that does not focus directly on success. Instead, testers try to break or destroy the software. Some consider this to be a cynical mentality and attempt to put a positive spin on the testing effort. However, in order to do the job properly, the focus needs to be kept on finding failures. If the errors are not found, they cannot be fixed, and they will make it to the field [6,7].

By its nature, testing is a destructive process. Good testers hunt errors and use *creative destruction* to break the software, rather than show its correctness. Edward Kit writes:

> Testing is a positive and creative effort of destruction. It takes imagination, persistence, and a strong sense of mission to systematically locate the weaknesses in a complex structure and to demonstrate its failures. This is one reason why it is particularly hard to test our own work. There is a natural real sense in which we don't want to find error in our own material [8].

Not everyone is effective at testing. In fact, in my experience, good testers tend to be in the minority. Most people want to make objects work rather than rip them apart [7]. However, Alan Page et al. point out that effective testers have a different *DNA* than most developers. Tester DNA includes (1) a natural ability for systems level thinking, (2) skills in problem decomposition, (3) a passion for quality, and (4) a love to discover how something works and how to break it [9]. Even though testers are often viewed as a pessimistic lot, their mission is critical. Testers look for the flaws in the software so the problems can be fixed and a high quality and safe product can be delivered.

DO-178C focuses on requirements-based tests to ensure that the requirements are met and that only the requirements are met. However, it is important to keep in mind that the requirements tell us how the program is supposed to behave when it is coded correctly. "They don't tell us what mistakes to anticipate, nor how to design tests to find them" [6]. Good testers anticipate the errors and write tests to find them. "Software errors are human errors. Since all human activity, especially complex activity, involves error, testing accepts this fact and concentrates on detecting errors in the most productive and efficient way it can devise" [8]. The harsher the testing, the more confidence we can have in the quality of the product.

Some believe that software can be *completely tested*. However, this is a fallacy. *Complete testing* would indicate that every aspect of the software has been tested, every scenario has been exercised, and every bug has been discovered. However, in their book *Testing Computer Software*, Cem Kaner et al. point out that it is impossible to completely test software for three reasons [10]:

- The domain of possible inputs is too large to test.
- There are too many possible paths through the program to test.
- The user interface issues (and thus the design issues) are too complex to completely test.

Kit's book *Software Testing in the Real World* includes several test axioms, which are summarized here [8]:

- Testing is used to show the presence of errors but not their absence.
- One of the most challenging aspects of testing is knowing when to stop.
- Test cases should include a definition of the expected output or result.

- Test cases must be written for invalid and unexpected, as well as valid and expected, input conditions.
- Test cases must be written to generate desired output conditions—not just the input. That is, output space as well as input space should be exercised through testing.
- To find the maximum number of errors, testing should be an independent activity.

9.6.2 Overview of DO-178C's Software Testing Guidance

Now that the overall purpose of software testing has been reviewed, let us consider what DO-178C says. DO-178C focuses on *requirements-based* testing: tests are written and executed to show that the requirements are met and to ensure that there is no unintended functionality. The more critical the software, the more rigorous the test effort.

9.6.2.1 Requirements-Based Test Methods

DO-178C section 6.4.3 promotes three requirements-based test methods.

1. *Requirements-based software/hardware integration testing* [1]: This test method executes tests on the target computer to reveal errors when the software runs in its operating environment, since many software errors can only be identified in the target environment. Some of the functional areas verified during software/hardware integration testing are interrupt handling, timing, response to hardware transients or failures, databus or other problems with resource contention, built-in-test, hardware/software interfaces, control loop behavior, hardware devices controlled by software, absence of stack overflow, field-loading mechanisms, and software partitioning. This method of testing is normally performed by running tests against the high-level requirements, with both normal and abnormal (robustness) inputs, on the target computer.

2. *Requirements-based software integration testing* [1]: This method of testing focuses on the interrelationships of the software to ensure that the software *components* (typically, functions, procedures, or modules) interact properly and satisfy the requirements and architecture. This testing focuses on errors in the integration process and component interfaces, such as corrupted data, variable and constant initialization errors, event or operation sequencing errors, etc. This testing is typically accomplished starting with the high-level requirements-based tests for hardware/software integration testing. However, requirements-based tests may need to be added to exercise the software architecture; likewise, lower level testing may also be needed to supplement the software/hardware integration testing.

3. *Requirements-based low-level testing* [1]: This method generally focuses on compliance with the low-level requirements. It examines low-level functionality, such as algorithm compliance and accuracy, loop operation, correct logic, combinations of input conditions, proper response to corrupt or missing input data, handling of exceptions, computation sequences, etc.

9.6.2.2 Normal and Robustness Tests

DO-178C also promotes the development of two kinds of test cases: normal and robustness.*

9.6.2.2.1 Normal Test Cases

Normal test cases look for errors in the software with normal/expected conditions and inputs. DO-178C requires normal test cases for high- and low-level requirements for levels A, B, and C, and for high-level requirements for level D. Normal test cases are written against the requirements to exercise valid variables (using equivalence class and boundary value tests, which are discussed in Sections 9.6.3.1 and 9.6.3.2), verify operation of time-related functions under normal conditions, exercise state transitions during normal operation, and verify correctness of normal range variable usage and Boolean operators (when requirements are expressed as logic equations) [1].

9.6.2.2.2 Robustness Test Cases

Robustness test cases are generated to show how the software behaves when exposed to abnormal or unexpected conditions and inputs. DO-178C requires robustness test cases for high- and low-level requirements for levels A, B, and C, and for high-level requirements for level D. Robustness test cases consider invalid variables, incorrect state transitions, out-of-range computed loop counts, protection mechanisms for partitioning and arithmetic overflow, abnormal system initialization, and failure modes of incoming data [1].

DO-178C places strong emphasis on using the requirements as the means to determine the test cases. Requirements specify intended behavior; such behavior needs to be explored. Although the behavior is specified, testers should strive to explore different or additional behaviors when writing tests cases using their understanding of what the system is intended to do and trying to check that different behaviors are not present in the implementation (i.e., apply the *break it mentality*). This is an open-ended process as it is impossible to determine every additional behavior that may exist. The initial set of requirements serves as a guide and any new tests cases added to check the behavior beyond that specified may lead to additional requirements, such as robustness requirements (since all tests used for certification credit need to trace to requirements).

---

* Per DO-178C section 6.4.2.

Testing serves two purposes: (1) to ensure that the requirements are satisfied and (2) to show absence of errors. The second purpose is the most challenging and is where experience is extremely beneficial. Keep in mind that the requirements proposed initially are the minimal set. Missing or inadequate requirements, particularly robustness requirements, may be identified when developing test cases. Requirements changes should be proposed and tracked through a formal change process to ensure that they are properly considered.

Many projects apply *requirements blinders* when writing tests (i.e., the tests are written just using the written requirements). This may result in faster testing and fewer test failures, but it is not an effective way to find errors. I have seen numerous cases where a the software testers have blessed a software load only to have an experienced and somewhat devious systems engineer come to the lab and break the software within 10 minutes. A good test engineer knows the requirements, how a designer/coder is likely to work, and the typical error sources. Something as simple as holding a button down for an extended period of time or pressing it for a very short time has led to some interesting problem reports.

James A. Whittaker's book *How to Break Software* provides some thought-provoking strategies for software testing. Although his approach is a bit unconventional for the safety-critical realm, he does have some good suggestions that can be applied to robustness testing. He emphasizes that there are four fundamental capabilities of software [11]:

1. Software accepts input from its environment.
2. Software produces output and transmits it to its environment.
3. Software stores data internally in one or more data structures.
4. Software performs computations using input and stored data.

He then focuses on attacking the four software capabilities. He provides suggestions for testing input, testing output, testing data, and testing computations. The input and output testing require detailed knowledge of the software functionality (high-level functionality). The data and computation tests concentrate on the design (low-level functionality) [11].

9.6.3 Survey of Testing Strategies

Having read dozens of books on software testing, I am astounded at the width of the chasm between what is in the books and what happens in real safety-critical software projects. For example, software engineering literature uses the terms *black-box testing* and *white-box testing* extensively. Black-box testing is functional and behavioral testing performed without knowledge of the code. White-box testing (sometimes called glass-box testing) uses knowledge of the program's internal structure and the code to test the software. These terms (especially *white-box testing*) tend to obscure the necessary connection to requirements. Therefore, these terms do not appear in DO-178C. In fact, DO-178C tends to stay clear of white-box testing strategies. Instead, DO-178C concentrates on testing the high- and low-level requirements to ensure that the code properly implements the requirements.

Despite the chasm between the literature and DO-178C, many of the concepts of black-box testing, and to some degree white-box testing, can be applied to safety-critical software. In my experience, black-box testing approaches are applicable to both high- and low-level requirement-based testing. White-box testing strategies are a little more challenging to apply. However, if the strategies are applied to the low-level requirements and architecture, rather than the code itself, they do provide some useful practices. Of course, this requires the low-level requirements to be written at an appropriate level. The DO-178C glossary defines low-level requirements as: "Software requirements developed from high-level requirements, derived requirements, and design constraints from which Source Code can be directly implemented without further information" [1]. If the low-level requirements are treated as design details just above the code, many of the white-box testing concepts can be applied. But, care must be taken to ensure that the tests are written against the requirements and not the code.

This section briefly examines several testing strategies that may be employed on a project. The information is provided at a high level to begin to bridge the chasm between DO-178C testing concepts and what is presented in software test textbooks.

9.6.3.1 Equivalence Class Partitioning

Pressman writes:

> Equivalence partitioning is a black-box testing method that divides the input domain of a program into classes of data from which test cases can be derived. An ideal test case single-handedly uncovers a class of errors (e.g., incorrect processing of all character data) that might otherwise require many test cases to be executed before the general error is observed [12].

Equivalence partitioning requires an evaluation of equivalence classes. The DO-178C glossary defines *equivalence class* as: "The partition of the input domain of a program such that a test of a representative value of the class is equivalent to a test of other values of the class" [1]. Equivalence class testing considers classes of input conditions to be tested, where each class covers a large set of other possible tests. It "enables the tester to evaluate input or output variables systematically for each parameter in a feature" [9]. It is typically applied to situations where valid (normal) and invalid (robustness) inputs over a range are identified. Rather than testing every value of the range, only representative values are needed (typically on both sides of the boundaries). For large ranges of data, a value in the middle and at the extremes of each end are also typical. In addition to ranges of values, equivalence classes consider similar groups of variables, unique values that require different handling, or specific values that must or must not be present.

Even though equivalence class partitioning may seem trivial, it can be quite challenging to apply. Its effectiveness relies on "the ability of the tester to decompose the variable data for a given parameter accurately into well-defined subsets in which any element from a specific subset would logically produce the same expected result as any other element from that subset" [9]. If the tester does not understand the system and the domain space, he or she may miss critical defects and/or execute redundant tests [9].

The following recommendations may help when looking for equivalence classes [7,10]:

- Consider the invalid inputs. These are often where the vulnerabilities in the software and the bugs are found. These also serve as robustness tests.

- Consider organizing classifications into a table with the following columns: input or output event, valid equivalence classes, and invalid equivalence classes.

- Identify ranges of numbers. Typically for a range of numbers, the following are tested: values within the range (at both ends and in the middle), a value below the smallest number in the range, a value above the largest number in the range, and a non-number.

- Identify membership in a group.

- Consider variables that must be equal.

- Consider equivalent output events. This can sometimes be challenging because it requires one to determine the inputs that generate the outputs.

9.6.3.2 Boundary Value Testing

Boundary value analysis (BVA) is a test-case design technique that complements equivalence partitioning. Rather than selecting any element of an equivalence class, BVA leads to the selection of test cases at the "edges" of the class. Rather than focusing solely on input conditions, BVA derives test cases from the output domain as well [12].

BVA is closely related to equivalence class partitioning with two differences: (1) BVA tests each edge of the equivalence class and (2) BVA explores the outputs of equivalence classes [8]. Boundary conditions can be subtle and difficult to identify [8].

Boundary values are the biggest, smallest, soonest, shortest, loudest, fastest, ugliest members of the equivalence class (that is, the most extreme values). Incorrect equalities (e.g., > instead of >=) are also considered [10]. Programmers may accidently create incorrect boundaries for linear variables; therefore, the boundaries are one of the best places to find errors. Boundary testing is especially effective for detecting looping errors, off-by-one errors, and erroneous relational operators [9]. It is important to consider outputs as well as inputs. Midrange values should also be examined.

A classical programming error is to be off by one in a loop count (e.g., the programmer may check for < when he or she should check for <=). Therefore, testing the boundaries of loops can be a good place to find errors. When testing the loop, one first bypasses the loop; then one iterates through the loop once, twice, maximum number of times, maximum minus one times, and maximum plus one times. This type of testing is often performed as part of low-level testing, since looping is normally a design detail rather than a functional requirements detail.

9.6.3.3 State Transition Testing

When state transitions are used, state transition testing considers the valid and invalid transitions between each state. Invalid state transitions that could impact safety or functionality should be tested in addition to the valid transitions.

9.6.3.4 Decision Table Testing

Requirements sometimes include decision tables to represent complex logical relationships. Since the tables function as requirements, they need to be tested. Typically, conditions in the table are interpreted as inputs and actions are interpreted as outputs. Equivalence classes may be present in the table as well. Testers need to show that they have thoroughly exercised the requirements presented in the table. Testing may reveal deficiencies in the table itself (such as, incorrect logic or incomplete data), as well as incorrect implementation in the code.

9.6.3.5 Integration Testing

Integration is the process of putting the software components together. The components may work beautifully when performing the low-level testing on each component; however, the process of integrating the components may result in unexpected results. When components are integrated, several things can go wrong, including data may be lost across an interface, one component may adversely affect another component, subfunctions may not properly contribute to the higher level functionality desired, global data may become corrupted, etc. [12]. To avoid such integration hurdles, an integration strategy is typically applied. Some common integration approaches are discussed in following.

Big bang integration: This approach throws the software components together to see if it works (it usually does not). Even though conventional wisdom tells us that big bang is a bad idea, it happens. I have seen multiple companies unsuccessfully attempt it. One of the main problems with big bang is that when a problem exists, it is difficult to track it down, because the integrity of the modules or their interaction is unknown. Some more effective integration strategies are now discussed.

Top-down integration: With this approach components are integrated moving downward through the hierarchy—typically starting with the main control module. Stubs are used for components that are not yet integrated. Stubs are replaced with one component at a time. Tests are conducted as each component is integrated.

Bottom-up integration: With this approach, the lower level components are tested using drivers. Drivers emulate components at the next level up the tree. Then the components are integrated up the hierarchy.

Combination of top-down and bottom-up integration (sometimes referred to as *sandwich integration* or *hybrid integration*): Oftentimes one branch at a time is integrated. With this approach not as many stubs and drivers are needed.

The preferred approach depends on the architecture and other programmatic details. Typically, it is best to test the most critical and high-risk modules first, as well as key functionality. Then functionality is added in an organized manner.

9.6.3.6 Performance Testing

Performance testing evaluates the run-time performance of the software within the integrated system and identifies bottlenecks in the system. Performance tests are used to demonstrate that performance requirements (such as, timing and throughput) are satisfied. Race conditions and other timing-related issues are also examined. Performance testing typically occurs throughout the testing process and may lead to redesign or optimization setting changes, if results are inadequate. As noted in Section 9.5, these tests are often also used to help determine the margins of the software in the target environment.

9.6.3.7 Other Strategies

There are several other strategies for testing that may be employed, depending on the system architecture and the requirements organization. For example, loop testing may be applied for simple loops, nested loops, and concatenated loops. Also, if a model is used in the requirements, model-based testing techniques may be applied (model-based development and verification are discussed in Chapter 14). Additionally, for algorithms, special testing may be needed, depending on the nature of the algorithm. For example, when testing the digital implementation of a first order lag filter, enough iterations need to be run to be able to show that the first order response actually curves instead of just being a linear ramp.

9.6.3.8 Complexity Measurements

Experience shows that complex code and interactions of complex portions of code are generally more prone to mistakes [9]. Complex code has a tendency to be buggy and difficult to maintain. Complexity measurements are often applied to code during the development phase to alert coders of the risky code constructs. When possible, the software is redesigned or refactored to make it less complex. Coding standards often include a guideline to measure and minimize the complexity (e.g., McCabe's cyclomatic complexity measure is often applied). However, if the code complexity is not resolved, it can be a good indicator to testers of where the code vulnerabilities exist. Complex software should be pounded hard (that is, more normal and robustness tests), because complex code will typically have more errors. The errors often creep in when the code is modified, because the code is difficult to understand and maintain.

One project that I lived through had a particularly complex feature. It was called the *fragile code*. Any time it was modified, the entire subsystem had to be completely retested, because the impact of the change was unpredictable. It ended up being an unacceptable situation for a safety-critical system and had to be redesigned late in the project. Highly complex or fragile code should be redesigned as early as possible. It is less agonizing to fix it up front than to try to track down phantom symptoms at the 11th hour.

9.6.3.9 Summary and Characteristics of a Good Test

Section 9.6.3 has surveyed several testing approaches. Most projects use a combination of many or all of the aforementioned approaches. There are others we could cover. Kaner et al. explain that regardless of the test approach, there are some common characteristics of a good test. Good test cases satisfy the following criteria [10]:

- "It has a reasonable probability of catching an error." Tests are designed to find errors—not merely to prove functionality. When writing a test one must consider how the program might fail.
- "It is not redundant." Redundant tests offer little value. If two tests are looking for the same error, why run both?
- "It is the best of its breed." Tests that are most likely to find errors are most effective.
- "It is neither too simple nor too complex." Overly-complex tests are difficult to maintain and to identify the error. Overly-simple tests are often ineffective and add little value.

9.6.4 Test Planning

Chapter 5 discussed the Software Verification Plan (SVP). The SVP provides the plan for how verification on the overall project will be performed. It also explains the test strategy and documentation. However, in addition to the SVP, most companies also develop a detailed test plan, which is often included in a Software Verification Cases and Procedures (SVCP) document. The SVCP document typically focuses on more than just testing; it also includes the procedures for reviews and analyses; however, this section concentrates on the test aspects of the SVCP.

The test planning portion of the SVCP (sometimes referred to as the *test plan*) goes into more detail than the SVP on the test environment (identifies equipment needed, test-related tools, test-station assembly), test case and procedure organization and layout, categories of tests, test naming conventions, test groupings, test tracing strategy, test review process and checklists, responsibilities, general test set-up procedures, test build procedures, test readiness review (TRR) criteria, and test station configuration audit or conformity plan.

The SVCP is normally a living document that gets updated as the testing program matures.* As the program matures, a listing of all requirements is normally added, along with a notation of the test technique that will be applied. As tests are written, the SVCP is updated to include a summary of all the test cases and procedures that will be executed during the test execution, as well as traceability matrices showing traceability between the test cases and requirements and between the test cases and test procedures. Test time estimations and test execution plan (including order of execution) are also added prior to test execution. The SVCP may also include the regression test approach.

The test planning in the SVCP is beneficial because it helps ensure the following:

- Everyone is on the same page and knows the expectations, including schedule, responsibilities, and assigned tasks.
- The test development and execution effort is organized.
- All necessary tests are developed and executed.
- There's no unnecessary redundancy.
- Schedules are accurately developed (the plan allows the test manager to develop an accurate schedule rather than aim at a fictitious date mandated by project management).
- Priorities are identified.
- All tasks are staffed (the plan may help identify staff shortages).
- Necessary equipment and procedures are identified.

---

* Rather than having one living document, some projects have a test plan which is completed up front, as well as a verification cases and procedures document that evolves throughout the program.

- Expected test layout is communicated.
- Test cases and procedures are developed, reviewed, and dry run.
- The TRR criteria are established ahead of time, so everyone knows what to expect.
- The order of test execution is planned to efficiently execute the program.
- The version of test cases and test procedures to be used for score are identified.
- Test set-up procedures are clear.
- All requirements are tested.
- Test groupings are logical.
- The necessary test equipment is acquired and properly configured.
- All test tools are properly controlled.
- Test files are built in such a way that they represent the actual software that will be fielded.
- Test stations are properly configured to represent the target environment.
- The approach for auditing test station configuration is identified.
- Risks are identified and mitigated.

The SVCP should be written to serve the project, not just to be a completed data item. It should be organized so that it is easy to update throughout the project. It provides evidence of an organized and complete test program. Otherwise, it is difficult to confirm that everything was properly executed.

Also note that Chapter 12 (Section 12.5) explains the expected maturity of software before flight testing for certification credit. This maturity criterion should be considered during test planning, since it could impact the test schedule and approach.

9.6.5 Test Development

The DO-178C expectations for testing and the typical test strategies employed have been examined. In this section more details of the test cases and procedures development from a DO-178C perspective are explored.

9.6.5.1 Test Cases

DO-178C defines a test case as: "A set of test inputs, execution conditions, and expected results developed for a particular objective, such as to exercise a particular program path or to verify compliance with a specific requirement" [1]. Typically, a template is provided for the test team to ensure that the tests cases are laid out in a common manner. If the tests are automated, the format is particularly important. Additionally, a tool is sometimes used to extract information from the test case headers to generate a test summary or traceability report. Whether the tests are manual or automated, each test case usually includes the following:

- Test case identification
- Test case revision history
- Test author
- Identification of software under test
- Test description
- Requirement(s) tested
- Test category (high-level, low-level, or both high- and low-level)
- Test type (normal or robustness)
- Test inputs
- Test steps or scenarios (the actual test)
- Test outputs
- Expected results
- Pass/fail criteria

9.6.5.2 Test Procedures

In addition to test cases, DO-178C mentions the need for test procedures. A test procedure is "detailed instructions for the set-up and execution of a given set of test cases, and instructions for the evaluation of results of executing the test cases" [1]. Test procedures come in different shapes and sizes. Some are actually embedded in the test cases themselves; others are separate high-level documents that explain how to execute the test cases. The test procedures are the steps used to execute the test cases and obtain the test results. Sometimes they are manual and sometimes they are automated.

Each test step and how to verify pass/fail should be clear and repeatable. The person who executes the test is often not the one who wrote the test. In fact some companies and certification authorities insist on independent test execution. Vague test procedures are a common problem; the steps may be clear to the test author, but someone who is not familiar with the test or functionality may not be able to execute the test successfully. To ensure clarity and repeatability, it is a good practice to have someone who did not write the tests perform a dry run of the tests as early as possible. This may actually be part of the test review process, which is discussed later.

9.6.5.3 DO-178C Requirements

Table 9.1 summarizes the DO-178C minimum requirements for high- and low-level requirements testing, as well as normal and robustness testing.

TABLE 9.1

Summary of DO-178C Test Requirements

	High-Level Normal	High-Level Robust	Low-Level Normal	Low-Level Robust
Level A	Required	Required	Required with independence	Required with independence
Level B	Required	Required	Required with independence	Required with independence
Level C	Required	Required	Required	Required
Level D	Required	Required	Not required	Not required

Source: RTCA DO-178C, *Software Considerations in Airborne Systems and Equipment Certification*, RTCA, Inc., Washington, DC, December 2011.

9.6.5.4 Low-Level Requirements Testing versus Unit Testing

Before exploring test execution, I want to briefly discuss *unit testing*. Some companies use *unit testing* for their low-level requirements testing, but this needs to be handled with care. Traditional unit testing is performed by writing tests against the code. This certainly is not prohibited and is a good practice for the code developers to perform to ensure their code does what they want (that is, debugging). However, DO-178C emphasizes that testing for certification credit is to be against the requirements. The concepts of unit testing can still apply when each module (or group of modules) is tested against the low-level requirements, but care must be taken to write tests against the requirements and not the code.* Additionally, testing module by module may create some issues for integration. If low-level testing is performed at the module or function level, a concrete plan for how the modules are integrated and tested together is needed. As discussed earlier, big bang integration is not advised and is normally unacceptable. Also, at the highest level of integration it may be difficult to confirm the correctness of interfaces between software modules or functions. When module-level testing is used, there is generally some additional software/software integration testing required in addition to the high-level software/hardware integration testing. When data coupling and control coupling analyses are discussed later (in Section 9.7), it should become more apparent why the software/software integration is needed.

9.6.5.5 Handling Requirements That Cannot Be Tested

During the test development phase, it may be discovered that some requirements are not testable. In some cases, the requirement may need to be re-written (e.g., negative requirements, requirements without tolerances,

---

* The ability to write low-level tests module by module will depend on how the low-level requirements are organized.

or ambiguous requirements). In other cases, an analysis or code inspection might be used in lieu of a test. Such situations should be handled carefully, because in general, analysis and code inspection are not considered as effective as testing.* If an analysis or code inspection is used instead of a test, it must be justified why the analysis or code inspection will find the same kinds of errors as a test and why the analysis or code inspection is equivalent to (or better) than a test. I recommend including the justification in the analysis or code inspection itself, so it is recorded. The analysis should also be well documented and repeatable (as noted in Section 9.5). An analysis or code inspection that is used in lieu of a test needs the same level of independence as a test would.

9.6.5.6 Obtaining Credit for Multiple Levels of Testing

In some situations a project may be able to write some tests so that they exercise both high- and low-level requirements. Generally, it is more likely that a high-level test to can satisfy low-level requirements, rather than low-level tests satisfying high-level requirements, but it depends on the integration approach and the granularity of the requirements. If a single test or group of tests is used to claim credit for multiple levels of requirements, there needs to be documented evidence that both levels of requirements are satisfied, including trace data and review records. In some cases an analysis may be needed demonstrate how the test(s) fully exercise both levels of requirements, particularly if the tests were not originally written to cover multiple levels of requirements. Such an analysis should be included as part of the test data.

9.6.5.7 Testing Additional Levels of Requirements

In some projects, in addition to high- and low-level requirements, there may be other intermediate levels of requirements. These requirements will also need to be tested, with the possible exception of level D software. As noted in the previous section, it might be possible to use one set of requirements-based tests to exercise multiple levels of requirements.

9.6.6 Test Execution

Test cases and procedures are written to be executed. This section discusses some of the things to consider when preparing for and executing tests.

---

* Code inspections are particularly subjective when used to confirm that executable object code satisfies the requirements. When used to satisfy DO-178C Table A-6 objectives, the focus of the code inspection is on how the executable object code meets the requirements, not a repeat of the code review (DO-178C Table A-5 objectives).

9.6.6.1 Performing Dry Runs

Most programs do a dry run of the tests, in order to work out any issues and resolve problems prior to *for-score* or *formal* runs that are used for certification credit. A dry run of tests is highly recommended. Trying to go straight from a paper review to formal execution typically results in some unexpected and undesired surprises. For higher software levels (levels A and B), it is a good practice to have an independent person who did not write the procedures execute the dry run. This helps work out issues with the procedures, as well as uncovers unexpected failures.

9.6.6.2 Reviewing Test Cases and Procedures

Prior to test execution, the test cases and procedures should be reviewed (for levels A, B, and C) and put under change control. Oftentimes, the tests are informally run prior to the review, so that the results can be referenced during the review. The review process is further discussed in Section 9.7.

9.6.6.3 Using Target Computer versus Emulator or Simulator

Tests are typically executed on the target computer—particularly the hardware/software integration tests. Sometimes a target computer emulator or a host-based computer simulator is used to execute the tests. If this is the case, the differences between the emulator/simulator and the target computer need to be assessed to ensure that the ability to detect errors is the same as on the target computer. Showing the equivalence may be achieved by a difference analysis* or by qualifying the emulator or simulator (using the tool qualification approach described in DO-178C and DO-330, which will be discussed more in Chapter 13). It should be noted that some tests will likely need to be rerun on the target computer, even if an emulator or simulator is used, since there are some types of errors that may only be detected in the target computer environment.[†,‡]

9.6.6.4 Documenting the Verification Environment

The Software Life Cycle Environment Configuration Index (SLECI) (or equivalent) should be up-to-date prior to executing the tests for certification credit. The SLECI documents the details of the verification environment and is normally used to configure the test stations before tests are run for certification credit. Chapter 10 provides information on the SLECI contents.

---

* DO-178C section 4.4.3.b.

† DO-178C section 6.4.1.a.

‡ FAA Order 8110.49 (change 1) chapter 16 provides some guidance on managing the development and verification environments when an emulator or simulator is used. The Order can be found on FAA's website: www.faa.gov

9.6.6.5 Test Readiness Review

DO-178C does not discuss the TRR; however, most projects perform a TRR prior to executing the tests to ensure the following:

- The requirements, design, source code, and executable object code have been baselined.
- The tests are aligned with the baselined version of the requirements, design, and code.
- The test cases and procedures have been reviewed and baselined.
- Any issues from dry runs have been addressed.
- The test builds have been completed and the part-number of the software under test is documented.
- The approved test environment is used.
- The test stations have been configured per the approved SLECI (or equivalent).
- The test trace data is up-to-date (traceability is discussed later in this chapter).
- The test schedule has been provided to those who are required to witness or support the test (for example, software quality assurance [SQA], customer, and/or certification liaison personnel).
- Any known problem reports against the software under test have been resolved or agreed upon for deferral (if there are tests that are expected to fail, it is important to know the expected test failures going into the formal test run).

It should be noted that for some larger projects, the tests are executed in groups. In this situation, there may be a TRR for each group of tests. When this occurs, care must be taken to ensure that all requirements are fully evaluated and that any dependencies between the test groups are considered.

9.6.6.6 Running Tests for Certification Credit

Test execution is the process of executing test cases and procedures, observing the responses, and recording the results. Test execution is often run by someone who did not write the tests and is frequently witnessed by SQA and/or the customer.

During the formal test execution, the test procedures are followed and checked off as they are completed (they should already have been reviewed and baselined).* A test log is typically kept to record who performed the test, when it was performed, who witnessed the test, the test and software

---

* Even automated tests require procedures and need a test log.

version information, and what the test results are. Oftentimes pass/fail will also be determined during the test execution. However, if the data needs to be analyzed, the pass/fail determination might occur later.

As the tests are run for certification credit (formal test execution) there may be some minor *redlines** to the procedures. If a good dry run is performed, redlines should be minimal. Too many redlines tends to be an indicator of immature test procedures. When redlines are minimal, they are typically recorded on a hard copy or on a test log and will be approved by engineering and SQA prior to release of the results. The redlined data are normally approved by SQA and are included in a problem report, so the tests are updated in the future.

When executed for certification credit, the tests may uncover some unexpected failures. When this happens, the failures are analyzed and appropriate action taken. Oftentimes, the tests and/or requirements are updated and tests are rerun. Test failures and resulting updates should be handled via the problem reporting process. For tests that are rerun, a regression analysis may be needed. See Section 9.6.9 for additional discussion on the regression test process.

9.6.7 Test Reporting

After the tests are run, the results are analyzed and summarized. Any test failures should result in a problem report and must be analyzed. The review of test failures is discussed in Section 9.7.

Typically, a Software Verification Report (SVR) is generated to summarize the results of verification activities. The SVR summarizes all verification activities, including reviews, analyses, and tests, as well as the results of the verification activities. The SVR often includes the final traceability data as well.

For any verification activities that were not successful, the SVR typically summarizes the analysis from the problem report and justifies why the failures are acceptable.

The verification report is normally reviewed by the certification authority and/or their designees, as well as the customer. Therefore, the report should be written for an audience that may not be familiar with the development and verification details.

9.6.8 Test Traceability

DO-178C section 6.5 emphasizes that trace data should exist to demonstrate bidirectional traces between all of the following [1]:

- Software requirements and test cases
- Test cases and test procedures
- Test procedures and test results

---

* A *redline* is a modification to the procedure. It is normally documented using a red pen on a hard copy or using change tracking for an electronic copy.

These trace data are often included in the SVR, but might be included in the SVCP or even in a stand-alone document. The trace data accuracy should be verified during the review of test cases, test procedures, and test results. In some cases, the tracing is implicit because the documents are combined; for example, the test cases and procedures may be in the same document. In other cases, a naming convention may be used; for example, each test procedure may generate a test result with the same name (such as TEST1.CMD and TEST1.LOG). If this is the case, the strategy should be explained in the SVP and summarized in the SVR.

9.6.9 Regression Testing

Regression testing is a re-execution of some or all of the tests developed for a specific test activity [8]. After the tests are executed for certification credit, there may be some additional tests that need to run. These may be tests that failed during the initial run and were modified, or there may be new tests added or modified because of new or modified functionality.

Pressman explains that a regression test run typically includes the following types of tests [12]:

- A representative sample of tests that will exercise all software functions (to support side-effect/stability regression).
- Tests that focus on the change(s).
- Tests that focus on the potential impacts from the software changes (that is, tests of impacted data items).

Before such tests are run, a change impact analysis is needed. As a minimum, the change impact analysis considers traceability data and data and control flows to identify all changed and impacted data items (such as requirements, design, code, test cases, and test procedures). The change impact analysis will also identify what reverification is required. Reverification will typically involve reviews of the updated and impacted data; rework of any impacted analyses; rerun of new, changed, and impacted tests; and execution of regression tests (explained later). Chapter 10 provides some additional insight on what is typically considered during a change impact analysis.

Depending on the extent of the change, it is sometimes advisable to rerun the all of the tests rather than try to justify a partial regression approach. This is particularly the case for automated tests. A thorough change impact analysis can take considerable time. It is sometimes easier to rerun the tests than to justify why they do not need to be rerun.

If a regression approach is taken and a small number of tests are run to address the change, it is advisable to run a subset of the overall test suite to confirm that there were no unintended changes. Some projects run a standard subset of tests every time a change is made, in addition to the tests of the changed or impacted software. The subset typically includes safety-related tests and tests that prove critical functionality. This is sometimes referred to as *side-effect regression* or *stability regression*, because it proves that the change did not introduce a side effect or impact the stability of the code base. These regression tests help to ensure that the changes do not introduce unintended behavior or additional errors.

9.6.10 Testability

Testability should be considered during the development. Chapter 7 explains that testable software is operable, observable, controllable, decomposable, simple, stable, and understandable. Chapter 7 also identifies features that help make software testable, including error or fault logging, diagnostics, test points, and accessible interfaces. Having testers involved during the requirements and design reviews can help make the software more testable.

9.6.11 Automation in the Verification Processes

Automation is commonly used during the verification activities. Tools do not replace the need to think, and users need to understand how the tools work in order to use them effectively; however, there are some tasks that tools can do more effectively and accurately than humans. Some examples of automation used during the verification processes are as follows:

- *Test templates* are commonly used for the test schedule and tasks, test plan, test cases, test procedures, and the test report. The amount of automation for these varies greatly.
- *Test script tools* provide a language to write automated tests, as well as a template to promote consistent test layout.
- *Traceability tools* are used to help capture and verify the trace data. The data in the tool still need to be verified, but trace tools can help to identify missing test cases or procedures, missing results, untested requirements, etc.
- *Test execution tools* execute test scripts and may be used to determine test pass or failure results. If the test output is not reviewed, the tool may need to be qualified, since it automates DO-178C objectives. Tool qualification criteria are discussed in Chapter 13.
- *Debugger tools* are sometimes used during the test process to set breakpoints in order to examine or manipulate the software. These should be used cautiously during formal testing to ensure they do not incorrectly change the software. Also, an overabundance of breakpoints may lead to difficulties when proving the completeness of integration. Depending how the debug tool is used, its functionality may need to be qualified.

- *Memory tools* are used to detect memory problems, memory over-writing, memory that has been allocated but not freed, and uninitialized memory use [8].
- *Emulators or simulators* are sometimes used during the test process to access internal data that may be difficult to access on the actual target.
- *Coverage tools* are used to help with the assessment of structural coverage analysis. These tools frequently instrument the code in order to gather metrics on the code execution. Therefore, it may be necessary to run the tests with and without the instrumentation, compare the results, and ensure that the instrumentation does not impact the test results.
- *Static analysis tools* are used to evaluate code complexity, compliance to rules, worst-case path, worst-case execution timing, worst-case stack usage, etc. Static analysis tools may also evaluate data and control flow to identify undefined variables, inconsistent interfaces among modules, and unused variables [7].
- *Test vector generators* are documented in a formal manner or with models. Such tools generate test vectors to verify the accuracy of the implementation. Depending how these tools are used, they may require qualification. It is also recommended that human-generated tests be used to supplement the tool-generated test vectors to thoroughly test the functionality. It is important that test vector generators use requirements as inputs and not the code. There are some tools that generate *requirements* from the source code, and then generate test vectors for those requirements. These should be avoided.

9.7 Verification of Verification

DO-178C Table A-7 is entitled "Verification of Verification Process Results." It is probably the most discussed and debated table in DO-178C, primarily because it includes several objectives that are unique to the aviation industry and are not well documented in other software engineering literature (e.g., modified condition/decision coverage). The DO-178C Table A-7 objectives essentially require a project to evaluate the adequacy and completeness of their testing program. That is, it calls for the *verification of the testing* effort. The objectives of DO-178C Table A-7 verify the test procedures and results, the coverage of the requirements, and the coverage of the code structure. Each of the major subjects of *verification of verification* is discussed in this section. Figure 9.1 provides an overview of the typical process.

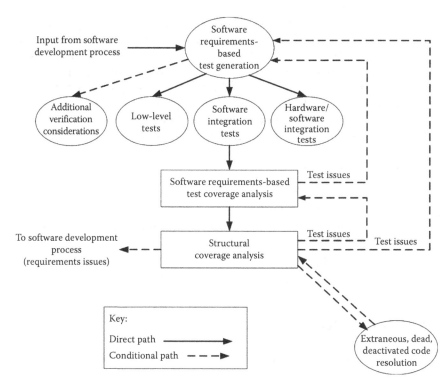

FIGURE 9.1
Overview of DO-178C software testing. (Adapted from RTCA DO-178C, *Software Considerations in Airborne Systems and Equipment Certification*, RTCA, Inc., Washington, DC, December 2011, DO-178C Figure 6-1. Used with permission from RTCA, Inc.)

9.7.1 Review of Test Procedures

DO-178C Table A-7 objective 1 is summarized as: "Test procedures are correct" [1]. The objective is to ensure that the test cases are correctly developed into test procedures. Typically, this is satisfied by a review of both the test cases and test procedures. Some argue that the objective only covers the test procedure review, since the test cases are addressed as part of objectives 3 and 4 (of DO-178C Table A-7). However, in my experience, this objective has been satisfied concurrently with objectives 3 and 4 to ensure that all of the requirements are tested, that the requirements are fully tested, and that the test cases and procedures are accurate, consistent, and compliant with the SVP.

DO-178C Table A-7 objective 1 applies to both high- and low-level testing and is required for levels A, B, and C. Level A requires independence. Typically, a peer review, similar to what is used for reviewing the requirements, design, and code is used to satisfy the objective. (See Chapter 6 for information on the peer review process.) The typical inputs to the review are the requirements and design, SVP, SVCP, preliminary test results (not same as dry run), test cases, test procedures, and trace data. If analyses or code inspections are used in lieu of any tests, they are also provided. Oftentimes, someone on the peer review team runs the test to verify that it produces the same results that are presented in the review, is repeatable, and is accurate. For manual tests, this is particularly effective at working out ambiguities in the test procedures and the pass/fail criteria. Clearly, the peer review should take place before the tests are executed for certification credit.

The most common issues identified during these reviews are noted here:

- Some requirements are not be tested.
- Robustness testing is not adequate.
- The tests do not completely exercise the requirements.
- Some tests are not well explained or commented (which makes maintenance difficult).
- Some steps in the procedures are missing or are not clear.
- The pass/fail criteria are not clear for some tests.
- Tests have not been run to ensure that they work.
- Traceability from requirements to test is not documented (only the test to requirements).
- Test environment is not documented well.

9.7.2 Review of Test Results

After the tests are executed, the results are reviewed. DO-178C Table A-7 objective 2 is summarized as: "Test results are correct and discrepancies explained" [1]. The objective is to ensure the correctness of the test results and to confirm that any failed tests are analyzed and properly addressed. This objective is typically satisfied by a review of the test results and the software verification/test report. The review includes an evaluation of explanations for any failed tests (typically in problem reports and the software verification/ test report). In some situations, test failures will require a fix to the software or the test cases/procedures. In general, if a dry run has been performed, the number of unexpected test failures is minimized. In an effort to shorten the schedule, some teams try to bypass the dry runs and go straight to formal testing. However, projects that attempt to go to formal testing without a thorough dry run normally end up doing the formal testing at least two times. In any case, all failures need to be documented in problem reports and addressed appropriately. Analysis of failures is also normally included in the SVR.

9.7.3 Requirements Coverage Analysis

Objectives 3 and 4 of DO-178C Table A-7 are referred to as the *requirements coverage objectives*. Objective 3 is summarized as: "Test coverage of high-level requirements is achieved" [1]. Similarly, objective 4 is summarized as: "Test coverage of low-level requirements is achieved" [1]. The purpose of these objectives is to confirm that all requirements have been tested (or verified by another means if properly justified). Typically, these objectives are evaluated during the test cases and procedures review. However, just before or after the final test run, there is often a final coverage analysis to ensure that all requirements are covered and nothing has been missed during the final stages of the test program (for example, a missing test case due a late-breaking requirement change). The trace data between test cases and requirements are typically used to confirm this coverage. However, there must also be a technical evaluation to ensure that the test cases fully cover the requirements (this is normally evaluated during the test case review). Basically, if the test cases have been adequately reviewed for complete coverage of the requirements, this coverage analysis is rather trivial. If an automated tool is used to run the tests for certification credit, then it is important to check that the tool actually ran all of the tests and none were skipped. This can be checked manually or automated as part of the tool that runs the tests. If credit is taken for the tool performing the check, then the tool needs to be qualified. Since low-level requirements do not have to be tested for level D, the requirements coverage of low-level requirements does not apply for level D. In my experience, the coverage analysis is either included in or summarized in the SVR.

9.7.4 Structural Coverage Analysis

The remaining five objectives of DO-178C Table A-7 address structural coverage. Objectives 5–7 and 9 pertain to the test coverage of the software structure, to ensure that the code has been adequately exercised during the requirements-based testing. Objective 8 is similar, except it focuses on the coverage of the data and control coupling between code components (for example, modules).

Structural coverage has the following purposes:

- Ensures that all code has been executed at least once.
- Finds unintended functionality and untested functionality.
- Identifies dead code* or extraneous code.†

---

* Dead code is: "Executable Object Code (or data) which exists as a result of a software development error but cannot be executed (code) or used (data) in any operational configuration of the target computer environment. It is not traceable to a system or software requirement. The following exceptions are often mistakenly categorized as dead code but are necessary for implementation of the requirements/design: embedded identifiers, defensive programming structures to improve robustness, and deactivated code such as unused library functions" [1].

† Extraneous code is: "Code (or data) that is not traceable to any system or software requirement. An example of extraneous code is legacy code that was incorrectly retained although its requirements and test cases were removed. Another example of extraneous code is dead code" [1].

- Helps confirm that deactivated code is truly deactivated.
- Identifies a minimal set of combinations for testing (i.e., it does not require exhaustive testing).
- Helps identify incorrect logic.
- Serves as an objective completion criteria for the testing effort (although it may not address completeness of the robustness testing).

It must be noted that *structural coverage analysis* in DO-178C is not equivalent to *structural testing* that is mentioned in much of the software engineering literature. Structural coverage *analysis* is performed to identify any code structure that was not exercised during the requirements-based testing. Structural *testing* is the process of writing tests from the code to exercise the software. Structural testing is not adequate for DO-178C because it is not requirements based and, hence, does not satisfy the purpose of structural coverage analysis.

A brief warning is in order before examining the structural coverage criteria in more detail. Some teams proactively run their structural coverage tools throughout the test cases and procedures development. This is a good practice; however, some teams use the coverage data to identify the tests needed—that is, they write tests to satisfy the tool rather than allowing the requirements to drive the testing. This is not the intent of structural coverage and can result in inadequate testing. Some software results in over 50% coverage just by turning on the system—without running a single test. Therefore, it must be emphasized that tests are to be written and reviewed against the requirements before structural coverage data is collected. To avoid the temptation to just *make the tool happy*, a different engineer is generally responsible for the coverage analysis (this also helps satisfy the independence required for level A and B structural coverage objectives).

Each DO-178C Table A-7 structural coverage objective is briefly discussed in the following subsections.

9.7.4.1 Statement Coverage (DO-178C Table A-7 Objective 7)

Statement coverage is required for levels A, B, and C; it ensures that "every statement in the program has been invoked at least once" [1]. This is achieved by evaluating the coverage of the code during requirements-based testing and ensuring that every executable statement has been invoked at least once. Statement coverage is considered a relatively weak criterion because it does not evaluate some control structures and it does not detect certain types of logic errors. An *if-then* construct only requires the decision to be evaluated to true to cover the statements. An *if-then-else* requires the decision to be evaluated both true and false. Basically, statement coverage ensures that the code has been covered, but it might not be covered for the right reason.

9.7.4.2 Decision Coverage (DO-178C Table A-7 Objective 6)

Decision coverage is required for levels A and B and is commonly equated with *branch* or *path* coverage. However, decision coverage is slightly different. DO-178C defines *decision* and *decision coverage* as follows [1]:

- "Decision – A Boolean expression composed of conditions and zero or more Boolean operators. If a condition appears more than once in a decision, each occurrence is a distinct condition."
- "Decision coverage – Every point of entry and exit in the program has been invoked at least once and every decision in the program has taken on all possible outcomes at least once."

The confusion centers around the traditional industry understanding of *branch point* (which is typically an *if-then-else, do-while,* or *case* statement) versus the DO-178C literal definition of *decision*. Basically, using the literal definition of DO-178C, a decision is not synonymous with a branch point. The main difference is that a Boolean value that is output on an input/output (I/O) port may have a binary effect on the behavior of the system (e.g., *wheels up* when TRUE and *wheels down* otherwise). It is important to make sure that this value is evaluated both TRUE and FALSE during test. This could be treated as a boundary value problem or a decision coverage problem.

In general decision coverage ensures that all paths in the code are taken and if there are any Boolean assignments, both the true and false conditions are exercised either when the value is used directly externally or when the value is used in a construct that results in a branch.

9.7.4.3 Modified Condition/Decision Coverage (DO-178C Table A-7 Objective 5)

Modified condition/decision coverage (MC/DC) is required for level A software and has been the subject of much debate most of my career and will probably continue to generate heated discussions after my retirement.

DO-178C defines *condition, decision,* and *modified condition/decision coverage* as follows [1]:

- "Condition – A Boolean expression containing no Boolean operators except for the unary operator (NOT)."
- "Decision – A Boolean expression composed of conditions and zero or more Boolean operators. If a condition appears more than once in a decision, each occurrence is a distinct condition."
- "Modified condition/decision coverage – Every point of entry and exit in the program has been invoked at least once, every condition in a decision in the program has taken all possible outcomes at least

once, every decision in the program has taken all possible outcomes at least once, and each condition in a decision has been shown to independently affect that decision's outcome. A condition is shown to independently affect a decision's outcome by: (1) varying just that condition while holding fixed all other possible conditions, or (2) varying just that condition while holding fixed all other possible conditions that could affect the outcome."

Even though an MC/DC criterion has been applied to aviation projects since the late-1980s/early-1990s, it is still rarely discussed in the mainstream software engineering literature. It was developed for the aviation industry to allow comprehensive criteria to evaluate test completion, without requiring that each possible combination of inputs to a decision be executed at least once (i.e., exhaustive testing of the input combinations to a decision, known as *multiple condition coverage*). While multiple condition coverage might provide the most extensive structural coverage measure, it is not feasible for many cases, because a decision with n inputs, requires 2^n tests [13]. MC/DC, on the other hand, generally requires a minimum of $n + 1$ test cases for a decision with n inputs.

Most projects apply MC/DC at the source code level, because the source code coverage tools are more available. However, there have been a few projects that have applied the structural coverage criteria at the object code or executable object code levels. There has been some debate on this approach, as noted in the Certification Authorities Software Team (CAST) position paper CAST-17, entitled *Structural coverage of object code*. The position paper provides a summary of the motivations behind object code coverage and some of the issues to be addressed [14]. The CAST-17 paper has served as the foundation for project-specific Federal Aviation Administration (FAA) issue papers. DO-178C section 6.4.4.2.b modified the wording from DO-178B to make it clear that object code or executable object code coverage are acceptable approaches: "Structural coverage analysis may be performed on the Source Code, object code, or Executable Object Code" [1]. DO-248C frequently asked question #42, entitled "What needs to be considered when performing structural coverage at the object code level?" also clarifies this topic [3].

It is tempting to write more about structural coverage—especially MC/DC, since I have spent considerable time investigating, debating, and evaluating it over the last several years. However, because it has been so widely discussed and debated in the aviation community, there are publicly available documents that cover the topic well. Items 1–5 in the "Recommended Reading" section of this chapter identify some particularly helpful resources.

9.7.4.4 Additional Code Verification (DO-178C Table A-7 Objective 9)

DO-178C Table A-7 objective 9 is summarized as follows: "Verification of additional code, that cannot be traced to Source Code, is achieved" [1].

This objective was added for DO-178C, although the text in DO-178B also required it. Most projects perform structural coverage on the source code (although a few do perform object code coverage or machine code coverage). However, the executable object code is what actually flies. For level A software, there needs to be some kind of analysis to ensure that the compiler does not generate untraceable or nondeterministic code.

For projects that perform their structural coverage on the source code, a source-to-object code traceability analysis is normally performed. In my experience, this includes a comparison between source code and object code to ensure that the compiler is not adding, deleting, or morphing the code. The analysis is usually applied using a sample of the actual code, rather than 100% of the code.* The sample used should include all constructs that are allowed in the source code and comprise at least 10% of the actual code base. It is also recommended that combinations of the constructs be evaluated. The analysis should be performed using the compiler settings that will be used for the actual code generation. The analysis normally involves a line-by-line comparison of the source code and the object code (or machine code). Any object code (or machine code) that is not directly traceable to the source code is analyzed and explained. In some situations, the analysis may identify unacceptable compiler behavior, which requires action. Compiler features such as register tracking, instruction scheduling, and branch optimization can lead to problems with the analysis. Highly optimized compiler settings and certain languages (e.g., languages with object-oriented features) can also result in untraceable code, and hence would not be certifiable. The analysis requires an engineer with knowledge of the specific language, assembly, machine code, and compilers. If the source code changes, there will need to be an assessment to ensure that no additional code constructs are added and that the analysis remains valid. Once this analysis is performed, it may be used for multiple projects, as long as the compiler and its settings are unchanged and the code uses the same constructs.

For the majority of projects that I have worked on, the source-to-object code analysis has not uncovered significant issues, as long as the optimization is kept minimal and mature languages and compilers are used. However, the analysis is still needed and should be done thoroughly.

9.7.4.5 Data Coupling and Control Coupling Analyses (DO-178C Table A-7 Objective 8)

I have saved the best for last in this section on structural coverage. The data coupling and control coupling (DC/CC) objective (Table A-7 objective 8) was included in DO-178B but was not clear. In particular, DO-178B varied from the overall software engineering definition of data coupling and control

---

* CAST-12, entitled *Guidelines for approving source code to object code traceability* provides additional information on this topic [15].

coupling but did not clearly explain the intent. According to members from the RTCA Special Committee #167 that developed DO-178B, the data and control coupling objective was added late in the committee deliberations and did not have extensive discussion. Consequently, there has been confusion about what was actually meant by the objective. Thankfully, DO-178C has clarified it. However, the clarification will likely cause some challenges throughout the industry, because many companies do not comply with the clarified objective (i.e., their interpretation of the original DO-178B objective is not consistent with the DO-178C clarification).

DO-178C defines *data coupling*, *control coupling*, and *component* as follows [1]:

- "Data coupling—The dependence of a software component on data not exclusively under the control of that software component."

- "Control coupling—The manner or degree by which one software component influences the execution of another software component."

- "Component—A self-contained part, combination of parts, subassemblies, or units that performs a distinct function of a system."

According to DO-178C, the intent of Table A-7 objective 8 is to perform an "analysis to confirm that the requirements-based testing has exercised the data and control coupling between code components" [1]. CAST-19 states that the purpose of data and control coupling analyses is as follows:

> To provide a measurement and assurance of the correctness of these modules/components' interactions and dependencies. That is, the intent is to show that the software modules/components affect one another in the ways in which the software designer intended and do not affect one another in ways in which they were not intended, thus resulting in unplanned, anomalous, or erroneous behavior. Typically, the measurements and assurance should be conducted on R-BT [requirements-based tests] of the integrated components (that is, on the final software program build) in order to ensure that the interactions and dependencies are correct, the coverage is complete, and the objective is satisfied [16].*

Unfortunately, this is not how many developers have interpreted the criteria. Many have applied it as a design and code review activity to ensure that the code accurately implements the control and data flow of the design. This activity during development is important to ensure that the final DC/CC analysis will be successful and to satisfy DO-178C Tables A-4 and A-5 objectives; however, this is not the intent of the Table A-7 objective. In addition to design and code reviews, those who comply with DO-178C also need to ensure that their tests cover the data couples and control couples between components (such as, modules or functions). Some organizations have applied the DC/CC

---

* Brackets added for clarification.

criteria as intended, and it can be quite challenging—particularly, if design and code reviews were inadequate or if the integration approach is not sound. If the requirements-based testing and structural coverage measurements are performed on an integrated system (rather than module by module) without instrumenting the code, this helps improve confidence.

In order to successfully perform DC/CC analysis using requirements-based tests, it is important to ensure during development that the architecture is consistent with requirements and that code complies with architecture. Satisfying DO-178C's guidance on DC/CC analysis normally involves four steps as described in the following.

1. First, the software architecture must be documented (DO-178C Table A-2 objective 3). DO-178C section 11.10 provides guidance on what to document in the software design description, including the following related to data or control flow: data structures, software architecture, internal and external input/output, data flow and control flow of the design, scheduling procedures and inter-processor/inter-task communication mechanisms, partitioning methods, and descriptions of software components [1]. Chapter 7 discussed the design process.

2. Second, the software architecture and code are reviewed and/or analyzed for consistency. DO-178C Table A-4 objective 9 references DO-178C section 6.3.3., which explains that one purpose of the design review and/or analysis is to ensure the correct relationship between "components of the software architecture. This relationship exists via data flow and control flow..." [1]. DO-178C Table A-5 objective 2 references DO-178C section 6.3.4.b, which explains that one purpose of the code review and/or analysis is to "ensure that the Source Code matches the data flow and control flow defined in the software architecture" [1]. Normally, both the design and code review/analysis activities involve a detailed checklist or questionnaire to consider common data coupling and control coupling issues. Table 9.2 provides some example issues to consider during the reviews and analyses; the specifics will vary depending on the architecture, language, environment, etc. These are merely examples.

3. Third, requirements-based integration tests are developed (DO-178C Table A-5). DO-178C section 6.4.3.b explains that requirements-based software integration testing is performed to ensure that "the software components interact correctly with each other and satisfy the software requirements and software architecture" [1]. The integration testing is intended to identify the following kinds of errors [1]: incorrect initialization of variables and constants, parameter passing errors, data corruptions (especially global data), incorrect sequencing of events and operations, and inadequate end-to-end

TABLE 9.2

Example Data and Control Coupling Items to Consider During Design and Code Review/Analysis

Example Data Coupling Items to Consider	Example Control Coupling Items to Consider
• All external inputs and outputs are defined and are correct • All internal inputs and outputs are defined and are correct • Data is typed correctly/consistently • Units are consistent and agree with data dictionary • Data dictionary and code agree and are both complete • Data is sent and received in the right order • Data is used consistently • Data corruption is prevented or detected • Data is initialized or read-in before being used • Stale or invalid data is prevented or detected • Data miscompares or data dropouts are prevented or detected • Unexpected floating point values are prevented or detected • Parameters are passed properly • Global data and data elements within global data constructs are correct • I/O is properly accessed from external sources • All variables are set (or initialized) before being used • All variables are used • Overflow and underflow is identified and correct • Local and global data are used • Arrays are properly indexed • Code is consistent with the design	• Order of execution is identified and correct • Rate of execution is identified and correct • Conditional execution is identified and correct • Execution dependencies are identified and correct • Execution sequence, rate, and conditions satisfy the requirements • Interrupts are identified and correct • Exceptions are identified and correct • Resets are identified and correct • Responses to power interrupts are identified and correct • Foreground schedulers execute in proper order and at the right rate • Background schedulers are executed and not stuck in infinite loop • Code is consistent with the design

numerical resolution. Some argue that including the *requirements-based* emphasis in the DO-178C explanation of DC/CC analysis weakens the criteria, because the architecture also needs to be exercised. However, during design review the consistency between architecture and requirements should be verified. If that occurs, then the requirements-based testing should also exercise the architecture. Also, as noted in DO-178C section 6.4.3.b the architecture should be considered during the DC/CC analysis.

4. Fourth, tests are analyzed to confirm that the data and control coupling between components are exercised by the requirements-based tests (DO-178C Table A-7 objective 8). If Steps 1–3 are not performed adequately, it will be difficult (probably impossible) to adequately complete Step 4. As far as I know, there are no commercially available tools to perform the data coupling and control coupling analysis. Most companies either development their own tool(s) or perform the analyses manually. Hopefully, there will be some expansion of the better commercially available tools in the future to help with the data and control coupling analyses.

Item 6 in the "Recommended Readings" section of this chapter provides some additional insight into data coupling and control coupling analyses. Additionally, since partitioning is closely related to data and control coupling, Chapter 21 discusses some additional items worth considering when developing an approach to satisfy DO-178C Table A-7 objective 8.

9.7.4.6 Addressing Structural Coverage Gaps

When measuring code coverage obtained through requirements-based tests, some gaps in coverage may be identified. An analysis is performed to determine the reason for the gaps. The following actions may be taken:

- If the coverage gap is caused because of missing tests, additional test cases are added and executed.
- If the gap is caused because of a missing requirements (an undocumented feature), the requirements are updated and tests are added and executed.
- If the gap identifies dead code or extraneous code, the dead or extraneous code is removed and an analysis is performed to assess the effect and the need for reverification. Typically, some reverification is needed. See Chapter 17 for information on dead and extraneous code.
- If the gap is caused by deactivated code, the deactivation approach is analyzed to ensure it is adequate and consistent with requirements and design. See Chapter 17 for discussion of deactivated code.
- If the gap is not addressed by any of the aforementioned, an additional analysis is needed to ensure that the code is adequately verified and works as intended. Typically, this analysis is either included in or summarized in the SVR.

9.7.4.7 Final Thoughts on Structural Coverage Analysis

Before we leave the subject of structural coverage, I want to summarize four important things.

First, when structural coverage credit is claimed, it should be based on requirements-based tests that pass (and satisfy the requirements they are traced to). If the test fails, the coverage may not be valid.

Second, structural coverage is typically measured by instrumenting the source code to identify where the source code has been exercised during the test execution. As the source code is changed it is important to assess if the changes introduced by the instrumentation are benign (as they should be). The resultant code optimization is also affected by the instrumentation because the information recording coverage features changes information flow. Therefore, it is necessary to rerun the tests without the instrumentation and compare the instrumented and non-instrumented results before claiming structural coverage credit.

Third, not all structural coverage tools fully satisfy DO-178C's criteria. In particular, be cautious of tools used to measure coverage in concurrent code that have not been designed to measure coverage in the presence of tasking constructs. Tools should be carefully evaluated and thoroughly understood before investing time and money. The FAA research report, entitled *Software Verification Tools Assessment Study*, provides some interesting insight. The FAA-sponsored research proposed a test suite to assess commercially available structural coverage tools. Anomalies were found in each of the three evaluated tools. This research demonstrates that care must be taken when selecting a tool, since some tools may not meet the DO-178C definition of structural coverage [17].

Lastly, the structural coverage analysis results are typically either included in or referenced in the SVR. Any structural coverage gaps and their analysis will be evaluated closely by certification authorities and/or their designees; therefore, they should be clearly stated and thoroughly justified.

9.8 Problem Reporting

Although problem reporting is included as part of the configuration management section in DO-178C and is further explored in Chapter 10 of this book, it is discussed in this chapter on verification, because many of the software problems are discovered during verification. Problem reports can be generated at any phase of the project and are usually initiated once a data item (e.g., requirements, design, code, test cases, or test procedures) has been reviewed and baselined. The problem reports are used to manage change between baselines and to address known errors with the software and its life cycle data.

During the verification process, it is common to discover problems with systems requirements, software requirements, design, and source code. Also, problems with the plans, the life cycle processes, tools, or even the hardware may be discovered. Problem reports are used to document issues with processes, data, and product. Problem reporting typically does not occur until after a life cycle data item has been reviewed, updated (if needed), and baselined. Once that has occurred, it is typical for all changes to be documented in a problem report. Some companies have two classes of problem reports— one for code and one for process and documentation errors. They sometimes have yet another class for future product enhancements. I tend to prefer one problem reporting system to document all kinds of issues or possible changes to the software, with a good classification system for each problem report (i.e., the problem report may be classified as code issues, documentation issue, process issues, future enhancement, etc.).

Each problem report (PR) typically contains the following information:

- PR number
- Date PR written
- Problem summary
- Affected life cycle data items and their versions
- Problem type (e.g., code, design, requirements, documentation, enhancement, process, tool, other)
- Problem severity (impact or consequence) from aircraft safety and compliance perspective (e.g., safety impact, certification/compliance impact, major functional/operational impact, minor functional/operational impact, documentation only)
- Problem priority from the software developer's perspective (e.g., fix immediately, fix as soon as possible, fix before next release, fix before certification, fix if possible, do not fix)
- PR author name and problem discovery date
- How the problem was found
- Functional area of the problem
- Problem description and how to reproduce it
- Suggested fix (usually an optional field)
- Supporting data (typically attached to the PR)
- Person assigned to investigate the problem
- Problem analysis and comments (i.e., where the problem investigation is documented)
- Status (e.g., open, in-work, fixed but not yet verified, verification in progress, fixed and verified, cannot be reproduced, deferred, duplicate, or canceled [if it is not a valid problem])

It should be noted that there are not always severity *and* priority classifications. Some problem reporting systems just have one classification and some have two. It will depend on the nature of the project. The PR classification scheme should be explained in the Software Configuration Management Plan (SCMP) to ensure it is consistent and clear to everyone involved. Chapter 10 provides additional information on problem reporting, including some of the certification authorities' recommended classifications.

Writing good PRs is challenging for many engineers. However, failure to write them well makes it difficult to confirm that changes are properly assessed, implemented, and verified. Following are some practical suggestions for writing and addressing PRs:

Suggestion 1: Document all problems in a PR, unless the document, data, or product has not yet been reviewed, in which case the problem should be noted in the review record. Verbal and e-mail reporting of problems are not trackable or actionable.

Suggestion 2: Write a descriptive, concise, accurate, and unique problem summary. The problem summary may end up in the SAS, if the problem is deferred; therefore, it could have high visibility. Even nondeferred problem reports may be read by managers, customers, systems engineers, safety personnel, SQA, certification liaison personnel, and occasionally, certification authorities. Therefore, the problem summary should be written in such a way that it makes sense even for someone not knowledgeable of the software details. Likewise, it should be value added for those who do know the software details. Essentially, extra time should be taken to write a good problem summary, since it could be around for a while.

Suggestion 3: Number and classify each PR. The classification may change as the problem is investigated, but it is important to have some idea of the severity or priority of the issue. Chapter 10 provides more discussion on the classification approach.

Suggestion 4: Ensure that the process for classifying PRs is clearly documented in the SCMP. As noted earlier, some projects will have multiple classification schemes (e.g., one for severity and one for priority). The classification approach must be clear to the developers, customer, change control board (CCB), certification liaison personnel, management, and anyone else who will read the PRs and make decisions using the PRs.

Suggestion 5: Only identify one problem in each PR. It is sometimes tempting to group related problems together; however, these are difficult to close if some are fixed and some are not.

Suggestion 6: Ensure that the PR is legible and understandable. A clear and complete description of the problem is needed, as well as specific information on how the problem was noted, how to reproduce the problem, and what the problem is. Using attachments with excerpts from the data item or screen shots of the problem can be very helpful. If the problem is not clearly described, proper action may not be taken.

Suggestion 7: Immediately document problems when they are discovered. Even obvious bugs should be reported; otherwise, they may not be fixed.

Suggestion 8: Document nonreproducible bugs in a PR. Such bugs may become more apparent as the project progresses.

Suggestion 9: Generally, the PR author should spend time explaining the problem but should not try to solve the problem. Some suggestions for resolution may be included, but the solutions will come during the investigation.

Suggestion 10: Ensure that the PRs are professional and nonjudgmental. It does not help to blame the coder, management, or the architect. Finger pointing is not productive. The PR should focus on the technical issue, not the people.

Suggestion 11: Once a problem is reported, it should be reviewed by one or more people to determine the next steps. A CCB (or a PR review board) is generally utilized to evaluate new problems and to make assignments. Once all necessary data is gathered and a recommendation for a solution is established, the CCB decides to approve the solution or defer the solution.

Suggestion 12: During the investigation phase of the problem, the engineer should consider related issues. The reported problem may just be a symptom of a bigger issue. The investigator looks for the root cause, not just the symptoms.

Suggestion 13: Remember that problem investigation and debugging can take time. Some problems are obvious, but some may take weeks to evaluate. Investigations should be assigned to experienced engineers who have a good understanding of the overall system and the development process.

Suggestion 14: Ensure that proposed solutions clearly identify what items will be changed and impacted and what the expected change will be to each item. Basically, a simplified change impact analysis is needed for each change. The better this analysis is documented, the easier it will be to support the consolidated change impact analysis for the software baseline.

Suggestion 15: Once a data item has been reviewed and baselined, it is only changed with a PR (or equivalent change authorization vehicle). This applies to CC1 (control category #1) data items (such as, requirements, design, code, verification cases/procedures). CC1/CC2 is explained in Chapter 10. It can be tempting to change obvious errors, when fixing a data item for a documented problem. It is not wrong to fix such errors; however, the issues and solutions need to be documented in a PR and not just fixed without a record.

Suggestion 16: Evaluate the problems together. Throughout the project, developers and management (as well as customers, quality, and certification liaison personnel) should have an understanding of the overall group of PRs.

Sometimes problems are related, but if the technical people are not in tune with the bigger picture, they may miss the connections.

Suggestion 17: Be aware of the tendency for bug morphing [9]. This occurs when a bug is reported and the investigator gets off track (perhaps he or she finds some other issues). If other issues are identified during a PR investigation, additional PRs should be created to address them.

Suggestion 18: Do not be afraid to report duplicate issues. Some developers hesitate to enter a PR because they think it might be a duplicate. They should quickly review the list of existing PRs and submit their problem if they do not clearly see the problem reported. It should also be noted that when there is a troubled area in the software, there may be multiple PRs that are related but are slightly different (oftentimes, there are multiple symptoms of an underlying issue). Therefore, it is better to write a potentially duplicate PR than to not report the issue. The CCB (or PR review board) will do the more thorough assessment to determine duplication.

Suggestion 19: To ensure quality and consistent PRs, assign an engineer to review PRs for readability, accuracy, completeness, etc. Oftentimes, this person also leads the CCB (or PR review board) and ensures the PR is mature before being evaluated by the board.

Suggestion 20: Verify the resolution(s) to the problem prior to closing the PR. The verification process may include re-review of requirements, design, code, and test data. Typically, an independent reviewer is preferred. The data item should go through the same level of rigor as it initially did (that is, the same processes, checklists, etc.).

Suggestion 21: If a PR is deferred, write a justification for deferral in the PR. This justification should explain why there is no safety, operation, performance, or compliance issues if the problem is not fixed prior to certification. The justification will have considerable scrutiny, so it must be clearly stated and well supported by the data. The justification will be evaluated by the CCB (and/or the PR board), the customer, safety personnel, and certification liaison to determine if deferral is feasible. Once the decision to defer is agreed, the justification is included in the SAS for agreement with certification authorities and/or their designees.

Suggestion 22: Include SQA in the PR deferral and closure process. SQA may be a member of the CCB or may have a separate approval process.

9.9 Recommendations for the Verification Processes

Some claim that there is no such thing as a *best practice* when it comes to verification, since so much depends on the specific situation. However, there are several practices that make the project go smoother and increase the probability of success. This section provides a summary of recommendations based on my involvement in numerous projects. Some of these are a summary of the aforementioned material and others have not yet been covered.

Recommendation 1: Plan early and adjust frequently. No project can predict all of the issues that will arise. However, planning is still important and can reduce some of the greater issues. In addition to the planning, continual adjustments must be made to address the actual issues.

Recommendation 2: Strive for a realistic schedule. Fictitious schedules are one of my pet peeves. Dream-based planning and head-in-the-sand management are just not effective. One time I was working on a schedule and told the manager that I needed to consider the tasks before I could project the schedule. He looked at me like I was crazy. He said: "I have my management hat on." I do not understand why a *management hat* cannot be based on reality. As one of my colleagues likes to say: "Good leaders don't ignore reality." Far too much scheduling is based on business agendas, rather than reality. Most engineers will work like maniacs to meet a realistic, even aggressive, schedule. However, too many impossible or dream-based schedules cause them to become demoralized and demotivated.

Recommendation 3: Design for robustness. It is nearly impossible to test in robustness (i.e., to implement robustness during the test phase). It is better to anticipate abnormal input and to design the system to address that input.

Recommendation 4: Design for testability. As noted earlier, it is helpful for developers to build the system so that it will support testing. For example, making some data structures visible at an outer level makes testing easier than verifying with data that is difficult to access. It is helpful to have a tester involved in the requirements and design reviews, because he or she will be thinking ahead about the test aspects.

Recommendation 5: Start verification early and continue it throughout the project. Verification should start as early as possible in the development effort. Reviews of requirements, design, and code should take place as the data is generated. It is extremely beneficial to have informal technical reviews of the data prior to holding the formal review. Putting verification off to the end is not effective, can be very expensive, and can harm the relationship with the certification authority and the customer.

Recommendation 6: Define roles during reviews. During peer reviews, it is helpful to define roles. For example, one person may focus on how well the tests exercise the requirements; one person may concentrate on the accuracy of the traces; one may evaluate correctness and accuracy of the test cases; and one may execute the test procedures to determine repeatability.

Recommendation 7: Involve the testers during the review phase. If testers are involved in the review of requirements and design, they can help ensure that the software is testable and can become familiar with the requirements that they will be testing.

Recommendation 8: Start testing activities early. As the requirements mature, someone should begin the test planning and test case development.

Recommendation 9: Give testers time and opportunity to understand the system. The better they understand the system's intent and its actual functionality, the better they can identify the errors. When the requirements are split among the testers and no one really has an overall understanding of the system, testing is not as effective.

Recommendation 10: Know the environment. Testers should understand the software's operational environment. Knowledge of the development environment, the hardware, and the interfaces is also important for effective testing.

Recommendation 11: Testers should be encouraged to and allowed to question everything. Obviously, the questions do not need to be out loud, but the best way to find errors is to continually be considering "what if?" or "how does it work?" or "why does it work?" A non-curious verifier tends to be ineffective.

Recommendation 12: Use experienced people and allow them to apply their experience. Obviously, the pool of experts is limited. However, experienced testers should be used for the key tasks. The junior engineers can play a role as well, and some of their work is incredibly ingenious. But, the senior and more experienced personnel should be allowed to lead. Some companies institute a form of *pair testing*—where two people work together on tests. This can be an effective way to train the junior engineer and utilize the experienced testers.

Recommendation 13: Encourage testers to think beyond the requirements—especially early in the project. If only the requirements are tested and the requirements are wrong or incomplete, some serious issues could be discovered late or may not be discovered at all.

Recommendation 14: Test critical and key functionality first. If the basic things do not work, the rest will not really matter.

Recommendation 15: If something is confusing, it should be tested more. If the requirements, design, and/or code are confusing, there is usually a reason. It might be because the developer did not understand the problem well (causing omissions or mistakes) or that he or she understood it too well (leading to oversimplification). If the requirements or design are confusing, it often leads to errors in the code. In many cases, the confusing area may need to be redesigned and recoded, or at least cleaned-up.

Recommendation 16: Expect to spend more time testing complex areas. As noted earlier, errors tend to hide in the caverns of complexity. If something is complex, it will likely have more issues and will take more time to verify and to fix when the problems are discovered.

Recommendation 17: Realize that quality needs to be built in—not tested in at the end. Testing is an indicator of the overall quality of the product, but waiting until the test phase to address quality is too late. Quality should be proactively built in. Involving testers in the development processes and developing and executing tests early (1) helps identify issues while they can still be efficiently fixed and (2) improves quality. Sometimes testing will uncover an issue that leads to redesign and reimplementation; it is better to find these errors early.

Recommendation 18: Testers should focus on the technical functionality—not the people. As mentioned, testers are often seen as a negative group. Their negativity helps identify problems that need to be resolved. However, they must take care to focus the negative energy on the software and not the people who wrote it.

Recommendation 19: Use as much independent verification as possible. Even though DO-178C does not require independence for many of the verification activities, some degree of independence is quite effective. It truly is difficult to find errors in your own work.

Recommendation 20: Write accurate PRs and keep abreast of the overall status of the program's problems. PRs should be generated as soon as problems are noted. Likewise, the PRs should be as accurate as possible—bad data does not help anyone. Additionally, management should regularly read the PRs in order to properly manage the project.

Recommendation 21: Automate wisely. Automation should only be used when it makes sense. Some projects automate just to be automating. Engineers love to build tools. Sometimes it can take longer to build and qualify a tool than to just test the product manually. Also, tools do not replace the need to think. If not used wisely, tools can give a false sense of security.

Recommendation 22: Set the team up for success. An effective manager fosters teamwork, encourages each member to do his or her best, utilizes diversity, encourages a nonhostile and productive work environment, rewards talent, encourages testers to find errors, deals with issues proactively, and makes decisions based on data rather than feelings (particularly, when it comes to schedules).

Recommendation 23: Foster creativity. Thorough testing requires creativity. Sometimes the testing requires more creativity than the development, because the testers have to think how to break the software rather than make it work. Creativity can be encouraged with free time to think (I call this *dream time*), fun activities, and competitions.

Recommendation 24: Invest in training for the testers. Good testers are always learning. Managers should provide opportunities for training and growth— even if it takes a day or two away from the project.

Recommendation 25: Implement continuous improvement. I once saw a comic strip where a guy was chopping down trees in the forest with an axe. He had a schedule to keep and did not have time to learn how to use the chainsaw in his truck. Sometimes we get so focused on the task at hand that we fail to look for opportunities to improve. Obviously, at the end of the project lessons learned should be captured and actions taken. However, it is also valuable to correct course during the project when something is not working or when there is a more effective way to do it.

Recommendation 26: Identify risks and issues and deal with them proactively. As issues arise, they should be addressed. They do not go away on their own.

Recommendation 27: When outsourcing some or all of the testing, do not just "throw it over the wall." Outsourcing or subcontracting will require close oversight, training, and continual communication. This is discussed more in Chapter 26.

Recommendation 28: Keep a master list of common issues and errors and make sure your team is aware of them. It is beneficial to compile a list of common issues based on input from experienced testers, problem reports, and literature. The list can be updated over time. Testers should be educated on the common mistakes. Obviously, they should not limit their verification to just these, but these will be a good start at shaking out problems in the system.

Recommendation 29: Be prepared for the challenges. One never knows what kind of challenges he or she will encounter when verifying software. Be ready to roll with the punches. Some of the most common challenges are: ambiguous requirements, missing requirements, determining how much robustness is enough, addressing schedule pressure (since testing is typically the last activity), exercising the architecture as well as the requirements, managing change (keeping up with changes to requirements, design, and code), keeping morale up, dealing with problems that are discovered late in the process, and maintaining an adequate tester-to-developer ratio. Every program has its own special set of issues. Be ready for anything.

References

1. RTCA DO-178C, *Software Considerations in Airborne Systems and Equipment Certification* (Washington, DC: RTCA, Inc., December 2011).
2. Certification Authorities Software Team (CAST), Verification independence, Position Paper CAST-26 (January 2006, Rev. 0).
3. RTCA DO-248C, *Supporting Information for DO-178C and DO-278A* (Washington, DC: RTCA, Inc., December 2011).
4. RTCA DO-254, *Design Assurance Guidance for Airborne Electronic Hardware* (Washington, DC: RTCA, Inc., April 2000).
5. Certification Authorities Software Team (CAST), Addressing cache in airborne systems and equipment, Position Paper CAST-20 (June 2003, Rev. 1).

6. C. Kaner, J. Bach, and B. Pettichord, *Lessons Learned in Software Testing* (New York: John Wiley & Sons, 2002).

7. G. J. Myers, *The Art of Software Testing* (New York: John Wiley & Sons, 1979).

8. E. Kit, *Software Testing in the Real World* (Harlow, England, U.K.: ACM Press, 1995).

9. A. Page, K. Johnston, and B. Rollison, *How We Test Software at Microsoft* (Redmond, WA: Microsoft Press, 2009).

10. C. Kaner, J. Falk, and H. Q. Nguyen, *Testing Computer Software*, 2nd edn. (New York: John Wiley & Sons, 1999).

11. J. A. Whittaker, *How to Break Software: A Practical Guide to Testing* (Boston, MA: Addison-Wesley, 2003).

12. R. S. Pressman, *Software Engineering: A Practitioner's Approach*, 7th edn. (New York: McGraw-Hill, 2010).

13. K. J. Hayhurst, D. S. Veerhusen, J. J. Chilenski, and L. K. Rierson, *A Practical Tutorial on Modified Condition/Decision Coverage*, NASA/TM-2001-210876 (Hampton, VA: Langley Research Center, May 2001).

14. Certification Authorities Software Team (CAST), Structural coverage of object code, Position Paper CAST-17 (June 2003, Rev. 3).

15. Certification Authorities Software Team (CAST), Guidelines for approving source code to object code traceability, Position Paper CAST-12 (December 2002).

16. Certification Authorities Software Team (CAST), Clarification of structural coverage analyses of data coupling and control coupling, Position Paper CAST-19 (January 2004, Rev. 2).

17. V. Santhanam, J. J. Chilenski, R. Waldrop, T. Leavitt, and K. J. Hayhurst, *Software Verification Tools Assessment Study*, DOT/FAA/AR-06/54 (Washington, DC: Office of Aviation Research, June 2007).

Recommended Readings

1. K. J. Hayhurst, D. S. Veerhusen, J. J. Chilenski, and L. K. Rierson, *A Practical Tutorial on Modified Condition/Decision Coverage*, NASA/TM-2001-210876 (Hampton, VA; Langley Research Center, May 2001). This tutorial provides a practical approach to assess aviation software for compliance the DO-178B (and DO-178C) objective for MC/DC (DO-178C Table A-7 objective 5). The tutorial presents a five-step approach to evaluate MC/DC coverage without a coverage tool. The tutorial also addresses factors to consider when selecting and/or qualifying a structural coverage analysis tool. Tips for reviewing MC/DC artifacts and pitfalls common to structural coverage analysis are also discussed.

2. J. J. Chilenski, *An Investigation of Three Forms of the Modified Condition Decision Coverage (MCDC) Criterion*, DOT/FAA/AR-01/18 (Washington, DC; Office of Aviation Research, April 2001). This report compares three forms of MC/DC and provides justification for why MC/DC should be part of the software system development process. The three forms of MC/DC are compared theoretically and empirically for minimum probability of error detection performance and ease of satisfaction.

3. V. Santhanam, J. J. Chilenski, R. Waldrop, T. Leavitt, and K. J. Hayhurst, *Software Verification Tools Assessment Study*, DOT/FAA/AR-06/54 (Washington, DC; Office of Aviation Research, June 2007). This report documents the investigation of criteria to effectively evaluate structural coverage analysis tools for use on projects intended to comply with DO-178B (and now DO-178C). The research effort proposed a test suite to increase objectivity and uniformity in the application of the structural coverage tool qualification criteria. The prototype test suite identified anomalies in each of the three coverage analysis tools evaluated, demonstrating the potential for a test suite to help evaluate a tool's compatibility with the DO-178B (and now DO-178C) objectives.

4. J. J. Chilenski and J. L. Kurtz, *Object-Oriented Technology Verification Phase 3 Handbook—Structural Coverage at the Source-Code and Object-Code Levels*, DOT/FAA/AR-07/17 (Washington, DC; Office of Aviation Research, June 2007). This handbook provides guidelines for meeting DO-178B (and now DO-178C) structural coverage analysis objectives at the source code versus object code or executable object-code levels when using object-oriented technology (OOT) in commercial aviation. The differences between source code and object code or executable object-code coverage analyses for the object-oriented features and MC/DC are identified. An approach for dealing with the differences is provided for each issue identified. While the focus is OOT, many of the concepts are applicable to non-OOT projects.

5. CAST-17, Structural coverage of object code (Rev 3, June 2003). This paper, written by the international Certification Authorities Software Team (CAST), explains some of the motivations behind structural coverage at the object code or executable object code level and identifies issues to be addressed when using such an approach.

6. J. J. Chilenski and J. L. Kurtz, *Object-Oriented Technology Verification Phase 2 Handbook—Data Coupling and Control Coupling*, DOT/FAA/AR-07/19 (Washington, DC; Office of Aviation Research, August 2007). This handbook provides guidelines for the verification (confirmation) of data coupling and control coupling within OOT in commercial aviation. Coverage of inter-component dependencies is identified as an acceptable measure of integration testing in both non-OOT and OOT software to satisfy DO-178B (and now DO-178C) Table A-7 objective 8. This approach is known as coupling-based integration testing.

10

Software Configuration Management

Acronyms

CC1	control category #1
CC2	control category #2
CCB	change control board
CD	compact disk
CIA	change impact analysis
CRC	cyclic redundancy check
DVD	digital video disk
EASA	European Aviation Safety Agency
FAA	Federal Aviation Administration
PR	problem report
PSAC	Plan for Software Aspects of Certification
SAS	Software Accomplishment Summary
SCI	Software Configuration Index
SCM	software configuration management
SLECI	Software Life Cycle Environment Configuration Index
SQA	software quality assurance

10.1 Introduction

10.1.1 What Is Software Configuration Management?

Software configuration management (SCM) is an integral process that goes from the cradle to the grave of the software life cycle. It spans all areas of the software life cycle and impacts all data and processes. Babich writes: "Configuration management is the art of identifying, organizing, and controlling modifications to the software being built by a programming team. The goal is to maximize productivity by minimizing mistakes" [1].

SCM is not just for source code as is commonly believed; it is needed for all software life cycle data. All data and documentation used to produce the software, verify the software, and show compliance of the software requires some level of configuration management. In other words, all of the software life cycle data listed in DO-178C section 11 requires SCM. The rigor of the SCM process applied depends on the software level and the nature of the artifact. DO-178C uses the concept of CC1/CC2 (control category #1 or control category #2) to identify how much configuration control applies to a data item. A data item classified as CC1 must apply all of the DO-178C SCM activities. A CC2 data item, however, may just apply a subset. CC1/CC2 is discussed later in this chapter.

This chapter examines the SCM activities that are required by DO-178C and that are considered best practices in any safety-critical domain. Special attention is given to the problem reporting, change impact analysis (CIA), and environment control processes, since they are important SCM processes that companies often struggle with in the real world.

10.1.2 Why Is Software Configuration Management Needed?

Software development in all domains, including the safety-critical domain, is a high pressure activity. Software engineers are required to develop complex systems with tight schedules and budgets. They are expected to make updates quickly and maintain high-quality software that complies with standards and regulations.

> To survive in this brutally competitive world, organizations need some sort of mechanism to keep things under control or total chaos and confusion will result, which could lead to product or project failures and put the company out of business. A properly implemented software configuration management (SCM) system is such a mechanism that can help software development teams create top-quality software without chaos and confusion. [2]

Good SCM helps to prevent problems such as: missing source code modules, inability to find the latest version of a file, reappearance of corrected mistakes, missing requirements, inability to determine what changed and when, two programmers overwriting each other's work when updating a shared file, and many more. SCM reduces these issues by coordinating the work and effort of multiple people working on the same project. When properly implemented, SCM "prevents technical anarchy, avoids the embarrassment of customer dissatisfaction, and maintains the consistency between the product and the information about the product" [3].

SCM also enables effective change management. Software-intensive systems will change—it's part of the nature of software. Pressman writes: "If you don't control change, it controls you. And that's never good. It's very easy for a stream of uncontrolled changes to turn a well-run software project into chaos" [4]. Because change happens so frequently, it must be effectively managed. Good SCM provides a means to manage change by: (1) identifying data items likely to change, (2) defining the relationship between the data items, (3) identifying the approach for controlling revisions of data, (4) controlling changes implemented, and (5) reporting any changes made.

Effective SCM provides many benefits to the software team, the overall organization, and the customer, including the following:

- Maintains integrity of the software by using organized tasks and activities.
- Builds confidence with certification authorities and customers.
- Supports higher quality, and hence safer, software.
- Enables the management of life cycle data required by DO-178C.
- Ensures the configuration of software is known and is correct.
- Supports the schedule and budget needs of the project.
- Provides the capability to baseline software and data items.
- Provides the ability to track changes to baselines.
- Avoids confusion and enhances communication among the development team members by providing a systematic approach to data management.
- Helps avoid, or at least reduce, surprises and wasted time.
- Provides a means to identify, record, and resolve problems in the code and the supporting life cycle data.
- Promotes a controlled environment for developing, verifying, testing, and reproducing software.
- Ensures software can be regenerated, even years after the initial development.
- Provides status accounting throughout the project.
- Provides a foundation for developers and customers to make decisions.
- Provides ability to reproduce issues during problem investigations.
- Provides a foundation for process improvement.
- Provides data to confirm when software is complete.
- Supports long-term maintenance.

The risks of not having good SCM are significant. Poor SCM leads to lost time, money, quality, and confidence. I recently consulted on a project with a company that had poor SCM. They had bright engineers, but they could not consistently reproduce the software or the avionics unit. This led to huge delays, increased oversight by the Federal Aviation Administration (FAA), and costly penalties due to the lateness of the product.

Although SCM does not make for exciting reading, it is extremely important. Projects get far too exciting when SCM is *not* applied.

10.1.3 Who Is Responsible for Implementing Software Configuration Management?

SCM is the responsibility of everyone involved in the software development process. Software development in all domains, including the safety-critical domain, is a high pressure activity. All developers need to be educated on the benefits of good SCM, the risks of poor SCM, SCM best practices, and the specific company SCM procedures. Such training helps engineers do an overall better job. When properly implemented and automated, SCM should not be difficult for the developers to perform on a daily basis.

In the past, SCM was a manual, time-consuming process; however, now, with the availability of good SCM tools, the SCM process can be implemented without a heavy burden on the developers.

> But thinking that SCM tools will take care of everything can be a recipe for disaster. Many SCM activities, including change management, build and release management, and configuration audits, require human intervention and judgment. Although SCM tools make these jobs easier, there is no substitute for human intelligence and decision making. [2]

SCM is everyone's responsibility. It requires communication and teamwork between those who want the software and data, those who produce the software and data, and those who use the software and data. Good SCM starts and ends with communication. Here are some suggestions to encourage communication that will enable effective SCM:

- Ensure the goals and objectives are clearly stated.
- Make sure that all stakeholders understand the goals and objectives.
- Ensure that all stakeholders are cooperating and address any issues that hinder cooperation.
- Ensure that all processes are documented and clearly understood by all stakeholders.
- Provide feedback frequently.
- Make sure data needed for decisions are available.
- Address issues as they arise.

10.1.4 What Does Software Configuration Management Involve?

In addition to the development of the SCM Plan, which is an output of DO-178C Table A-1 objectives, DO-178C identifies six objectives for the SCM process in Table A-8. Each objective is required for all software levels. The six objectives are the following [5]:

- *DO-178C, Table A-8, objective 1*: "Configuration items are identified."
- *DO-178C, Table A-8, objective 2*: "Baselines and traceability are established."
- *DO-178C, Table A-8, objective 3*: "Problem reporting, change control, change review, and configuration status accounting are established."
- *DO-178C, Table A-8, objective 4*: "Archive, retrieval, and release are established."
- *DO-178C, Table A-8, objective 5*: "Software load control is established."
- *DO-178C, Table A-8, objective 6*: "Software life cycle environment control is established."

These objectives are applied to ensure integrity, accountability, reproducibility, visibility, coordination, and control of the software life cycle data as it evolves. In order to satisfy the DO-178C objectives, several activities are required, including: configuration identification, baselining, traceability, problem reporting, change control, change review, status accounting, release, archival and retrieval, load control, and environment control. Each activity is discussed in the next section.

10.2 SCM Activities

This section describes the activities of the SCM process. Each activity should be detailed in the SCM Plan early in the project. The SCM Plan ensures that the SCM team and the project team are aware of the procedures, duties, and responsibilities they are to carry out during the project to support and maintain SCM. Both the SCM team and the project team should be trained on the SCM expectations and requirements. The SCM Plan is the foundation for training and equipping the SCM and project teams on the required SCM processes. See Chapter 5 for more details on the SCM Plan.

10.2.1 Configuration Identification

"Configuration identification is one of the cornerstones of configuration management, as it's impossible to control something whose identify you don't know" [6]. It is the first activity of configuration management. Configuration identification "identifies items to be controlled, establishes identification schemes for the items and their versions, and establishes the tools and techniques to be used in acquiring and managing controlled items" [2]. Configuration identification provides the starting point for other SCM activities; it is the first major SCM function that needs to be started in a project. It is essentially a prerequisite for other SCM activities, since all the other activities use the output of the configuration identification activity [2].

Configuration identification provides the ability to (1) identify the components of a system throughout its life cycle and (2) trace between the software and its life cycle data. Each configuration item must be uniquely identified. The identification method typically includes a naming convention along with version numbers and/or letters. The identification approach facilitates the "storage, retrieval, tracking, reproduction, and distribution of configuration items" [2].

10.2.2 Baselines

In software, a baseline is the software and its supporting life cycle data at a point in time. The baseline serves as a basis for further development. Once a baseline is established, changes should only be made through a change control process [2,7].

Baselines should be established early in the project; however, bringing all items under configuration control too early can impose unnecessary procedures and slow the developers' work [2]. "Before a software configuration becomes a baseline, changes may be made quickly and informally. However, once a baseline is established, change can be made, but a specific, formal procedure must be applied to evaluate and verify each change" [4]. So, the question arises, "When should a baseline be established?" There is no hard and fast rule. Usually baselines are established after each life cycle phase at the completion of the formal review that ends the phase. Thus, there is a requirements baseline, a design baseline, a code baseline, etc. The SCM Plan should identify the plan for baselining. It is also important to align the code baseline with the requirements and design that it implements.

Baselines must be immutable. This means the data item(s) is *locked down* so it cannot be altered. This immutability characteristic is important because a permanent record of the code version (and its supporting data) used to build the release for software testing, flight testing, production, etc. is necessary [8].

"The number and type of baselines depend on which life cycle model the project is implementing. Life cycle models, such as spiral, incremental development, and rapid prototyping, require more flexibility in the establishment of baselines" [9]. Baselines need to be established for configuration items that are used for certification credit [5].

10.2.3 Traceability

Traceability is closely related to baselining and problem reporting. Once a configuration item is baselined, changes are documented (typically through a problem report [PR] and/or change request). Therefore, when a new baseline is established, it must be traceable to the baseline from which it came.

10.2.4 Problem Reporting

Problem reporting is one of the most important SCM activities. PRs are used to identify issues with baselined data, process noncompliances, and anomalous behavior with the software. PRs may be generated to address problems as well as to add or enhance software functionality. Effective problem reporting is essential to managing change and fixing known issues in a timely manner. As noted previously, problem reporting typically starts once a data item has been baselined.

See Section 9.8 for a summary of what a PR contains and recommendations for writing and addressing PRs. Problem reporting is an ongoing process throughout the software development and verification. Problems should be investigated and addressed promptly after identification. Additionally, every effort should be made to write PRs that are understandable by a wide audience, including developers, managers, quality, safety engineers, systems personnel, certification liaison representatives, and certification authorities.

Following are a few topics related to PR management that are especially important to certification.

10.2.4.1 Problem Report Management with Multiple Stakeholders

Most software-intensive systems involve multiple stakeholders. For example, a typical avionics system may have the following stakeholders: aircraft manufacturer, systems integrator, avionics systems engineers, safety team, avionics software developers, operating system supplier, software verification team, and certification personnel. Each entity has a different focus and area of expertise. When multiple stakeholders are involved, several issues can arise, such as the following:

- The software and verification teams may not understand the system, safety, or aircraft implications of some of their problems.
- The aircraft manufacturer or systems integration team may not have access to the software problem data. If they do have access, they may not have adequate software background to fully understand the issues and to make the appropriate safety decisions regarding the risks to their aircraft and flight test crew.
- A large number of problems may mask other issues—both at software and systems levels.

In order to address these and other similar issues related to multiple stakeholders, the following recommendations are offered:

Recommendation 1: Coordinate and agree upon a PR classification approach. Typically, the final customer (e.g., the aircraft manufacturer) identifies their problem classification scheme to all of their suppliers. The PRs are categorized based on potential impact on safety, functionality, performance, operation, or development assurance [10]. For example, an aircraft manufacturer may have severity categories, such as the following:

1. Safety and compliance issues (must fix immediately)
2. Significant functionality, performance, or operation issues (must fix before certification)
3. Nonsignificant functionality, performance, or operation issues (fix if possible)
4. Nonfunctional issues (e.g., documentation only) (may be deferred)

In their Certification Memorandum CM-SWCEH-002, the European Aviation Safety Agency (EASA) proposes a four-type classification scheme, as summarized in Table 10.1. The terms *error*, *failure*, and *fault* are defined as follows [11]:

- *Error*: With respect to software, a mistake in requirements, design, or code.
- *Fault*: A manifestation of an error in software. A fault, if it occurs, may cause a failure.
- *Failure*: The inability of a system or system component to perform a required function within specified limits. A failure may be produced when a fault is encountered. But a fault may also remain hidden at system level and have no operational consequences.

Additionally, the software team may have an additional priority classification scheme to help with their PR management. These may include the following:

- Must fix immediately
- Must fix prior to flight (safety issue)
- Must fix prior to certification (functional issue)
- Document/noncode issue (fix if possible; okay to defer)
- Cancelled

The classification schemes for all entities need to be coordinated, agreed upon, and explained in the plans (including SCM Plan and system-level plans).

Recommendation 2: Coordinate PR implementation, cancellation, and closure. Action taken on PRs, cancellation of PRs, and closure of PRs should be agreed upon with the stakeholders.

Recommendation 3: Coordinate flow-down PRs. Any software-related PRs found at the higher level (e.g., aircraft or system level) should be flowed down to the software team to address. A separate software PR is typically opened.

Recommendation 4: Coordinate flow-up PRs. All software developers should supply their PRs to their customer in a timely manner and establish feedback.

TABLE 10.1

Example PR Classification Scheme

Type	Potential Impact	Description
0	Safety impact	A problem whose consequence is a failure, under certain conditions of the system, with an adverse safety impact.
1A	Significant functional failure	A problem whose consequence is a failure that has no adverse safety impact on the system/equipment, but the failure has a *significant* functional consequence. The meaning of *significant* needs to be defined in the context of the related system.
1B	Nonsignificant functional failure	A problem whose consequence is a failure that has no adverse safety impact on the system/equipment, and the failure has no *significant* functional consequence. The meaning of *significant* and *nonsignificant* needs to be defined in the context of the related system.
2	Nonfunctional fault	A problem that is a fault that does not result in a failure.
3A	Significant deviation from plans or standards	Any problem that is not Type 0, 1, or 2, but is a deviation to the software plans or standards. Type 3A is a *significant* deviation whose effects could be to lower the assurance that the software behaves as intended and has no unintended behavior.
3B	Nonsignificant deviation from plans or standards	Any problem that is not Type 0, 1, or 2, but is a deviation to the software plans or standards. Type 3B is a *nonsignificant* deviation that does not affect the assurance obtained.
4	All other types of problems	All other open PRs that do not fall into the earlier classification Types 0–3. Due to the mutually exclusive nature of these classifications, the problems of Type 4 are of no functional consequence. In many cases these problems might be described as typographical errors.

Source: European Aviation Safety Agency, Software aspects of certification, Certification Memorandum CM-SWCEH-002, Issue 1, August 2011.

Agreement will need to be obtained on the problem criticality and the plan to address it. There are several ways in which coordination may occur. The PRs may be provided to customers, the PR database may be made available to customers, and/or there may be a PR review board involving the customer. For complex systems with several stakeholders, a weekly or biweekly PR review meeting can be very effective. For an aircraft project, the stakeholders might include representatives from aircraft systems, avionics systems, flight test, software, safety, software quality assurance (SQA), and certification. This broad team of stakeholders is needed because what may seem to be an insignificant issue from one perspective may indeed be a safety issue when examined from another perspective. The multiple disciplines help to ensure all issues are thoroughly analyzed and understood, so they can be properly addressed. PRs should be provided to all stakeholders to review prior to the PR review meeting; this also allows those with schedule conflicts the opportunity to provide input or to send a delegate.

The certification authorities have raised issue papers and policy documents on management of multi-stakeholder PRs (e.g., FAA's Order 8110.49 [10] and EASA's CM-SWCEH-002 [11]). Some of the guidelines are still evolving. Be sure to address any specific issues that are noted by the certification authority on your particular program.

10.2.4.2 Managing Open/Deferred Problem Reports

It is highly desirable and recommended that PRs be addressed and closed prior to certification. Justifying open PRs and managing them during maintenance can be a large overhead for a project and a potential troublemaker for subsequent upgrade projects.

I recently consulted on a project that had 2000 test failures because of an ambiguous requirement. The developer wanted to defer the PR. However, justifying 2000 failures is not an easy task. After much debate, everyone agreed that it was easier and cleaner to fix the requirement and pass the tests than to try to justify why it is okay to certify with 2000 failed tests.

Most projects end up with at least a few PRs that are deferred. When this is the case, the deferred PRs need to document a thorough analysis and justification as to why the deferral does not impact safety or compliance. Problems that impact safety or compliance are not suitable for deferral. The decision to defer must be understood and agreed upon by all the stakeholders and summarized in the Software Accomplishment Summary (SAS) for certification authority agreement. Chapter 12 explains the SAS, as well as what is expected in the SAS for any deferred PRs.

A large number of deferred PRs is problematic for several reasons:

- It indicates that the system may not be mature.
- It indicates that development assurance may be questionable.
- It is difficult to assess the interaction of problems.
- It is difficult to adequately justify a large number of issues from a safety and compliance perspective.
- It can be difficult to claim that software performs its intended functionality, as required by the regulations.
- It is extremely difficult to maintain the PRs during maintenance. Each time the software changes or is used in another installation, the PRs need to be reevaluated.

All of these should be considered when deciding to fix or defer a PR.*

10.2.5 Change Control and Review

Change control and review are closely related to the problem reporting process because PRs are often used as the vehicle to manage change. Change may occur at any phase of the software development process and is actually a fundamental characteristic of software. Change may occur to add or modify functionality, fix a bug, improve performance, update hardware and the supporting interfaces, improve processes, change the environment (e.g., compiler enhancement), etc. It is relatively easy to change software, which is one of the main advantages of using software; however, it is not so easy to manage it. Without effective change management, chaos occurs. Change management is complex but essential. An effective change management process involves several tasks, including the following:

1. *Protection from unauthorized changes.* Once a baseline is established, it should be protected from inadvertent and unauthorized changes. All changes to a baseline should be planned, documented, and approved (typically through a change control board [CCB]).

2. *Change initiation.* All proposed changes to a baseline should be documented. A PR or a change request is the normal vehicle for documenting the proposed change.

3. *Change classification.* As noted earlier in Section 10.2.4, changes are classified by their impact on the overall system safety, functionality, and compliance.

4. *Change impact assessment.* In order to properly plan the resources that are needed to implement a change and the amount of reverification required, the impact of the change is documented in a PR or change request. It should be noted that if the software has already been through formal testing or certification, a more formal CIA is usually needed, as discussed later in this chapter.

5. *Change review.* DO-178C section 7.1.e states that the objective of change review is to ensure that "problems and changes are assessed, approved, or disapproved, approved changes are implemented, and feedback is provided to affected processes through problem reporting and change control methods defined during the software planning process" [5]. Change review involves the review of all changes to determine the potential impact of the change. Typically,

---

* Chapter 14 of FAA Order 8110.49 [10] and chapter 16 of EASA CM-SWCEH-002 [11] provide guidelines for managing open problem reports.

a CCB is established as the gatekeeper for the change process. The CCB reviews all PRs and change requests, approves or disapproves the proposed changes, confirms that approved changes are properly implemented and verified, and confirms that PRs or change requests are properly updated throughout the development and verification effort and closed upon completion. Therefore, the CCB should include members who know the software and system well and who have authority and knowledge to make changes. Oftentimes, the customer (e.g., aircraft manufacturer and/or avionics integration team) will be part of the CCB. Earlier, a PR review board was mentioned. In most projects the PR review board and the CCB are the same board. However, for some large projects, they may be separate.

6. *Change decision.* The CCB evaluates the proposed change and approves, denies, defers, or sends it back for more information.

7. *Change implementation.* Once a proposed change is approved by the CCB, the change is made as agreed upon in the PR or change request. If the scope of the change is modified during implementation, the PR or change request needs to be updated and may require additional review by the CCB. As noted earlier, change to a configuration item should result in a change to its identification (typically, the version or revision is updated; but sometimes a part number change is needed). Additionally, as previously noted, software changes should be traceable from their origin. Anne Hass writes: "For any configuration item, it must be possible to identify changes in it relative to its predecessor. Any change should be traceable to the item where the change was implemented" [6].

8. *Change verification.* Once the change is implemented, it is verified. This is usually carried out by an independent review of all changed or impacted artifacts, as well as a rerun of any necessary tests.

9. *Change request or PR closure.* Once the change has been implemented, verified, and accepted, it may be closed. The CCB is typically responsible for the PR or change request closure.

10. *Updated baseline release.* The baseline should not be released until all of the approved PRs or change requests have been verified.

10.2.6 Configuration Status Accounting

Configuration status accounting involves recording and reporting information that is needed to effectively manage the software development and verification effort. Reports are generated to inform managers, developers, and other stakeholders about the project's status. Configuration status accounting provides consistent, reliable, timely, and up-to-date status information that helps to enhance communication, avoid duplication, and prevent repeat mistakes [2]. It often includes reports that provide the following:

- Status of data items, including configuration identification.
- Status of PRs and change requests (including classification, impacted data items, root cause of problems, configuration identification of updated data items).
- Status of released data and files.
- List of baseline contents and differences from previous baseline.

SCM tools are often used to automate the configuration status accounting reports. For the tools to be successful, the data must be accurately and consistently entered. Additionally, to avoid erroneous usage, the tool functionality must be well understood by those using the tool.

Configuration status reports should be planned early in the project in order to properly capture the data. However, as a project progresses, it is sometimes desirable to modify or expand the metrics. For example, when summarizing PRs, one might decide to add fields to identify personnel information (who identified the problem, what team caused the problem, who fixed the problem, etc.) in order to help with staffing needs.

Status reports should be updated at a defined frequency. Most companies either automate them, so they are always current, or update weekly prior to their project team meetings. An out-of-date or erroneous report is useless and can even lead to bad decisions. I was recently involved in a project that frequently provided an out-of-date and inaccurate summary of the PRs. It was very hard for me, as the FAA designee, to determine what data to look at for safety impact. The erroneous data also made me question the accuracy of other data. Therefore, it is important to make every effort to generate accurate status reports; otherwise, they may do more harm than good.

10.2.7 Release

Once a data item is mature, it is released into the formal configuration management system (typically an enterprise-wide library). DO-178C defines *release* as: "The act of formally making available and authorizing the use of a retrievable configuration item" [5].

Not all data need to go through the formal release process. Some data may just be stored and protected. Typically, the data needed to rebuild and maintain the software (such as requirements, design, code, configuration files, and configuration index) need to be formally released. Supporting data (such as review records and SQA records) just need to be stored with the project records (typically on a secure server). DO-178C identifies the minimum set of data items to be released using the CC1/CC2 categorization. This will be discussed more when we examine control categories later in this chapter (Section 10.2.9).

The release process typically involves review and sign-off of the document/data by key stakeholders (e.g., author, reviewer, software quality engineer, certification liaison representative, and data quality). Prior to signing/approving a data item, it should actually be read. This probably seems obvious to many, but I have come across more than a few signed documents that had not been read.

Typically, data (including executable code) are released prior to providing to customer and prior to using for certification credit.

10.2.8 Archival and Retrieval

The archival and retrieval process involves the storage of data (both released data and other data used to support certification) so that it can be accessed by authorized personnel. The archival and retrieval process should consider the following [5]:

- *Accuracy and completeness.* There should be verification that the proper data are archived. This is often done when preparing and reviewing the Software Configuration Index (SCI) to make sure data in the SCI match what has been archived.

- *Protection from unauthorized change.* Data should only be updated by authorized personnel and only with the proper change authorization.

- *Quality of the storage media* to minimize regeneration errors in the short-term and deterioration errors in the long-term. Not all media is suitable for storing safety-critical software.

- *Protection during disasters.* This is typically carried out by using some kind of off-site storage.

- *Readability of data.* Type design data need to be available and readable as long as the equipment is in an aircraft. This may require data refresh on a periodic schedule, depending on how reliable the storage media is.

- *Archival of supporting tools.* If tools are required to read or generate the data, they too may need to be archived; this may include the development and verification environments. If the tools require licensing agreements, those should also be considered.

- *Accuracy of retrieval and duplication.* When the data are retrieved or duplicated, it must be free from corruption.

- *Ability to handle modification without losing data.* When data are modified, the previously released and archived data should not be affected.

10.2.9 Data Control Categories

In order to identify how much configuration control is required by data type, DO-178C uses the concept of *control category*. The concept of control categories is unique to aviation and is explained in DO-178C. There are two control categories defined: control category 1 (CC1) and control category 2 (CC2). The DO-178C Annex A tables identify the applicability of CC1 or CC2 for each data item.*

CC1 requires maximum control and applies to key certification data and data required for regeneration or accident investigation. Also, for levels A and B more data items are classified as CC1. CC1 requires application of the following SCM processes: configuration identification, baselines, traceability, problem reporting, change control (integrity, identification, and tracking), change review, configuration status accounting, retrieval, protection against unauthorized changes, media selection, refreshing, duplication, release, and data retention [5].

CC2 is a subset of CC1 and requires limited configuration management. It applies to data items that aren't as critical to executable object code regeneration, such as SCM records and verification records. CC2 requires application of the following SCM processes: configuration identification, traceability to source, change control (integrity and identification), retrieval, protection against unauthorized changes, and data retention [5].

Table 10.2 summarizes DO-178C's CC1 and CC2 requirements by data item. There are some data items that are always CC1: Plan for Software Aspects of Certification (PSAC), requirements, development trace data, source code, executable object code, parameter data item files, SCI, and SAS. Likewise, there are some data items that are always CC2: PRs, SQA records, verification results, and SCM records. The control category for other data items vary by software level.

The DO-178C control category assignments are considered a minimum. Many companies opt to go above and beyond and require more items to be CC1 than is required by DO-178C. For example, the Software Verification Report is often treated as CC1, even though DO-178C identifies software verification results as CC2.

10.2.10 Load Control

Executable code does nothing until it is actually loaded in the target hardware. Some software is loaded in the factory and some is loaded in the field. (See Chapter 18 for information on field-loadable software.) Following are the key components of a controlled software load process:

- *Approved load procedures*. Although the actual loading is often outside the software development and DO-178C scope (since loading is often part of the manufacturing process or aircraft maintenance), it needs to be considered during the software approval process. The loading

---

* DO-330, DO-331, DO-332, and DO-333 Annex A tables also identify CC1 or CC2 for each data item.

TABLE 10.2

Control Categories by Data Type

DO-178C Software Life Cycle Data	DO-178C Section	A	B	C	D
PSAC	11.1	CC1	CC1	CC1	CC1
Software Development Plan	11.2	CC1	CC1	CC2	CC2
Software Verification Plan	11.3	CC1	CC1	CC2	CC2
SCM Plan	11.4	CC1	CC1	CC2	CC2
SQA Plan	11.5	CC1	CC1	CC2	CC2
Software Requirements Standards	11.6	CC1	CC1	CC2	N/A
Software Design Standards	11.7	CC1	CC1	CC2	N/A
Software Code Standards	11.8	CC1	CC1	CC2	N/A
Software Requirements Data	11.9	CC1	CC1	CC1	CC1
Design Description	11.10	CC1	CC1	CC1	CC2[a]
Source Code	11.11	CC1	CC1	CC1	CC1
Executable Object Code	11.12	CC1	CC1	CC1	CC1
Software Verification Cases/Procedures	11.13	CC1	CC1	CC2	CC2[a]
Software Verification Results	11.14	CC2	CC2[a]	CC2[a]	CC2[a]
SLECI	11.15	CC1	CC1	CC1	CC2
SCI	11.16	CC1	CC1	CC1	CC1
PRs	11.17	CC2	CC2	CC2	CC2
SCM Records	11.18	CC2	CC2	CC2	CC2
SQA Records	11.19	CC2	CC2	CC2	CC2[a]
SAS	11.20	CC1	CC1	CC1	CC1
Trace Data (Development)	11.21	CC1	CC1	CC1	CC1
Trace Data (Verification)	11.21	CC1	CC1	CC2	CC2[a]
Parameter Data Item File	11.22	CC1	CC1	CC1	CC1

[a] Only applies when the software life cycle data are required (varies by objective).

procedures should be developed and verified as part of the DO-178C compliance effort. DO-178C section 11.16.k recommends that the loading procedures be included in the SCI, just as the build instructions are. This was added to DO-178C since DO-178B. In DO-178B, it was not clear where the load procedures were to be documented.

- *Load verification.* There should be some means to ensure that the software is completely loaded without corruption. This is often carried out by some kind of integrity check, for example, a cyclic redundancy check (CRC).

- *Part marking verification.* There needs to be some way to identify the loaded software to confirm that the part number and version loaded are consistent with what was approved. Once software is loaded, it should be verified that the identification agrees with the approved data.

* *Hardware compatibility.* Approved hardware and software compatibility should be documented and adhered to.

10.2.11 Software Life Cycle Environment Control

The software life cycle environment includes the methods, tools, procedures, programming languages, and hardware that will be used to develop, verify, control, and produce the software life cycle data and software product [5]. An uncontrolled environment can lead to numerous problems, including errors in the code, lost data items, elusive errors during testing, inadequate testing, nonrepeatable build process, etc. "Software life cycle environment control ensures that the tools used to produce the software are identified, controlled, and retrievable" [5].

In the planning phase, the software development environment is described in the Software Development Plan, the verification environment is described in the Software Verification Plan, the SCM environment is described in the SCM Plan, and the SQA environment is described in the SQA Plan. The plans describe the procedures, tools, methods, standards, and hardware used to implement processes. It is also recommended to summarize all of the tools in the PSAC.

However, during the planning phase, the details of the environment are often not known (e.g., the compiler may be known but the specific version, options/settings, and supporting libraries may not be known during planning). In order to have a deterministic environment, the details (including specific executable files) must be documented. The details of the software life cycle environment are identified in the Software Life Cycle Environment Configuration Index (SLECI) or an equivalent document. The typical contents of the SLECI are described later (see Section 10.4.3). The SLECI is used to control the environment. There should be processes in place to ensure that the environment identified in the SLECI is what is actually being used by the engineers. This is a commonly overlooked task. The SLECI is frequently not completed until the end of the program. However, in order for the environment to be controlled, it must be completed early in the process. Because the development, SCM, and SQA environments may be known earlier than the verification environment, there may be multiple iterations of the SLECI.

Additionally, the tools used to develop, verify, control, build, and load the software and its data need to be kept under configuration control. Therefore, the tools must be controlled as CC1 or CC2 data. DO-178C requires the executable object code for nonqualified tools to be controlled at CC2, as a minimum. The appropriate control category for qualified tools and their supporting qualification data are specified in DO-330, *Software Tool Qualification Considerations*. Tool qualification is discussed in Chapter 13; however, be forewarned about the environment used for qualified tools. If your project is using a qualified tool, it is important to verify that the environment in which you are using the tool (the operational environment) is representative of the environment in which the tool was qualified. Otherwise, the tool qualification credit may not be applicable.

10.3 Special SCM Skills

Experienced and qualified personnel are important for SCM. Configuration management is everyone's job, but the SCM manager or librarian and the CCB members play a distinct role in the SCM process.

The *SCM librarian* establishes the SCM library or libraries and ensures each library's integrity as well as the integrity between libraries [6]. The librarian's role may be supported by automation. The SCM librarian should possess the following skills: understanding of the company's overall configuration management needs, attention to details, ability to document detailed procedures, ability to ensure adherence to procedures, ability to communicate well with a variety of people types, and an understanding of any automation used.

The *CCB* is responsible for evaluating changes, approving/disapproving changes, and following up on all agreed actions. The CCB members must understand the product, the potential impact of changes, and the overall SCM system. The CCB leader or chair should be good at coordinating a variety of personalities, managing meetings, ensuring follow through, and communicating at multiple levels. This includes the ability to consider the needs of the various stakeholders when determining change impact, priorities, and safety-related issues.

10.4 SCM Data

DO-178C Table A-8 identifies the objectives for the SCM process and the data generated during the process. This section briefly explains each of the SCM life cycle data items.

10.4.1 SCM Plan

The contents of the SCM Plan were explained in Chapter 5. The SCM Plan should be developed early in the project, so that SCM is properly applied. The SCM Plan should address the items discussed in this chapter, as well as what was discussed in Chapter 5. In addition to the SCM Plan, there may need to be more detailed procedures for some SCM activities. For example, the problem reporting process and the CIA process often require specific details above and beyond what is in the SCM Plan.

10.4.2 Problem Reports

PRs were discussed earlier in this chapter and in Chapter 9. PRs should be identified and kept current throughout the project. It is also important to have a good status summary of PRs to support project management decisions, as well as certification and safety evaluations.

10.4.3 Software Life Cycle Environment Configuration Index

As noted previously and in DO-178C section 11.15, the SLECI is used to document and control the environment for development, build, load, verification, and control. It helps support software regeneration, reverification, or modification. It contains a list of the following [5]:

* Hardware and operating system of the development environment.
* Tools used to develop, build, and load the software (such as compilers, linkers, loaders, requirements management tools).
* Tools used to verify the software (including hardware for testing and operating systems).
* Tools used by SCM and SQA processes.
* Qualified tools and the data used to support the qualification.

Many projects package their SLECI with the SCI. When doing this, it should be noted that multiple iterations of the SCI will be needed, since the environment needs to be controlled before code is built and released.

10.4.4 Software Configuration Index

The SCI provides a listing of all the life cycle data for the software product with its specific configuration (including the source code and executable object code files). If the software contains multiple components, there may be a hierarchy of SCIs; for example, an SCI for each component and then a top-level SCI to pull all of the components together.

The SCI is an important data item because it essentially "identifies the configuration of the software product" [5] and is important for certification and maintenance. The SCI defines what the software is and what data were used to develop it. It is also one of the three data items that are required to be submitted to the certification authority (in addition to PSAC and SAS).

Per DO-178C section 11.16, the SCI identifies the software product (the part number(s) of the software loaded into the equipment), the executable object code, the source code files (with version information), archive and release media, the life cycle data, build instructions for generating the executable object code, reference to or inclusion of the SLECI, data integrity checks used (such as CRCs), loading procedures and methods, and procedures and methods for user-modifiable software (if used) [5]. If parameter data item files are used, they are also listed, along with any build instructions for creating them. (Parameter data item files are discussed in Chapter 22.)

Typically, an SCI is generated for each baseline of software. For the development baselines, the SCI often just includes a list of the source code files, build instructions, resolved PRs and change requests, and reference to or inclusion of SLECI data. However, for each baseline, it is also recommended to identify the version of plans, standards, requirements, and design that were used to generate the code. The other data can be added as the project progresses, but these items are important to know what drove the code.

I normally examine several SCIs each year. There are a couple of trends that I often see. These are noted in the following for your awareness:

- Oftentimes, the SCI includes the code listing but doesn't include the other software life cycle data. This seems to be particularly prevalent in companies that have a military background and are used to developing a Version Description Document (which was required by DOD-STD-2167A and other military standards).
- Many teams choose to combine the SLECI and SCI into a single document. This is fine and is even explicitly mentioned in DO-178C. However, teams often do not complete the document until the software is released. If the SLECI is part of the SCI, this means an early version of the document will be needed to define the development environment. Otherwise, it is difficult to prove that the environment was controlled.

10.4.5 SCM Records

SCM Records include additional data used in the SCM process, such as status accounting reports, release records, change records, software library records, and baseline records. The specific records generated vary from company to company. The SCM Plan should either identify the data to be generated or point to company procedures that provide the details.

10.5 SCM Pitfalls

There are several SCM pitfalls to avoid. Some pits are deeper than others, but all of them can be problematic and should be avoided. Certain issues were referred to earlier, but are mentioned here again, in order to provide a consolidated list:

Pitfall 1: Failure to address all of the activities required. All of the SCM activities mentioned in Section 10.2 are needed. If some are missing, it can quickly lead to havoc.

Pitfall 2: Improper planning. Good SCM doesn't just happen. It requires planning and detailed procedures. The plans and procedures should address the engineering or developmental configuration management, as well as the enterprise-wide or formal configuration management.

Pitfall 3: Lack of management understanding, commitment, or support. SCM requires resources, including tools, training, and staff. Without management support, the resources are typically limited and hence the SCM process falls short of what it needs to be.

Pitfall 4: Lack of qualified personnel. SCM requires a specialized skill set, as noted in Section 10.3. Failure to have people with these skills leads to ineffective SCM.

Pitfall 5: Improper use of automation. A tool alone will not solve an organization's SCM problems. As Jessica Keyes puts it, "automating a money-losing process allows you to lose more money faster and with greater accuracy" [9]. There are plenty of SCM tools available, ranging from free to very expensive. Some are tailored for large software teams, whereas others are more suitable for a smaller team. Some are good for formal SCM and long-term storage, whereas others work better for the day-to-day needs of engineering. Some projects may find it advantageous to utilize both a long-term and a day-to-day toolset. Great care should be taken when determining what to automate and when selecting the tool to perform the automation.

Pitfall 6: Lack of training. In order to consistently apply SCM, all parties need to understand the procedures. SCM training should be mandatory for all team members. It does not need to be time consuming. Computer-based training can be quite effective.

Pitfall 7: Environment isn't controlled. As mentioned earlier, the lack of controlled development and verification environments is a common problem for software projects. The SLECI should be developed in a timely manner and kept current, so that engineers all use the proper environment. Additionally, the settings of all tools should be documented somewhere (typically in the SLECI or in the build instructions section of the SCI) in order to ensure that the tools are properly set and used.

Pitfall 8: PR process not adequately defined or implemented. Frequently, the plans and procedures fail to explain when the problem reporting process will begin and how the team members are to carry it out. As a result, the situation arises where problems are not properly documented, are not fixed in a timely manner, are fixed without documentation, are not assessed for impact, are not coordinated among the stakeholders, etc.

Pitfall 9: PRs not written for a more general audience. Frequently, PRs are not written so they can be read by the wide range of people who will need to understand them, such as managers, the customer, and the certification authority.

Pitfall 10: Changes are made without authorization. Sometimes, while an engineer is doing his or her job, he or she sees another problem and decides to fix the problem. In and of itself, this isn't bad, unless no one else is apprised of the change and it isn't documented. Once data are baselined, all changes should be documented in a PR or a change request (or an equivalent process).

Pitfall 11: Failure to understand and control supplier's SCM. Some suppliers may have well-established SCM processes, while others may be at the infancy stage of SCM. It is important to understand the SCM processes of all suppliers (including subcontractors and offshore teams) and to deal with any limitations, weaknesses, or incompatibilities as soon as possible. Examples of things to consider when using suppliers are the following:*

- How does the supplier identify their data and is it compatible with your company's configuration identification scheme? It is important to understand how they name or number their data items. For supplier-delivered data it may be necessary to add your own company's number or store the data in a unique library or with a special attribute.
- What data will be delivered and what data will be maintained by the supplier?
- Once the data are delivered, who is responsible for maintaining it?
- Who is responsible for storing what data and where?
- What happens if the supplier is sold to another company or goes out of business?
- How is change control of the supplier's data managed? Will they have their own CCB? If so, what type of insight will be provided to their CCB's activity?
- What status accounting does the supplier provide?
- What problem reporting process will the supplier use?
- How does the supplier control their environment?

Pitfall 12: Not confirming the operational environment of qualified tools. When qualified tools are used to support the development or verification of the safety-critical software, it is important to confirm that the tool is used in the

---

* This is not an exhaustive list but is provided as a starting point.

operational environment for which it was qualified. Each tool is qualified for one or more operational environments. Tool users must confirm that they are using the tool in that environment. Otherwise, the tool may not operate properly. In some cases, the improper operation may not be obvious. Tool qualification is examined in Chapter 13.

Pitfall 13: Not archiving the environment. The environment (hardware, operating system, and supporting software tools) may need to be archived in addition to the requirements, design, source code, and executable object code in order to support regeneration, accident investigation, continued airworthiness, etc.

Pitfall 14: Procedures are not controlled. Many companies have enterprise-wide procedures on an intranet for easy access to the entire enterprise. Sometimes the plans will reference such procedures, but there is limited configuration control of the procedures. The procedures may not have revision status, may be updated without notice, and may not be archived (i.e., previous versions may not be retrievable). The software team must work to a known set of procedures. This may require them to capture the set of procedures in a separate work area or to use a different SCM process above and beyond the enterprise-wide intranet.

Pitfall 15: Failure to consider long-term readability of data. The media or data format should be considered for maintenance purposes. In addition to saving data on a dependable media (such as compact disks [CDs] or digital video disks [DVDs]), the equipment and software to read that media should be available. For this reason, it is recommended that data be stored in hard copy or .pdf format so that it can be read in the future. This is important when dedicated tools are needed to view the data, especially if the tool requires an annual contract extension or licensing.

Pitfall 16: Build and/or load procedures are not repeatable. As noted earlier, the procedures to build and load the software are documented in the SCI. Oftentimes, the only one who can perform the build or load is the person who performs those activities on a daily basis. Since that person may not be around every hour of every day for the lifetime of the equipment's use, the procedures need to be documented to be repeatable. They should be run by someone unfamiliar with the procedures to ensure that the procedures are understandable and the same results are obtained (repeatable).

10.6 Change Impact Analysis

Software change is a way of life for safety-critical systems. Several years ago at a conference, an engine control manufacturer claimed that over 90% of their software development activity consisted of changes to already flying software.

In the aviation industry, derivative products comprise the vast majority of our work. These derivative products tend to reuse as much software as possible, adding new features or corrections, as needed. Therefore, it is important to design the software to be maintainable and upgradable.

When software changes occur in safety-critical systems, great care must be taken. The changes to in-service software must be carefully planned, analyzed, implemented, controlled, and verified/tested. Additionally, steps must be taken to assure that a change will not negatively impact other functions in the system or on the aircraft.

The change process was discussed in Section 10.2.5. Each change during development is documented in a PR or change request. Each PR or change request considers the impact of that change. Each changed and impacted software component and its supporting life cycle data are then verified prior to accepting it into the software build and closing the PR or change request.

Once the software has been fielded, the CIA becomes a more formal process, requiring additional documentation. It should be noted that such formality may also be warranted during the initial development when formal testing has been completed and then regression testing is needed for a change. In 2000, the FAA first published policy on the CIA. The guidance has remained stable since that time. For software that has been approved on a certified product and is being modified (perhaps to fix a bug, add some functionality, or modify for a different user), a CIA is required. There are several purposes for the CIA that should be considered when doing the analysis:

- It guides the software development and verification team to determine the amount of rework.
- It helps the CCB determine which changes to approve or not.
- It provides a way to assess the potential impact of the change on safety.
- It provides insight to the customer to support their system-level or aircraft-level CIA.
- It serves as a vehicle to justify the categorization of the change as minor or major.

Typically, a preliminary CIA is performed early in the software modification effort in order to evaluate the amount of work needed to implement the change, determine the potential impact of the change on safety, and obtain the certification authority agreement on major or minor classification. The preliminary CIA is typically either documented as part of the PSAC for the upgraded software or is included in a separate document that is referenced in the software planning documents. At the end of the project, the CIA is updated to reflect the actual analysis and is often included in the SAS.

Per FAA Order 8110.49, software changes that do not have a potential impact on safety can be classified as minor; whereas, changes that could affect safety are typically classified as major [10]. Both major and minor changes go through the same process; however, major changes require involvement from the certification authority.

Chapter 11 of FAA Order 8110.49* identifies items that need to be assessed as part of the CIA, as a minimum [10]. Each item should be assessed even if it ends up having no impact (if it is not applicable, the CIA section can say "Not Applicable" and explain why). The following is a brief summary of the analyses summarized in the CIA [10,12]:

- *Traceability analysis*—identifies the requirements, design elements, code, and test cases and procedures that may be either directly or indirectly affected by the change. Forward traceability of changes identifies design components affected by the change. Backward traceability helps determine other design features and requirements that may be inadvertently affected by the change. Overall, the requirements traceability helps determine the impact of change on the software project. It is important to identify both changed and impacted software and data.

- *Memory margin analysis*—ensures that the memory allocation requirements are still satisfied, the original memory map is maintained, and adequate memory margins are maintained. This analysis typically cannot be completed until the change has been implemented. It can be estimated early on, but the actual impact assessment takes place later.

- *Timing margin analysis*—confirms that the timing requirements (including scheduling and interface requirements) are still satisfied, that resource contention characteristics are known, and that timing margins are still maintained. Like the memory margin analysis, the timing analysis typically cannot be completed until the change has been implemented.

- *Data flow analysis*—identifies any adverse effects due to changes in data flow, as well as coupling between and within components. In order to perform the data flow analysis, each variable and interface affected by the change should be analyzed to ensure that the original initialization of that variable remains valid, that the change was made consistently, and that the change does not affect any other usage of that data element. This can be a costly and difficult process, if large global data sections are used.

- *Control flow analysis*—identifies any adverse effects due to changes to the control flow and coupling of components. Task scheduling,

---

* With the publication of DO-178C and its supplements, the FAA policy will likely be updated. The CIA process will likely remain very similar, but if you are required to perform a CIA please confirm that you are using the latest policy.

execution flow, prioritization, and interrupt structure are examples
of the items that are considered.

- *Input/output analysis*—ensures that the changes have not adversely
  impacted the input and output requirements of the product. Things
  such as bus loading, memory access, throughput, hardware input,
  and output device interfaces are considered.

- *Development environment and process analyses*—identify any change(s)
  which may impact the software product (such as compiler options
  or versions and optimization change; linker, assembler, and loader
  instructions or options change; or software tool change). The target
  hardware should also be considered. If the processor or other hardware that interfaces with the software changes, it could impact the
  software's ability to perform its intended function.

- *Operational characteristics analysis*—considers changes that may
  impact system operation (such as evaluation of changes to displays
  and symbols, performance parameters, gains, filters, limits, data
  validation, interrupt and exception handling, and fault mitigation)
  to ensure that there are no adverse effects.

- *Certification maintenance requirements analysis*—determines whether
  new or changed certification maintenance requirements are neces-
  sitated by the software change. During the original certification of
  an aviation product, certification maintenance requirements are
  identified. For example, the brakes may need to be inspected after
  100 landings. If a change to software affects a certification maintenance requirement, it should be addressed during the change process.

- *Partitioning analysis*—ensures that the changes do not impact any
  protective mechanisms incorporated in the design. If architectural
  mitigations are employed as part of the partitioning scheme, the
  change must not affect those strategies. For example, data should
  not be passed from a less critical partition to a more critical partition,
  unless the more critical partition appropriately checks the data.

The CIA also documents the software life cycle data (such as requirements,
design, architecture, source and object code, test cases and procedures)
changed and affected by the change and verification activities (reviews,
analyses, inspections, tests) needed to ensure that no adverse effects on the
system are introduced during the change.

Having an organized, documented, and thorough CIA process is important for project planning and implementation, reverification and regression testing (see Chapter 9 for more information on regression testing), and
certification. Companies developing safety-critical software should have a
company-wide CIA process that meets the project needs, supports safety,
and meets the needs of the certification authority. In order for this to happen,

the analysis should identify any changed and impacted data (including the software itself), verification of the changed and impacted data, and analysis of safety impact. The CIA should be readable and comprehensive and should address the certification authority policy and guidance (e.g., FAA Order 8110.49 or EASA Certification Memorandum CM-SWCEH-002). The preliminary CIA may be higher level, but the final CIA needs to clearly identify what data were changed or impacted, what verification took place, what the final characteristics of the software are compared to the original characteristics (e.g., timing and memory margins), and how any safety impacts have been identified and verified. It should be noted that the preliminary CIA not only supports the certification effort but also provides a good project management tool to plan the resources, budget, and schedule for the software update.

The company-wide process should also identify the approach for documenting the preliminary and final CIAs. That is, explain if CIAs are part of the PSAC and SAS, stand alone, or some other arrangement. The process should also identify who approves the CIAs and whether they are submitted to the certification authority. For example, some organizations require that an FAA authorized designee review and approve the CIA. If their change is deemed to be major, the CIA is included as part of the PSAC and SAS and submitted to the FAA; however, if the change is minor, the CIA may be separate from the PSAC and SAS and is not submitted to the FAA.

Many companies perform a CIA for each PR or change request. This is a good practice; however, the combination of the PRs and/or change requests should also be considered in the CIA.

References

1. W.A. Babich, *Software Configuration Management* (Chichester, U.K.: John Wiley & Sons, 1991).
2. A. Leon, *Software Configuration Management Handbook*, 2nd edn. (Norwood, MA: Artech House, 2005).
3. A. Lager, The evolution of configuration management standards, *Logistics Spectrum*, Huntsville, AL, January–March 2002.
4. R.S. Pressman, *Software Engineering: A Practitioner's Approach*, 7th edn. (New York: McGraw-Hill, 2010).
5. RTCA DO-178C, *Software Considerations in Airborne Systems and Equipment Certification* (Washington, DC: RTCA, Inc., December 2011).
6. A.M.J. Haas, *Configuration Management Principles and Practice* (Boston, MA: Addison-Wesley, 2003).
7. IEEE, *IEEE Standard Glossary of Software Engineering Terminology*, IEEE Std-610–1990 (Los Alamitos, CA: IEEE Computer Society Press, 1990).

8. B. Aiello and L. Sachs, *Configuration Management Best Practices* (Boston, MA: Addison-Wesley, 2011).
9. J. Keyes, *Software Configuration Management* (Boca Raton, FL: CRC Press, 2004).
10. Federal Aviation Administration, *Software Approval Guidelines*, Order 8110.49 (Change 1, September 2011).
11. European Aviation Safety Agency, *Software Aspects of Certification*, Certification Memorandum CM-SWCEH-002 (Issue 1, August 2011).
12. L. Rierson, A systematic process for changing safety-critical software, *IEEE Digital Avionics Systems Conference* (Philadelphia, PA: 2000).

11

Software Quality Assurance

11.1 Introduction: Software Quality and Software Quality Assurance (SQA)

11.1.1 Defining Software Quality

To meet the needs of the safety-critical domain, software developers must be committed to quality. High-quality products don't just happen. They are the result of well-managed organizations with a commitment to quality, as well as talented, conscientious, and disciplined engineers.

The *American Heritage Dictionary* includes the following in the definition of *quality*: "an inherent or distinguishing character," an "essential character," and a "degree or grade of excellence" [1]. In software engineering, there are multiple views of what quality is—two tend to be pervasive. First is the developer's perspective: it is a quality product if the software meets their defined requirements. The second is the customer's perspective: it is a quality product if the software meets their needs. A product that meets its defined requirements but not the needs of the customer isn't considered high quality by the customer. The requirements are crucial to bridging the gap between the developer's and customer's view of quality. The requirements must be developed to meet the customer's needs, so that the developers can produce and verify software that meets those requirements. As noted in Chapter 6, it is important to get the customer involved in the requirements definition process. Without close customer and developer coordination during the requirements definition phase, quality is an elusive goal.

Roger Pressman writes: "You can do it right, or you can do it over again. If a software team stresses quality in all software engineering activities, it reduces the amount of rework that it must do. That results in lower costs, and more importantly, improved time-to-market" [2]. "The Quality Program is a framework for building quality into a product, doing the evaluations necessary to determine if the framework is working, and evaluating the quality actually achieved in the product" [3]. Quality is affected by activities throughout the life cycle, including requirements definition, design, coding, verification, and maintenance.

11.1.2 Characteristics of High-Quality Software

Quality attributes are often used to identify the goals of high-quality software. Such attributes include correctness, efficiency, flexibility, functionality, integrity, interoperability, maintainability, portability, reusability, testability, and usability. The International Standards Organization (ISO) and International Electrical Technical Commission (IEC) Standard 9126 (ISO/IEC 9126) define a set of quality attributes called characteristics and subcharacteristics. These characteristics and subcharacteristics are summarized in the following [4,5]:

- *Functionality*: The capability of the software product to provide functions which meet stated and implied needs when the software is used under specified conditions. Subcharacteristics of functionality include suitability, accuracy, interoperability, security, and functionality compliance.

- *Reliability*: The capability of the software product to maintain a specified level of performance when used under specified conditions. Subcharacteristics of reliability include maturity, fault tolerance, recoverability, and reliability compliance.

- *Usability*: The capability of the software product to be understood, learned, used, and attractive to the user when used under specified conditions. Subcharacteristics of usability include understandability, learnability, operability, attractiveness, and usability compliance.

- *Efficiency*: The capability of the software product to provide appropriate performance, relative to the amount of resources used, under stated conditions. Subcharacteristics of efficiency include time behavior, resource utilization, and efficiency compliance.

- *Maintainability*: The capability of the software product to be modified. Modifications may include corrections, improvements, or adaptation of the software to changes in the environment and in requirements and functional specifications. Subcharacteristics of maintainability include analyzability, changeability, stability, testability, and maintainability compliance.

- *Portability*: The capability of the software product to be transferred from one environment to another. Subcharacteristics of portability include adaptability, installability, coexistence, replaceability, and portability compliance.

Companies often implement metrics to measure a subset of these attributes, although many of them are actually quite difficult to quantify.

11.1.3 Software Quality Assurance

Most companies implement an SQA process to help ensure that the required quality attributes are satisfied and that the software meets its requirements. "A formal definition of Software Quality Assurance is that it is the systematic activities providing evidence of the fitness for use of the total software product" [6]. Pressman writes: "Quality assurance establishes the infrastructure that supports solid software engineering methods, rational project management, and quality control actions ...In addition, quality assurance consists of a set of auditing and reporting functions that assess the effectiveness and completeness of quality control actions" [2]. SQA activities provide confidence that adequate processes are established and followed in order to produce products that meet the requirements and the customer's needs.

The DO-178C SQA process is an integral process that runs continuously throughout the software planning, development, verification, and final compliance efforts. One or more software quality engineers (SQEs) ensure that the plans and standards are developed and followed throughout the implementation. The SQE(s) also performs a conformity review to confirm completeness and compliance.

Although DO-178C requires an SQA process, it should be noted that quality is not solely the responsibility of the SQA personnel. Additionally, software quality cannot just be assessed at the end of the project, as it is for some engineering disciplines. As William Lewis notes: "Quality cannot be achieved by assessing an already completed product. The aim, therefore, is to prevent quality defects or deficiencies in the first place, and to make the products assessable by quality assurance measures ...In addition to product assessments, process assessments are essential to a quality management program" [6]. Emanuel Baker and Matthew Fisher echo the sentiment: "While evaluation activities are essential activities, they alone will not achieve the specified quality. That is, product quality cannot be evaluated (tested, audited, analyzed, measured, or inspected) into the product. Quality can only be 'built in' during the development process" [3]. Quality is an ongoing process and is the responsibility of the entire software team. DO-178C encourages quality through the use of standards and verification activities throughout the software life cycle. Quality must be built in—it cannot be audited in, policed in, or tested in.

DO-178C requires that all SQA objectives be satisfied with independence. DO-178C provides a slightly different definition for independence for SQA than it does for verification. DO-178C defines independence as:

> Separation of responsibilities which ensures the accomplishment of objective evaluation. (1) For software verification process activities, independence is achieved when the verification activity is performed by a person(s) other than the developer of the item being verified, and a tool(s) may be used to achieve equivalence to the human verification activity. (2) For the software quality assurance process, independence also includes the authority to ensure corrective action [7].

The first half of the definition relates to verification independence which was discussed in Chapter 9. The second half applies to the SQA process and where it exists in the organizational structure. Although it isn't required, most companies meet the independence requirement by having a separate SQA organization which does not report to engineering. This section uses the term *SQA* to refer to the group that is responsible for performing the SQA activities. SQA may be a separate organization (which is preferred) containing one or more SQEs or just a separate function within the engineering organization (this approach tends to be limited to very small companies).

DO-178C encourages a process-oriented SQA approach. SQA is primarily responsible for ensuring that the processes in the approved plans and standards are followed. However, some organizations are beginning to realize that process alone is inadequate. Some companies are embracing the Software Engineering Institute (SEI) concept of product quality assurance (PQA), where an engineer is responsible for ensuring the quality of the product and not just the process. With the PQA approach, the product quality engineer (PQE) and SQE work closely together to ensure both product and process quality. Process evaluations demonstrate that the processes were followed; whereas product evaluations demonstrate the correctness of the process outputs. It should be noted that this has always been the concept of verification in DO-178B and now DO-178C; however, PQEs can help bridge the gap between engineering and SQA. The use of the PQE does require a PQE with domain knowledge and experience for the product being developed.

ISO has developed a number of documents to define good SQA practices, including a family of documents referred to as *ISO 9000** (which includes

---

* ISO-9000 is entitled *Quality Management Systems – Fundamentals and Vocabulary*; ISO-9001 is *Quality Management Systems – Requirements*; and ISO-9004 is *Managing for the Sustained Success of an Organization – A Quality Management Approach*.

ISO 9000, 9001, and 9004). ISO 9000 gives the vocabulary and basics of the series of standards on quality systems. ISO 9001 integrates earlier standards known as ISO 9001, 9002, and 9003. ISO 9001 is general and can be applied to any product. ISO 9001 includes the following basic elements [2]:

- Establish the elements of a quality management system.
- Document the quality system.
- Support quality control and assurance.
- Establish review mechanisms for the quality management system.
- Identify quality resources, including personnel, training, and infrastructure elements.
- Define methods for remediation.

Most developers of safety-critical software pursue and maintain an ISO 9000 registration. ISO 9000 registration is granted by an accredited third-party body. Surveillance occurs every 6 months. Reregistration is required every 3 years [5].

Other industry-wide organizations and standards, such as SEI's Capability Maturity Model Integration® (CMMI®) and the Six Sigma strategy, which was originally popularized by Motorola in the 1980s, also provide a framework for SQA.

11.1.4 Examples of Common Quality Process and Product Issues

There are a variety of quality issues that might be present in a software development effort. Table 11.1 summarizes some of the common process and product issues. Obviously, most of the process issues can lead to product issues if not proactively resolved. Therefore, both process quality and product quality are important to consider.

11.2 Characteristics of Effective and Ineffective SQA

11.2.1 Effective SQA

The purpose of SQA has been explained and SQA's activities will be explored shortly. Before examining what SQA does, let's consider some of the prerequisites for effective SQA. "Quality is never an accident; it is always the result of intelligent effort" [8]. The following characteristics are essential to effective SQA implementation.

Characteristic 1: Top-level management support. In order for SQA to be effective, they must have the buy-in, commitment, and support of top-level management. This needs to be a sincere commitment, not just lip service.

TABLE 11.1

Common Process and Product Issues

Common Process Issues	Common Product Issues
• Requirements and design reviews are skipped or are performed by people without the proper technical skills or understanding of the system.	• Defective software is shipped to the customer.
• Build process is not repeatable; every time the software is built the result is different.	• Software does not meet all of the requirements.
• Source code is added to fix some improper functionality, but the requirements and design are not updated to reflect the change.	• Software is not robust; it works okay in a normal situation but fails when something out of the ordinary happens.
• Configuration of the life cycle data, including the source code, is not maintained.	• Requirements are incomplete or ambiguous; therefore, the product does not meet the customer's expectations or needs.
• Plans and standards are not followed.	• Source code does not align with the requirements and design.
• Development and verification environments are not defined.	

Top-level management must be committed to the overall quality of their products and to providing the resources required to staff and train the SQA organization. This concept is strongly enforced by quality experts such as Kaoru Ishikawa, Joseph M. Juran, W. Edward Deming, and Watts S. Humphrey.

Characteristic 2: Independence. Effective SQA is independent of the development and verification organizations. Independence helps to reduce errors from extensive familiarity with the day-to-day process or product being evaluated. Additionally, independence helps to enforce corrective action when needed. Without independence, there is pressure (sometimes intentional and sometimes unintentional) to compromise when the budget and schedule demands arise.

Characteristic 3: Technical competence. In order for the SQA organization to be effective and to earn the respect of the development and verification teams, it must be staffed with well-qualified and experienced technical engineers. Without technical strength, SQA will not identify the real issues in a timely manner, and they will not be respected by the engineering organization.

Characteristic 4: Training. SQA personnel, as well as the overall engineering organization, must be properly trained regarding software quality expectations. Training should be ongoing to ensure that any improper mindset or behaviors are erased or at least minimized and that new personnel are properly indoctrinated. SQA should be trained in the following areas: software development and verification, auditing or assessment skills, and DO-178C. In teaching DO-178B (and now DO-178C) classes around the world for over 15 years, I've had no more than 10 SQA personnel attend. Unfortunately, most companies are not investing in training their SQA engineers, and their overall software quality reflects the lack of investment.

Characteristic 5: Continuous improvement. Organizations with good quality are always aiming to improve. Deming recommends the PDCA (plan, do, check, act) process, which is commonly referred to as the *Deming Cycle* or the *PDCA Circle* because it is a circular process. The concept is to develop a plan and then execute the plan. The activity is monitored (checked) and action is taken to improve the process. The PDCA process is ongoing in order to continuously improve the overall effectiveness of the process and the quality of the products produced.

11.2.2 Ineffective SQA

Over the years, I've been frustrated by the lack of good SQA processes in many organizations. In evaluating why SQA is not effective, I find it is nearly always related to a failure in one of the five areas noted earlier (i.e., top-level management support, independence, technical competence, training, or continuous improvement). Failure in one or more of these areas leads to the following issues:

- *Understaffed SQA organization.* If management doesn't support SQA, they are often spread too thin to carry out the job.
- *Unqualified SQEs.* It is challenging to attract good engineers to SQA. Many of the top engineers want to design software, rather than look at other people's design. Therefore, SQA managers need to get creative to attract good talent. Some companies use a rotation process (using 1 or 2 year terms) as a stepping stone to project management. Some companies also use an empowering process (using the SQA personnel to train and lead the company's commitment to quality). When people understand some of the benefits of being in SQA, they are more likely to join the team. Table 11.2 summarizes the qualifications and characteristics to look for when hiring an SQE.
- *Disregarded and hence ineffective SQA organization.* When SQA has unqualified or ineffective personnel, engineering doesn't respect or take action on SQA's input.

TABLE 11.2

Qualifications and Characteristics of Effective SQEs

SQE Background and Skill Set

- Three to five years of development or verification experience
- Educational background in engineering or computer science
- Experience in various aspects of development (e.g., coding, writing requirements, testing)
- Good written and spoken communication skills (since SQEs will interact with management, engineers, certification liaison personnel)
- Seeking advancement to management or program management position (SQA will give them good insight into the overall company)
- Capability of handling challenging situations
- Willing to stand up for what is right, even when pressured to do otherwise
- Trustworthy (someone who instills trust and gets along with multiple stakeholders without caving in to pressure)
- Independent but willing to take directions (someone who can work with little oversight but who also has a respect for authority)
- Passionate about quality and safety
- Teachable and eager to learn

11.3 SQA Activities

This section considers the typical tasks that the SQA organization (using one or more SQEs) performs in order to ensure that plans, standards, and company processes are followed.

Task 1: Review plans. SQA reviews the PSAC, the software plans, and the development standards, considering the compliance with DO-178C objectives and the consistency between the plans [7].* (Chapter 5 provides information on the plans and standards.) Most companies hold a peer review of the plans and SQA participates in that peer review. Checklists are normally completed by both the project and SQA as part of the review effort. In addition to the software plans, SQA should be knowledgeable of the system-level and safety plans since they may impact the software processes.

Task 2: Write the SQA Plan. The SQA team is responsible for authoring the SQA Plan and keeping it updated. Most companies have a company-wide SQA Plan. However, each project may have some unique considerations which are typically detailed in a project-specific SQA Plan or in the PSAC.

Chapter 5 provided a summary of what is typically included in the SQA Plan. Additionally, other resources may be used to help identify the

---

* This is identified in DO-178C Table A-9 objective 1. This objective was part of DO-178B Table A-1, but was moved to Table A-9 for DO-178C. It has always been the responsibility of SQA to review plans for consistency and compliance with DO-178B (and now DO-178C) objectives.

components of the SQA Plan. For example, the Institute of Electrical and Electronic Engineers (IEEE) Standard 730 identifies typical components for inclusion in the SQA Plan [9]. Table 11.3 summarizes the IEEE Standard 730 proposed sections.

Task 3: Approve key data items. SQA generally approves key data items, such as the plans, standards, requirements, design, verification cases and procedures document, verification report, Software Life Cycle Environment Configuration Index (SLECI), Software Configuration Index (SCI), tool qualification data, and Software Accomplishment Summary (SAS). Oftentimes, key data items cannot be released without SQA's approval.

TABLE 11.3

Summary of IEEE Standard 730

Section Name	Section Description
Purpose	Explains the purpose of the SQA Plan
Referenced documents	Lists all documents referenced in the plan
Management	Explains the organization structure, tasks, and responsibilities for the project
Documentation	Identifies the documentation used to define the development, verification, and configuration of the software
Standards, practices, conventions, and metrics	Identifies the standards, practices, conventions, and metrics to be applied and describes how compliance will be monitored and assured
Reviews and audits	Defines the reviews to be performed by the technical team, as well as quality assurance's plan for participating in reviews and auditing the processes
Test	Explains the test approach and techniques, as well as SQA's role in test witnessing and oversight
Problem reporting and corrective action	Describes the problem reporting process and SQA's role
Tools, techniques, and methodologies	Identifies and explains any tools, techniques, or methodologies that SQA uses to perform their role
Code control	Explains the process for documenting, maintaining, storing, releasing, and retrieving software
Media control	Defines the methods used to identify the media for each product
Supplier control	Explains SQA's role in supplier control and oversight
Records collection, maintenance, and retention	Explains the SQA documentation approach
Training	Explains necessary training for SQA personnel, as well as the project engineers
Risk management	Explains how risks are identified and managed throughout the project

Source: Software Engineering Standards Committee of the IEEE Computer Society, *IEEE Standard for Software Quality Assurance Plans*, IEEE Std. 730-2002, IEEE, New York, 2002.

Task 4: Audit life cycle data. SQA audits the software life cycle data to ensure that the processes are followed, standards are used, checklists are properly completed, and DO-178C objectives are satisfied.* Sometimes audits are performed during the peer reviews or in conjunction with the certification liaison reviews. However, SQA audits may be performed independent of other activities. Some SQA organizations use the plans and standards to develop a *software audit kit*; SQA then audits the processes and data using the audit kit. This approach provides detailed evidence that the project is complying with the approved plans and standards.

Task 5: Participate in reviews. At their discretion,[†] SQA participates in peer reviews throughout the software development and verification processes. SQA looks for technical issues, as well as the engineering team's compliance with the peer review process, completion of checklists, satisfaction of entry and exit criteria, etc. Participation in peer reviews is a good way for SQA to stay in tune with the project's activities and issues.

Task 6: Transition criteria assessment. DO-178C Table A-9 objective 4 states: "Assurance is obtained that transition criteria for the software life cycle processes are satisfied" [7]. This SQA objective is often addressed by SQA's participation in the peer review process. However, SQA may have a separate assessment activity to evaluate transition criteria.

Task 7: Witness tests. SQA often plays a key role in witnessing software tests. Some organizations require a certain percentage of SQA witnessing during test execution. The expectations should be identified in the SQA Plan and clearly communicated to the project.

Task 8: Audit the environment. SQA typically audits the development and verification environments to ensure that they are consistent with the approved configuration (usually documented in the SLECI). Prior to building the software for release, SQA may witness the build machine setup (as explained in Section 8.4.1) to ensure the approved procedures are followed and that the build machine complies with the approved configuration. Prior to formal testing (i.e., testing for certification credit), SQA normally inspects or witnesses the test station setup to ensure that it complies with the approved configuration.

Task 9: Witness build and load. Using the documented build and load procedures, SQA may perform the build and load themselves or witness someone else implementing the build and load procedures in order to confirm repeatability. As explained in Chapter 8, it is best to have someone who doesn't normally do the build and load (e.g., an SQE or another engineer) execute the procedures.

---

* Per DO-178C Table A-9 objectives 2 and 3, one of SQA's objectives is to ensure compliance with the plans and standards. Auditing is one of the ways that this is carried out.
† Frequently, the SQA Plan identifies the target percentage of peer reviews the SQA will participate in. The percentage may be driven by the overall risk of the project.

Task 10: Participate in change control board. SQA is often a key player on the change control board as they evaluate proposed changes and assess the implementation of the changes. SQA's approval is normally required for the closure of a problem report or change request.

Task 11: Track and evaluate problem reports. SQA tracks the problem reports throughout the software development and verification effort to make sure that the problem reporting processes defined in the Software Configuration Management (SCM) Plan are followed. The problem reports are also evaluated for completeness, accuracy, and adequacy. SQA also ensures that all changes driven by problem reports (or change requests) are verified prior to closure.

Task 12: Evaluate SCM processes and records. SQA audits SCM records and processes to ensure that they comply with the SCM Plan. SQA evaluates SCM during development and verification activities. SQA ensures that the data in the configuration management system align with the data identified in the SCI; that is, they verify that the configuration management library and the SCI are complete and accurate. SQA also audits the change control process throughout the software development and verification efforts to ensure that problems are properly documented, changes are agreed upon, changes are correctly implemented and verified, and all documentation is appropriately updated.

Task 13: Audit suppliers. If the project uses suppliers or subcontractors, SQA audits the suppliers' SQA processes, as well as their overall development and verification effort.

Task 14: Identify corrective actions. As SQA performs their tasks, they may identify required corrective actions. The corrective actions may be documented in a problem report or in a separate SQA record known as a corrective action report or something similar. In addition to identifying corrective actions, SQA needs to follow up to ensure that the actions are properly carried out. Corrective actions should be addressed prior to release of the SAS.

Task 15: Review and approve deviations and waivers. SQA reviews any deviations or waivers to approved processes to ensure DO-178C compliance and safety are not impacted. SQA also ensures that all such deviations or waivers are documented per the approved process (such as a problem report). Similarly, SQA reviews and approves any redlines to approved test procedures.

Task 16: Perform conformity review. Typically, the last step of the DO-178C compliance effort is the conformity review. Per DO-178C section 8.3: "The purpose of the software conformity review is to obtain assurances, for a software product submitted as part of a certification application, that the software life cycle processes are complete, software life cycle data is complete, and the Executable Object Code is controlled and can be regenerated" [7]. The review is typically documented in a conformity review report and ensures the following [7]:

- The planned software life cycle processes have been completed and software life cycle data have been generated.
- The life cycle data are traceable to the system and safety requirements that they came from.
- There is evidence that the software life cycle data were produced and controlled in accordance with the plans and standards.
- There is evidence that each problem report has been evaluated and has a recorded status.
- Any requirements deviations have been documented and approved (typically using the problem reporting process).
- The executable object code can be generated from the source code documented in the SCI, using the identified build instructions.
- The approved software can be loaded using the released load instructions.
- Any problem reports deferred from previous conformity reviews are reevaluated.
- If any previously developed software is used for certification credit, it is confirmed that the current software baseline is traceable to the previous baseline and all changes have been approved.

Task 17: Document activities in SQA records. SQA activities are documented in SQA records to provide evidence of SQA involvement throughout the software development and verification effort. SQA records typically include the following information: what was evaluated, date, name of evaluator, evaluation criteria, evaluation status (compliance or noncompliance), findings, person(s) required to respond, due date for response, severity of the finding(s), and updates relative to the response acceptance [10]. Because SQA records are considered control category #2 (CC2) data, they need unique configuration identification. (See Chapter 10 for information on CC2.) Configuration identification is typically accomplished by including the date and the subject in the title of the SQA file.

References

1. *The American Heritage Dictionary of the English Language*, 4th edn. (Boston, MA: Houghton Mifflin Company, 2009).
2. R. S. Pressman, *Software Engineering: A Practitioner's Approach*, 7th edn. (New York: McGraw-Hill, 2010).

3. E. R. Baker and M. J. Fisher, Chapter 1—Organizing for quality management, in *Software Quality Assurance Handbook*, 4th edn., G. Gordon Schulmeyer, Ed. (Norwood, MA: Artech House, 2008), pp. 1–34.

4. ISO/IEC 9126, *Software Engineering—Product Quality Int'l Standard* (International Standard, 2003).

5. H. Van Vliet, *Software Engineering: Principles and Practice*, 3rd edn. (Chichester, U.K.: Wiley, 2008).

6. W. E. Lewis, *Software Testing and Continuous Quality Improvement*, 2nd edn. (Boca Raton, FL: CRC Press, 2005).

7. DO-178C, *Software Considerations in Airborne Systems and Equipment Certification* (Washington, DC: RTCA, Inc., December 2011).

8. G. Gordon Schulmeyer, Chapter 2—Software quality lessons learned from the quality experts, in *Software Quality Assurance Handbook*, 4th edn., G. Gordon Schulmeyer, Ed. (Norwood, MA: Artech House, 2008), pp. 35–62.

9. Software Engineering Standards Committee of the IEEE Computer Society, *IEEE Standard for Software Quality Assurance Plans*, IEEE Std 730-2002 (New York: IEEE, 2002).

10. J. Meagher and G. Gordon Schulmeyer, Chapter 13—Development quality assurance, in *Software Quality Assurance Handbook*, 4th edn., G. Gordon Schulmeyer, Ed. (Norwood, MA: Artech House, 2008), pp. 311–330.

12

Certification Liaison

Acronyms

CAST	Certification Authorities Software Team
CC1	control category #1
CC2	control category #2
CM	configuration management
CRI	certification review item
DER	Designated Engineering Representative
EASA	European Aviation Safety Agency
FAA	Federal Aviation Administration
ODA	Organization Designation Authorization
PR	problem report
PSAC	Plan for Software Aspects of Certification
PSCP	Project-Specific Certification Plan
SAS	Software Accomplishment Summary
SCI	Software Configuration Index
SCM	software configuration management
SCMP	Software Configuration Management Plan
SDP	Software Development Plan
SLECI	Software Life Cycle Environment Configuration Index
SOI	stage of involvement
SQA	software quality assurance
SQAP	Software Quality Assurance Plan
SVP	Software Verification Plan
TQL	tool qualification level
TSO	Technical Standard Order
WCET	worst-case execution time

12.1 What Is Certification Liaison?

Certification liaison is the ongoing communication and coordination between the applicant* and the certification authority. It is an essential process for successful certification. Most industries have some kind of certification or approval process. For the aviation industry, certification is an onerous process that requires continual attention. Failure to address certification needs throughout the project results in failure to obtain certification or delays the certification and makes it overly expensive. This chapter is specific to the aircraft certification process for software required by the Federal Aviation Administration (FAA) and other civil aviation certification authorities. DO-178C identifies certification liaison as an *integral process*, which means it applies throughout the entire software life cycle. Certification liaison begins early with planning and ends with the final compliance substantiation. DO-178C Table A-10 identifies three objectives for the certification liaison process, which apply to all software levels. Following is a summary of the certification liaison objectives [1]:

- *DO-178C Table A-10 objective 1*: "Communication and understanding between the applicant and the certification authority is established." This is accomplished by the development and submittal of the Plan for Software Aspects of Certification (PSAC) (Chapter 5 discussed the PSAC). On most projects communication between the certification authority and the applicant continues throughout the project's life cycle.

- *DO-178C Table A-10 objective 2*: "The means of compliance is proposed and agreement with the Plan for Software Aspects of Certification is obtained." This is accomplished by the certification authority's approval of the PSAC, which is normally communicated in a letter.

- *DO-178C Table A-10 objective 3*: "Compliance substantiation is provided." The output of this objective is the Software Configuration Index (SCI) (discussed in Chapter 10) and the Software Accomplishment Summary (SAS) (discussed later in this chapter). Approval of these two documents signifies the completion of the DO-178C effort—at least for that version of the software.

---

* The applicant is the entity applying for certification or TSO authorization. The applicant is responsible for showing compliance to the regulations. When DO-178C is the selected means of compliance, the applicant is also responsible for showing compliance to the DO-178C objectives and guidance.

12.2 Communicating with the Certification Authorities

Most companies have a well-defined process for communicating with the certification authority; they may have a certification office (sometimes called an airworthiness office), use designees that have been authorized by the certification authority (e.g., Designated Engineering Representatives [DERs] or Organization Designation Authorization [ODA] unit members), or a combination of the two. Generally, the certification office handles the logistics of coordination with the FAA and the basic certification activities, while the designees (DERs or ODA unit members) focus on the technical aspects of finding compliance with the certification regulations and guidance. Larger companies might also have certification liaison engineers who prepare data and coordinate with the designees for DO-178C compliance activities. The specifics of how the overall certification liaison process is organized vary depending on the location of the applicant and certification authority, the type of approval sought (e.g., an aircraft, engine, or propeller type certification versus an appliance Technical Standard Order [TSO] authorization), the relationship of the software developer to the applicant, etc. More information on the FAA's certification process is available in FAA Order 8110.4[ ] *Type Certification* and/or Order 8150.1[ ] *Technical Standard Order Program.**

As noted in Chapter 5, it is important to secure the certification authority's agreement on the PSAC. Communication with the certification authority begins even before the submittal of the PSAC. Regular technical meetings from the beginning to the end of the project are an effective means of coordination. Meetings do not replace the need for the data, but they do provide a forum to discuss and resolve issues throughout the project with the goal of *no surprises* to either party. Detailed meeting minutes and an ongoing action item list are recommended to record interactions with the certification authority and to ensure that agreed actions are completed.

The certification authority or designee typically performs a level of involvement assessment early in the project (either formally or informally). This assessment considers the experience of the team with certification and DO-178C (or DO-178B) compliance, the newness and novelty of the proposed technology, the criticality of the software to the aircraft functionality, the track record of the software and avionics or electrical systems team, the quality of the certification liaison personnel (e.g., the designees), etc. FAA Order 8110.49 chapter 3 and appendix 1 provide insight into the certification authority's assessment process and criteria [2].† For projects that are assessed as high risk and that may impact safety, the certification authority

---

* The brackets ([ ]) indicate the latest revision. These documents are updated from time to time. The latest version can be found on the FAA's website at www.faa.gov.
† Section 4 of EASA certification memo CM-SWCEH-002 includes similar information for projects certified in Europe [3].

involvement is high. For lower risk and criticality, the certification authorities may delegate much of the compliance activity to a designee. The level of involvement assessment determines the following:

- How much data need to be submitted to the certification authority (higher risk projects often need to submit all of their plans and the test results, in addition to the PSAC, SCI, and SAS).
- How many audits the certification authority will perform.
- How much of the compliance activity will be delegated to the designees.
- How frequently the applicant and the certification authority will need to meet.

12.2.1 Best Practices for Coordinating with Certification Authorities

As noted earlier, the communication approach between the applicant and certification authority varies depending on the project details and the authority's preferences. A majority of the compliance findings are generally delegated to the designees or the delegated organization, with oversight by the certification authority. Following are some suggestions for effective coordination with the certification authorities.

Suggestion 1: Submit a project-level plan. Many organizations submit a Project-Specific Certification Plan (PSCP) (or equivalent) that outlines the overall project plans. This plan includes an overview of the project (including the aircraft and systems), identifies the planned software and aircraft electronic hardware activities (with the software and hardware levels and the selected suppliers/developers), defines the proposed means of compliance to the applicable regulations, identifies points of contact, provides a project schedule, etc. Such a plan establishes the framework for the software effort and initiates the communication with the certification authority.

Suggestion 2: Hold an early familiarization meeting. The sooner the certification authority becomes acquainted with the project, the better. The initial familiarization meeting is generally scheduled about the same time the PSCP is ready to submit. The kickoff meeting provides an overview of the project, system, software, schedule, and certification plans. Although the primary purpose of the meeting is to familiarize the certification authority with the planned project, the meeting also provides valuable information to the applicant. The certification authority usually identifies concerns and anticipated certification issues, discusses expected roles and responsibilities, explains communication expectations, and proposes next steps.

Suggestion 3: Identify potential certification and technical issues. Be forthright about any anticipated certification issues or technical challenges. It is important to inform authorities if any new technology or novel approaches are planned. In general, certification authorities don't like surprises. Surprises can give the impression that someone is hiding something and can reduce the certification authority's trust level. Therefore, it is important to be honest about potential challenges. It is also important to identify the plans to mitigate or address the challenges. Early disclosure of potential issues will allow the authorities to officially document any concerns. The FAA generates issue papers to identify unique compliance issues. The applicant then provides a response to the issue papers to explain their intended approach for dealing with the issue. Other certification authorities have a similar vehicle for identifying certification issues (e.g., the European Aviation Safety Agency [EASA] uses certification review items [CRIs] and the Transport Canada Certification Agency uses certification memorandums).

Suggestion 4: Provide timely response to issue papers. While issue papers are not pleasant, the earlier they are issued and resolved, the better. Receiving an issue paper late in a program can be extremely disruptive to the project. Once an issue paper is received, it is important to develop a thorough response as quickly as possible. Most issue papers can be anticipated. The FAA has a standard set of software issue papers that are issued on most new certification projects (examples of common topics for issue papers include object-oriented technology, model-based development, object code coverage).* These general issue papers are sometimes called *generic issue papers* and serve as a starting point for project-specific issue papers. Additionally, there may be some other project-specific issue papers based on the challenges of the project (e.g., if a large amount of the work is outsourced or offshored, there may be an issue paper to address oversight and problem reporting). The applicant's response to each issue paper should clarify any of the issues (sometimes the issue paper might present a slightly incorrect understanding of the details) and explain how the issue will be addressed on the specific project. It is important to follow up with the certification authority until the issue paper is closed (i.e., to obtain the certification authority's approval of the project position). Sometimes, it takes a few iterations to obtain agreement. An unresolved issue paper can be just as risky as a not-yet-issued issue paper.

Suggestion 5: Prepare, brief, and submit PSAC early. As noted in Chapter 5, it is best to prepare and submit the PSAC early. The longer a project waits to submit a PSAC, the higher the risk if the certification authority rejects the plan. For high-risk projects it is strongly recommended to brief the certification authority on the PSAC before submitting it. This allows some informal feedback to adjust the plan if needed or to go ahead and submit it. (See Chapter 5 for details on the PSAC.)

---

* Once the certification authorities formally recognize DO-178C, DO-330, and the supplements, several of the typical software issue papers should no longer be needed.

Suggestion 6: Follow up quickly on any questions or issues on the PSAC. The certification authority will often have some questions on the PSAC. Be sure to answer the questions quickly, thoroughly, and accurately. Certification authorities are typically overworked and underpaid, so when they are working on your data, it's important to be responsive. Otherwise, it could delay the PSAC acceptance.

Sometimes the certification authority will request some updates to the PSAC. It is important to understand the issues; a meeting or phone call can be helpful, if clarification is needed. Oftentimes, the issues are resolved by a brief discussion and possibly an agreement to make a note in the SAS. Such agreements should be documented in writing; an email or meeting minutes are usually sufficient to document such agreements. If the issues still exist after a discussion, present intended updates to the certification authority and get general agreement before making the updates to the PSAC and resubmitting it.

Suggestion 7: Meet with certification authorities throughout the project. For projects that require a high level of certification authority involvement, a periodic meeting with the certification authority should be established. The frequency may vary throughout the project, depending on how things progress. There may be times when a monthly meeting is needed. At other times quarterly or as-needed meetings may be more appropriate. For high-risk projects, designees should be consulted throughout the project and involved in all coordination with the certification authorities.

If the project is a medium or low level of certification authority involvement, most of the coordination will likely be with the designees or company certification liaison group.

The interactions with the certification authorities should focus on the status of the project, any issues that have come up, and the plan for addressing the issues. Action items and due dates should be established and included in the official meeting minutes. It is important to follow through on agreed upon actions.

Suggestion 8: Deal with issues as they arise. Problems rarely go away when they are ignored. They should be addressed promptly before they get out of control. Be sure to inform the designees or certification liaison personnel of any issues that might affect compliance. They will help to determine the preferred approach for informing the certification authorities. When divulging issues to the certification authorities, always have a proposed plan for addressing the issues.

Suggestion 9: Submit any changes or deviations to the certification authority for approval or acceptance. As noted in Chapter 5, the PSAC should identify the plan for handling changes to the approved processes. The certification authorities will likely want to be informed of any changes to processes throughout the software life cycle.

Suggestion 10: Support audits by certification authorities and/or their designees. Software projects are typically audited by the certification authorities or their designees to ensure compliance to DO-178C and the identified issue papers (or equivalent). Audits require preparation and support. More information on what to expect during an audit and how to prepare for it is provided later in this chapter.

Suggestion 11: Submit software compliance data in a timely manner. The SCI and SAS are the two life cycle data items that are typically submitted to the certification authority to demonstrate compliance with the DO-178C and the regulations. These items are normally submitted a few months before the final issuance of the aircraft or engine type certificate. For TSO authorization projects, the SCI and SAS are submitted with the TSO package. Be sure to coordinate the submittal of the data so that the certification authority has adequate time to review and approve it.

In some projects, the certification authority may also request that the software verification results be submitted as part of the compliance data. The expected submittals are usually discussed with the certification authority during the planning phase and documented in the PSAC.

12.3 Software Accomplishment Summary

DO-178C Table A-10 identifies three types of data to support the certification liaison process: PSAC, SCI, and SAS. As noted earlier, the software verification results are sometimes submitted as well. The PSAC contents were discussed in Chapter 5; the SCI was described in Chapter 10; and the software verification results were explained in Chapter 9. The expected contents of the SAS are now covered.

The SAS summarizes the DO-178C compliance effort. DO-178C section 11.20 provides guidance for its contents. The PSAC and SAS are like bookends. The PSAC is submitted early in the project and explains what will be done. The SAS is submitted at the end of the project and explains what actually occurred. The SAS contains much of the same material as the PSAC (including the system overview, software overview, certification considerations, life cycle summary, life cycle data summary or reference), but the SAS is written in past tense instead of future tense. Additionally, the SAS identifies the following [1]:

1. *Deviations* from the approved plans and processes (this is often noted throughout the SAS and summarized in an appendix). For example, if the actual tools used differed from what was planned, this should be explained.

2. *Configuration* of the software to be approved.

3. *Open or deferred problem reports* (PRs), including a justification of why they do not impact safety, functionality, operation, or compliance. Normally, a table is included to summarize the following information for all PRs that have not been resolved at the time of certification or TSO authorization [1–3]:

   a. PR number

   b. PR title

   c. Date opened

   d. Description of the problem, including root case (if known) and impacted data

   e. Classification (see Chapter 10 for discussion on PR classifications)

   f. Justification for deferral, including why the problem does not negatively impact safety, functionality, operation, or compliance to the regulations (including XX.1301 and XX.1309)*

   g. Mitigation means, if applicable (e.g., operational limitations or functional restrictions)

   h. Relationship to other open PRs

   Additionally, some authorities may request that the plan for resolving the PR be included in the SAS. This is a relatively new expectation from the certification authorities and is being requested in order to promote the closure of PRs rather than keeping them open for years.

   It is also a good practice to explain in the SAS how any post-certification problems will be documented, evaluated, and managed.

4. *Resolved PRs*: If the SAS is for a follow-on certification or based on previously developed software, all PRs or change requests since the last approval are identified (either in the SAS or SCI).

5. The *software characteristics*, such as size of executable object code, timing margins, and memory margins. Chapter 9 discusses the analyses to determine these characteristics.

6. A *compliance statement* stating that the software complies with DO-178C and any other certification requirements. Frequently, the SAS includes or references a compliance matrix summarizing how each DO-178C objective is satisfied and the data that proves the compliance.

---

* XX may be Parts 23, 25, 27, or 29 of Title 14 of the Code of Federal Regulations.

12.4 Stage of Involvement (SOI) Audits

12.4.1 Overview of SOI Audits

In the mid-1990s, the certification authorities in both the United States and Europe began auditing software projects to assess compliance to DO-178B. Unfortunately, it was discovered that many projects were not complying with the objectives. Also, it was noted that there was a huge variance in how certification authorities assessed a project. Because of this, the international Certification Authorities Software Team (CAST) coordinated and identified a software compliance assessment approach. The CAST paper identified four intervention points throughout the project: (1) planning, (2) development, (3) test, and (4) final. The FAA further documented this approach in Order 8110.49 chapter 2 and in a Job Aid entitled *Conducting Software Reviews prior to Certification.** Order 8110.49 explains what the FAA or authorized designees will do and what data will be examined.[†] The Job Aid provides a process for how the certification authorities will perform the *reviews* (also called *Stage of Involvement* [SOI] reviews). The Job Aid was intended to be a training tool for FAA engineers and their designees to standardize their review process. However, for many projects the FAA now requires that the questions in the Job Aid be completed by the applicant or their designees prior to submittal of certification data. Table 12.1 provides an overview of the SOI number and type, data examined, and the DO-178C objectives assessed.

The Job Aid and FAA Order 8110.49 refer to SOI *reviews*; however, the term *audit* distinguishes it from the verification reviews; therefore, the terms *audit*, *auditor*, and *auditee*[‡] are used in this chapter. More information is provided later about conducting a SOI audit (for auditors) and preparing for a SOI audit (for auditees).

12.4.2 Overview of the Software Job Aid

DO-178C sections 9.2 and 10.3 explain that the certification authority may perform reviews; however, no further explanation is given for what these reviews include. Order 8110.49 and the FAA's Software Review Job Aid provide additional explanation of *what* will be assessed (Order 8110.49) and *how* to assess it (the Job Aid). Table 12.1 provides an overview of the four SOI reviews, including the data and DO-178C objectives

---

* The original Job Aid was released in June of 1998. Rev 1 was released in January of 2004. It is anticipated that the Job Aid will be updated to be consistent with DO-178C, DO-330, and the supplements. See the FAA's website (www.faa.gov) for the latest version of the Software Review Job Aid.
† Chapter 4 of EASA's Certification Memorandum CM-SWCEH-002 is very similar to FAA Order 8110.49 Chapter 2.
‡ An *auditee* is the organization being audited.

TABLE 12.1

Summary of the Four SOIs

SOI # and Type	Data Examined	DO-178C Objectives Assessed
1. Planning	PSAC, SDP, SVP, SCMP, SQAP, development standards, verification results (planning review records), SQA records, SCM records, and tool qualification plans (if separate from the PSAC).	• Primarily Table A-1 • Tables A-8 through A-10 as they apply to planning
2. Development	Any planning data not completed in SOI 1 or that changed since SOI 1, system requirements allocated to software, software requirements (high-level and derived high-level), software design description (low-level requirements, derived low-level requirements, architecture), source code, build procedures, Software Life Cycle Environment Configuration Index (SLECI), verification records (for requirements, design, and code verification), trace data, PRs and/or change requests, SQA records, and SCM records.	• Primarily Tables A-2 through A-5 • Tables A-8 through A-10 as they apply to development
3. Verification (Test)	Data not examined or not resolved in previous SOIs, object code, verification cases and procedures, verification results, SLECI, SCI (with test baseline), trace data, PRs and/or change requests, SQA records, and SCM records.	• Primarily Tables A-6, A-7 • Tables A-8 through A-10 as they apply to testing
4. Final	Any data not completed or examined in previous SOIs, verification results (often packaged as a Software Verification Report), SLECI, SCI, SAS, PRs and/or change requests, trace data, SQA records, software conformity review report, and SCM records.	• Primarily Table A-10 • Objectives that were not assessed in previous SOIs • Objectives that had issues noted in previous SOIs

that are assessed. The Job Aid provides recommendations for how to conduct a SOI audit and provides activities and questions to be assessed during the audit. The Job Aid provides insight into the certification authorities' thought process and expectations; this insight can help developers prepare for audits and better plan their projects from the beginning. Some companies use the Job Aid to perform self-audits.

The Job Aid is divided into four parts. Part 1 provides an overview, including an introduction to the four SOI audits and the definition of some key terms. Following are the terms that are particularly important to understand [4]:*

- *Compliance* is the satisfaction of a DO-178B objective.
- A *finding* is the identification of a failure to show compliance to one or more of the RTCA/DO-178B objectives.
- An *observation* is the identification of a potential software life cycle process improvement. An observation is not an RTCA/DO-178B compliance issue and does not need to be addressed before software approval.
- An *action* is an assignment to an organization or person with a date for completion to correct a finding, error, or deficiency identified when conducting a software review.
- An *issue* is a concern not specific to software compliance or process improvement but may be a safety, system, program management, organizational, or other concern that is detected during a software review.

Part 2 explains the typical audit tasks. Regardless of the SOI type (planning, development, verification/test, or final), the auditor must prepare for the audit, conduct the audit, document the audit results, summarize the audit in an exit briefing (or sometimes a written executive summary), and conduct follow-up activities (such as preparing the SOI report and ensuring that the findings, observations, and actions are addressed).

Part 3 comprises the bulk of the Job Aid. It summarizes the activities and questions for each SOI audit. A table is included for each SOI audit, summarizing the SOI activities (identified in bold font) and questions that are used to complete the activity. Each question is mapped to the DO-178B objective(s) that the question assesses.* Table 12.2 provides an example of one SOI 1 activity and some of the supporting questions. Typically, during a SOI audit, a column is added to the table to provide a response to each of the questions (based on the evaluation of the project's data) and is included in the SOI audit report.

Part 4 of the Job Aid provides some examples of how to summarize the SOI audit results. The typical SOI audit report contents are discussed later in this chapter.

---

* At time of this writing, the Job Aid references DO-178B rather than DO-178C objectives. The majority of the Job Aid still applies to DO-178C. It is anticipated that the FAA will update the Job Aid to align with DO-178C.

TABLE 12.2

Excerpt from the Software Job Aid SOI 1 Activities/Questions

Item #	SOI #1 Evaluation Activity/Question	DO-178B objective(s)
1.1	Review all plans (PSAC, SCMP, SQAP, SDP, SVP, software tool qualification plans, etc.) and standards. Based on your review of all the plans, consider the following questions:	
1.1.1	Have the planning data been signed and put under CM (CC1 or CC2 as appropriate for the software level)? Verify there is objective evidence of coordination (e.g., authorized signatures) from all organizations controlled and affected by the software plans and standards.	• A-1, #1–7
1.1.2	Have the plans and standards been reviewed and approved by a software designee (as authorized)?	• A-1, #1–7
1.1.3	Are plans and standards cited complete, clear, and consistent?	• A-1, #1, 7
1.1.4	Were reviews of the plans and standards conducted and review results retained? And, were review deficiencies corrected? (See section 4.6 of DO-178B.)	• A-1, #6, 7
1.1.5	Do the plans and standards comply with the content as specified in DO-178B section 11 (i.e., sections 11.1 through 11.8)? Note: The plans and standards are not required to be packaged as identified in 11.1 through 11.8; however, the items specified in 11.1 through 11.8 should be documented somewhere in the plans and standards.	• A-1, #1–7
1.1.6	Do the plans and standards address the software change process and procedures for the airborne software and tools (if tools are used)?	• A-1, #1, 2
1.1.7	Are all software tools identified in the plans and is rationale included for why each does or does not need to be qualified?	• A-1, #4
1.1.8	Are the inputs, activities, transition criteria, and outputs specified for each process?	• A-1, #1
1.1.9	Are the development and verification life cycle activities defined in sufficient detail (reference DO-178B sections 11.1 through 11.3) to satisfy section 4.2 of DO-178B?	• A-1, #1–7
1.1.10	Do the plans and standards meet the DO-178B planning objectives in Table A-1? (i.e., Is each plan and standard internally consistent? Are the plans and standards consistent with each other? Is the software life cycle defined? Are the transition criteria defined?)	• A-1, #7
1.1.11	If the plans and standards are followed, would this ensure that all applicable DO-178B objectives in Tables A-2 through A-10 are met? (i.e., do the plans and standards address how each of the applicable DO-178B objectives will be satisfied?)	• A-2–A-10 (all objectives)

Source: Federal Aviation Administration, *Conducting Software Reviews prior to Certification,* Aircraft Certification Service, Rev. 1, January 2004.

The Job Aid also includes four supplements to help the auditors and the project teams being audited (the auditee) including the following:

- Supplement 1: "Typical Roles and Responsibilities of the FAA Software Team." This supplement gives a summary of what to expect from the various FAA offices during a software project. Most of the SOI audits are performed by designees, but the certification authorities may be involved in some of the high-risk projects.
- Supplement 2: "Typical Roles and Responsibilities of the Software Designee." This supplement provides a summary of what a software designee does. This information can be quite helpful to properly utilize designees.
- Supplement 3: "Example Letters, Agendas, and Report." This supplement provides some example materials for auditors to use when notifying applicants of audits. A sample report is also included.
- Supplement 4: "Optional Worksheets for Reviewers." This supplement includes some worksheets that can help auditors with their record keeping tasks. These are considered optional and are provided primarily for training purposes.

12.4.3 Using the Software Job Aid

The following sections provide recommendations for auditors and auditees. General recommendations are provided first, followed by a summary of what to expect during a SOI audit and how to prepare for a SOI audit. Having performed or supported hundreds of audits, the subsequent sections are based on my personal experience and lessons learned. Because this information is presented from my personal experience, it is presented using a more interactive tone than other sections of this book.* Please note that the term *applicant/developer* is used throughout the upcoming sections to refer to *applicant and/or developer*. Sometimes the applicant is the software developer; however, many times the software developer is a supplier to the applicant. When the applicant and developer are separate organizations, both should be present at the audit.

12.4.4 General Recommendations for the Auditor

Following are some recommendations for those of you who find yourselves in the auditing role. You may be a designee, a certification authority, a certification liaison engineer, a software quality engineer, or a project engineer.

---

* Some personal and somewhat humorous stories are included in boxed text.

Regardless of why you're required to perform the SOI audit, these recommendations are intended to help you do a better job and avoid common mistakes.

Auditor Recommendation 1: Communicate and coordinate the audit plan in advance. Communicate with the applicant/developer's point of contact a few weeks prior to the audit to address the following:

- Ensure that you understand the applicant/developer's status and confirm that they meet or will meet the agreed entry criteria for the SOI audit (typical entry criteria are discussed later).
- Coordinate the audit date.
- Prepare the agenda for the on-site or desktop audit* (the Job Aid Supplement 3 provides some sample agendas).
- Ensure that expectations are clearly communicated.

Let the applicant/developer know who will be on the audit team, how long the audit will take, what kind of meeting space will be needed (one or more rooms), etc. The better the audit plan is communicated ahead of time, the smoother things tend to go on-site. Detailed planning can help make the on-site time more productive and enjoyable.

Auditor Recommendation 2: Get a helper. It is extremely beneficial to have a teammate when assessing compliance, particularly for more formal audits (e.g., SOI audits performed for certification credit). There are several advantages of the team approach. First, it doesn't take as long; the team can divide the tasks in order to examine more data faster. Second, it provides a more thorough review. Multiple eyes and brains identify more issues, faster. I especially find a team helpful for SOIs 2 and 3 where there is a large amount of data to examine. Third, teamwork provides a witness. I've been involved in a few audits where the project fabricated stories about what happened. With a witness, it's harder to do that.

Obviously, it's great to have a technically experienced reviewer as a teammate. However, sometimes that is not possible because of the resource limitations. It is still helpful to have someone to act as support even if they are not experienced enough to lead an audit. In fact, it can be a great way to train future auditors. They can help with note taking, reviewing configuration management (CM) and quality assurance data, examining review records, witnessing build and loads, etc.

---

* An on-site audit happens at the developer's facility or sometimes at the applicant's facility (if the applicant and developer are separate). A desktop audit is performed remotely by examining data. Many times, SOIs 1 and 4 can be performed remotely but SOIs 2 and 3 are performed on-site.

On one of my first audits, I went alone. The project was a mess. There was no traceability; requirements, design, and code didn't match up; and quality assurance was nonexistent. I conducted the audit, gave the exit briefing, and left for the airport. Later, I learned that the applicant (who was partly responsible for the mess, because they failed to perform adequate oversight) used the audit results to hammer the supplier. The applicant had selectively heard the exit briefing and used it to place the blame solely on the supplier. The audit findings were both misinterpreted and misquoted. With a witness, it would have been harder for the applicant to manipulate the results.

On the flip side, I recently went through a challenging project that had far too many compliance issues. This time I had two teammates. One represented the applicant (who was not the developer) and one was a trainee. During the audits we divided the workload. It allowed us to examine more data and to examine it from different perspectives. It was not pleasant to find so many issues, but the assessment was much more thorough. Later, the project was assessed by multiple certification authorities. Because we had assessed it so thoroughly during the initial audits, the follow-on issues noted were very minor.

Auditor Recommendation 3: Be prepared. It is important to prepare for the audit. This may include reading or rereading the plans (if they changed or if it has been considerable time since the last audit), reviewing responses to previous SOI audits, and closing out issues from the previous SOI audits. Unprepared auditors tend to be ineffective.

Auditor Recommendation 4: Be considerate and kind. I find that professionalism and respectfulness are far more productive than power games and fear tactics. As I've heard it said: "Honey gets better results than vinegar." People respond better when they are treated with respect and don't feel threatened.

I once worked with a guy who didn't follow this advice. He was extremely confrontational to the developers. When the engineers or programmers met with him, they were shaking. When he asked a question, they didn't know how to answer, so they just stared at him. We nicknamed this auditor *Headlight John* because everyone looked like a deer in the headlights in his presence. Interestingly, John never really got accurate information from the developers; they were too intimidated to share freely with him. His attitude distracted their thought process, so they couldn't accurately explain what they knew.

Auditor Recommendation 5: Strive to really understand the system, process, and implementation. During the first part of an audit, the applicant/developer will provide a high-level view of their system, software, processes, and status. It's important to understand the project framework, processes, and philosophy. Read the plans ahead of time and ask questions during the presentations and demonstrations to make sure you understand what is being developed and how it is being developed. The big picture (the forest) is really important to understand before diving into the details (the trees).

Auditor Recommendation 6: Communicate the intent up front and throughout the SOI audit. Each day, be sure to communicate your intentions with the applicant/developer's team lead. If plans change, let them know. I often schedule a few minutes at the beginning and end of each day to communicate where we are and where we are going. Most applicants/developers are quite responsive if you just let them know what you're thinking.

Auditor Recommendation 7: Have the applicant/developer walk you through the data initially. If the company is one you are not familiar with or the project is still relatively new to you, it is good to have the applicant/developer walk you through their data at first. Let them show you their requirements, design, code, verification cases, verification procedures, and verification results. Have them demonstrate how their traceability works. You might request that they do one top-down thread (system requirement to high-level software requirement[s] to low-level requirement[s] to code) and one bottom-up thread (code to low-level requirement[s] to high-level software requirement[s] to system requirement[s]). After this, you may feel comfortable examining the data on your own, or you may just have the applicant/developers serve as your *driver* (i.e., the person who operates the computer per your instructions). I tend to prefer driving on my own, but some auditors like to have the applicant/developer do the driving, so they are free to take notes and ask questions. Either is fine; just be sure to communicate your preference with the applicant/developer.

Auditor Recommendation 8: Be persistent in getting answers. Sometimes it can take multiple attempts to get to the bottom of an issue. Not all issues are obvious. I find that looking at additional data and interviewing multiple sources often helps to get a clearer perspective.

Auditor Recommendation 9: Document as you go. It's important to keep notes as you perform the audit. If you wait, the details get hazy and may be incomplete. I find it helpful to keep draft notes as I go along and then clean them up each evening. If I wait more than one or two evenings, it gets harder to remember the details.

Auditor Recommendation 10: Don't lose credibility. When assessing data you will see a lot of potential issues—some of them may end up being major showstoppers and some may just be minor speed bumps. It normally takes a while to calibrate the significance of the issues. If you start picking at the minor details, you may lose the applicant/developer's ear when you find some truly significant issues. When I'm assessing a relatively new project, I generally keep my thoughts to myself for at least a day or two until

I'm more familiar with the data. Sometimes what initially appears to be a big deal dulls in comparison to what I discover as time goes on.

Auditor Recommendation 11: Beware of personalities. One of the challenges of auditing is the *interesting* personalities that present themselves. Audits are not really pleasant for anyone. Sometimes the stress of the situation brings out the worst in people. Some people are confrontational. Some totally avoid providing any useful information. Some are nervous and high strung. Others make promises and don't keep them. The personality factor is another reason it's helpful to have a teammate. Sometimes when if you can't crack the code to communicate with a person, maybe your teammate can.

Personality Story #1—Nasty-cons

I recently performed a desktop pre-SOI 1 audit. It started out as a SOI 1 audit, but the plans were so poorly prepared, that I opted to downgrade it to a *pre-SOI* to give the team the opportunity to get their act together before doing the *real SOI*. Despite my kindness, the team was confrontational. They informed me that "this was not their first time in the ball park" and that they had multiple FAA approvals "under their belt." They acted as if I had never seen a software plan or read DO-178B. I was dumbfounded and more than a little annoyed. However, I found that persistence paid off. After two frustrating teleconferences (I called them *nasty-cons*), I was able to gain the team's ear so that we could focus on the technical issues.

Personality Story #2—Afraid for My Life

Several years ago, I performed an on-site audit by myself. The results of the audit were awful, and there wasn't enough honey in the world to sweeten up the exit briefing. I was straightforward about the issues. I even issued a finding against SQA for their lack of activity (they had three SQA records to show for the 5-year project). Afterward, the software quality engineer (whom I will call *Mister SQA*) asked if he could meet me for a side discussion. I agreed and he proceeded to walk me to his office. It was dark by this time, and his office was on the far side of the facility. We walked across the empty factory, and I began to get nervous. When we finally got to his office, away from everyone else, he started yelling at me. "How can you write us up for this?" he demanded, "I could lose my job!" My imagination ran wild. I envisioned him pulling out a baseball bat, whacking me over the head, cutting my body into small pieces, and burying me under the calibration machine. I immediately began looking for the quickest means of escape. I made some parting comments and said I needed to get to the airport for my flight. I left *Mister SQA* with steam flowing from his ears, happy to escape with my life. Yet another reason to have a witness!

Auditor Recommendation 12: Use the objectives as the measuring stick. Throughout the audit, it's important to use the objectives (either DO-178B or DO-178C, whichever is the means of compliance, and any of the supplements that are applied) as the evaluation criteria. The Job Aid questions and your experience are helpful, but the objectives are what you are evaluating.

Auditor Recommendation 13: Stay focused and follow through. Most audits have at least a few distractions: the data may be unclear, the people may be peculiar, or things may not go according to plan. In these situations, it's important to stay focused and to get to the bottom of the issue(s). You may need to request additional time to look at the data alone or have someone else walk you through the data.

Auditor Recommendation 14: Communicate potential issues throughout. Once you are familiar with the project and the company's processes and you've gained the respect of the team, it is good to mention issues as you identify them. This gives the applicant/developer the chance to develop a strategy for addressing the issue and possibly discuss it with you before the audit ends. I used to wait until the exit briefing and drop all of the issues in the applicant/developer's lap at once, but I soon learned that that the brief-and-run approach was not effective. There are a variety of ways to inform the team of noted issues. One option is to hold a brief meeting at the end of the day where you share *preliminary* issues noted. It's best to emphasize them as *preliminary*, since the audit is still under way. Another option is to verbally mention issues as you see them, so the applicant/developer can keep their own list of issues. The important thing is to share the information with the right people throughout the audit, so that the exit briefing and SOI report are merely a summary of what you've already discussed.

Auditor Recommendation 15: Reschedule or downgrade, if needed. Sometimes, no matter how much planning and preparation you perform, you will begin an audit and realize that the data just are not ready (stated another way, it does not comply). There are several options in this situation, for example,

- You may continue with the audit and write-up the findings and observations for the applicant/developer to address. You will normally have to redo the audit later.
- You might downgrade the audit to a *pre-SOI* or an informal assessment.
- You may choose to stop the audit and come back later.

There are other options. The decision will depend on several factors, including personal preference.

Auditor Recommendation 16: Provide the report quickly. Once you finish the SOI audit, it's important to provide the report as soon as possible, so the applicant/developer can take immediate action. I normally provide a preliminary list of findings, observations, and actions as part of the exit briefing and then send the official report within 1 week. In the list of findings, I find it helpful to distinguish between *systemic findings* and *isolated findings*. *Systemic findings* are findings that were noted in several places and, therefore, require a project-wide action. *Isolated findings* are noted and need to be fixed but are not widespread; they just require that particular instance to be fixed.

I typically document the findings, observations, actions, and comments in a tabular format. The table is usually presented in landscape view in order to have adequate space for the applicant/developer to respond. Each column is described in the following:

- *Issue #*—This column identifies the number of the table entry. Like requirements, if a number is assigned, it is best not to renumber; that way cross-references can be made elsewhere in the report. (For example, if a finding is deleted, the number stays and a note is made explaining the deletion.)
- *FOCA*—This column classifies the issue as a finding (F), observation (O), comment (C), or action (A). Findings and actions require the application/developer to take action. An observation needs a response, but does not necessarily require action. A comment is typically a note that may be useful for compliance assessment. Comments do not require an applicant/developer response or action. When project feedback is needed in order to determine a classification, an "F?" or "?" can be used. Once the project response is received, the classification can be updated.
- *Objective*—This column notes the DO-178C (or DO-178B) objective(s) related to the issue. Most of the time, DO-178C (or DO-178B) Annex A table number is adequate. If the issue is somewhat controversial, the specific DO-178C (or DO-178B) objective number and section reference may be needed.
- *Systemic?*—A *"Yes"* in this column identifies systemic issues.
- *Data*—This column identifies the document or data item that the issue was noted against. If it is a general issue, and not document-specific, the word *"General"* can be used. Somewhere in the report, the data version should be identified as well.
- *Issue Description*—This column identifies the issue. It should be specific. Include the section, requirement number, code line, excerpt, etc., and clearly explain the issue noted.
- *Applicant/Developer Response*—This column is used by the applicant/ developer to respond to the issue. It is a good practice to date and initial the responses because sometimes there will be multiple updates before the issue is resolved.
- *Evaluation*—This column summarizes the SOI team lead's evaluation of the issue. It is recommended to date and initial the evaluation

comment. If the applicant/developer's response requires them to take additional action, the expected action is explained. If the applicant/developer's response is acceptable, state this and close the issue. If the applicant/developer's response is acceptable but requires some future action (perhaps in a later SOI), this should be noted. I normally identify responses that require future action with blue font, so I remember to follow up at the next SOI.

- *Status*—The status column is also updated by the SOI team lead. Typical entries include: *Open, In-Work, Response Acceptable, Closed.* The goal is to get all entries to the *Closed* state.

Table 12.3 provides a summary of the sections that I normally include in a SOI report.

12.4.5 General Recommendations for the Auditee (the Applicant/Developer)

If you are the auditee (the applicant or software developer being audited), here are some suggestions.

Auditee Recommendation 1: Assign a single point of contact to interface with the SOI audit team lead. This is usually the software project lead. If the SOI audit will be by the certification authorities, the point of contact might be the designee

TABLE 12.3

Sample SOI Report Outline

1.0 Introduction

    1.1 Purpose—includes a brief summary of the project and the purpose of the SOI audit and the report.

    1.2 Report overview—provides an overview of the report.

    1.3 Date(s) of the audit—identifies when the SOI audit was performed.

    1.4 Participants—identifies both the audit team and the applicant/developer's team members who participated.

    1.5 Data examined—lists the data examined along with the configuration identification (document numbers and versions).

2.0 Summary of findings, observations, actions, and comments—this section includes the FOCA table that was described earlier.

3.0 Job Aid questions—includes a response to each of the Job Aid questions for the specific SOI audit. This is usually presented in table format, using the Job Aid tables and adding an evaluation column. Some Job Aid questions may be evaluated in the next SOI. If this is the case, it should be noted (I highlight the text in blue, so I do not forget it during the next SOI).

4.0 Objectives compliance assessment—includes an evaluation of each of the DO-178C (or DO-178B) objectives for compliance.

5.0 Supporting information—includes details that may be used to support the previous sections (e.g., notes from test witnessing or details of requirements examined).

(e.g., a DER) for the project. The point of contact will handle the coordination with the SOI audit team lead and the project team.

Auditee Recommendation 2: Coordinate with the SOI audit team lead to understand the plan and agenda. The assigned point of contact coordinates with the SOI audit team lead to ensure that the SOI audit logistics, agenda, and expectations are well understood. Most of the coordination can be handled via email; however, a telephone call or teleconference may be preferred.

Auditee Recommendation 3: Perform a self-audit and be honest about any known problems (and how they are being resolved). Before having a formal SOI audit, it is highly recommended to perform a dry run (pre-SOI) ahead of time. If the formal SOI audit will be performed by a designee, the dry run may be done by SQA and a project team member. If the formal SOI audit will be performed by the certification authority, the designee or certification liaison personnel will likely perform the pre-SOI audit.

Auditee Recommendation 4: Be prepared. Have the data and personnel ready when the SOI audit team arrives. It is frustrating to the auditors if they have to wait for the auditee to gather people and data. First impressions are important in audits and are rarely erased. You don't want *lack of preparation* to be the first thing the auditors note. If you are not sure what data are expected, get that clarified ahead of time. Not all software team members need to be in the room for the full audit, but they should be available throughout the event. If a key developer or verifier or the lead engineer will be gone during the audit, be sure to let the audit team lead know ahead of time. The auditing team might prefer to postpone the audit. I joke at how *coincidental* it is that the key personnel have all-day dentist appointments, vacation, or grandma's funeral (for the third time) when I arrive for an audit. Obviously, scheduling conflicts arise, but do everything you can to have the key people there. Normally, SQA and team leads participate in the entire audit.

Auditee Recommendation 5: Present the project accurately and concisely. In a routine audit, the applicant/developer has a half day to tell the project's story. Use the time wisely, since it will set the tone for the rest of the audit. Provide an accurate and concise overview of your system, software architecture, processes, and known issues. If it is a SOI 3, explain the testing approach. Provide a brief walkthrough of requirements, design, code, and testing data to help the SOI audit team understand the project's approach and data layout. Provide the background and overview that will enable the auditors to effectively do their job. Be as concise but complete as possible. Auditors may get annoyed if they suspect someone is wasting their time. One company that I audited tried to stretch their presentation out over the entire 3-day period, even though the agenda was quite clear that they needed to finish at noon on the first day. They acted *surprised* when I explained that I wanted to see the data after lunch.

Auditee Recommendation 6: Be cordial, cooperative, and positive. Audits are not pleasant for anyone, but they can be handled with professionalism.

Applicant/developer teams that are cordial, cooperative, and positive will almost always end up with a better SOI report. Attitude *does* count. An uncooperative attitude draws suspicion and scrutiny, but a cooperative attitude and behavior develops trust.

Auditee Recommendation 7: Look at the experience as an opportunity to improve. As in every field, there are a few loose cannons out there who perform SOI audits. However, most of the SOI auditors have vast experience. They have seen numerous projects and can be a wealth of knowledge. It is best to look at the SOI audit as a learning opportunity, rather than a torture chamber. The best companies I've worked with are eager to learn how to improve.

Auditee Recommendation 8: Make sure the team thoroughly understands any findings, including the DO-178C (or DO-178B) foundation. It is important to understand the findings and actions documented by the SOI audit team. Occasionally, there will be a *finding* that is really an opinion. If you suspect this is the case, tactfully inquire about the DO-178C (or DO-178B) objective that is not being satisfied. Also, be sure you understand which findings are *systemic* (apply across the project) and which ones are *isolated*, since this significantly impacts the responses. When in doubt, ask for clarification.

Auditee Recommendation 9: Take findings seriously and follow up. By definition, a *finding* is a noncompliance that must be addressed prior to certification. Therefore, action is required.

Years ago, when I was a relatively new FAA engineer, I performed the equivalent of a SOI 2 audit (this was before the Job Aid, so the *SOI* term did not yet exist) on a flight control system at a company that was doing their first DO-178B project. They had done multiple military projects but were just learning DO-178B and the interactions with the FAA. I performed the audit and provided a report with numerous systemic findings. The project had a DER, so I relied on the DER to perform the follow-up work. A year later, I came back for the equivalent of SOI 3. I was dumbfounded to learn that they had not yet addressed any of the SOI 2 audit findings. It was a fiasco for everyone and impacted the overall certification schedule. Everyone learned lessons through this situation, including myself. I now follow up, even if it's not technically my job, to ensure action is being taken.

Auditee Recommendation 10: Document how the team will address all findings, actions, and observations. As soon as possible, begin working on a response to the SOI issues. Typically, a SOI audit report will be generated with a list of findings, actions, and observations. Oftentimes, the list will be in a table format, so you can just add a column with the project's response. If the list is not provided in a table format, I recommend putting it in a table format (some companies use a spreadsheet because it provides a convenient way to sort and filter). Be sure to respond to all findings, actions, and observations. Observations don't have to be addressed but they do need a response. If you have any questions, get clarification from the SOI audit team lead or your designee. Once a response to all the findings, actions, and observations has been prepared, provide the response to the SOI audit team lead to get feedback. Many times, it takes a few iterations to reach agreement on all issues. In some situations it may be more expedient to schedule a teleconference or a meeting to discuss the responses rather than going back and forth on email.

12.4.6 SOI Review Specifics

This section examines more details on each of the four SOI audits. The following topics are discussed:

- Typical entry criteria for a SOI audit (items to be done before a SOI audit can be performed).
- What the SOI audit team typically does during the audit.
- How to prepare for a SOI review.

12.4.6.1 SOI 1 Entry Criteria, Expectations, and Preparation Recommendations

12.4.6.1.1 SOI 1: When It Occurs

The SOI 1 audit occurs after the plans and standards have been reviewed and baselined by the applicant/developer. If the SOI audit is performed by a certification authority, they will typically want to review released data. If the SOI audit is performed by a designee, they may evaluate prereleased data, but it will still need to be baselined. The expectations on data release should be clarified with the SOI audit leader.

12.4.6.1.2 SOI 1: What to Expect

The SOI 1 audit is often performed remotely. If this is the case, the SOI audit team lead will request that the plans and standards be provided. It will typically take the SOI audit team at least a month to examine the data. The summary of issues will then be presented in writing—generally in a SOI audit report. There may also be a teleconference or meeting to discuss the noted issues. Following are things that the SOI audit team generally does:

- Ensure that plans and standards are under configuration control before evaluating them. It can be frustrating to spend 40 hours reviewing a set of data and then discover that everything has changed (without change tracking), creating more work for the auditors.

- Use DO-178C sections 11.1–11.8 to ensure all of the expected content is included in the plans and standards. The data may be packaged differently than DO-178C suggests (e.g., the standards may be in the Software Development Plan (SDP), or the SDP and Software Verification Plan (SVP) may be combined); however, the basic contents of DO-178C sections 11.1–11.8 need to be included in the plans and standards.

- Use DO-178C Annex A objectives to make sure all applicable objectives are addressed by the plans. The auditors will usually make sure that the PSAC addresses all applicable objectives. The PSAC may not go into detail but should provide some coverage of how each objective will be addressed. The auditors will also ensure that the SDP, SVP, Software Configuration Management Plan (SCMP), and Software Quality Assurance Plan (SQAP) provide the details for how the objectives will be addressed.

- Examine details of *additional considerations* (e.g., tool qualification). The auditors will ensure that the additional considerations are adequately explained and that the approach described is acceptable.

- Ensure that the plans are consistent. The auditors will make sure that the plans are both internally consistent (each plan's content is consistent) and externally consistent (all plans agree).

- Make sure issue papers (or equivalent) are addressed in the plans. In some cases (such as TSO projects) the issue paper numbers might not be included, but there should still be evidence that the issues are being addressed. For example, even though the issue paper for model-based development isn't mentioned in the PSAC, there may be a section that explains how model-based development is being carried out. For software that is specific to a certain aircraft or engine, the software-related issue papers (or equivalent) and the project's response are normally summarized in the PSAC.

- Examine the standards to ensure they exist, are usable, are being used, and are appropriate for the project.

- Evaluate any tool qualification plans, if the plans are separate from the PSAC, using DO-330. If the tool is developed before DO-178C and DO-330 recognition, the DO-178B and FAA Order 8110.49 or EASA CM-SWCEH-002 criteria will be used.

- Ask about how the team is using the plans and standards. The auditors may want to know how the project is ensuring that the plans are being followed. Many companies have mandatory training and reading. Training and reading records may be examined during the audit.

12.4.6.1.3 *SOI 1: How to Prepare*

Following are some suggestions for how to prepare for SOI 1:

- Complete all the plans and standards using the DO-178C section 11.1–11.8 guidelines.
- Complete tool qualification plans, if needed, using DO-330 guidelines.
- Perform a peer review of the plans and standards, including a review of the plans and standards together.
- Ensure consistency between plans. This is normally assessed during the planning peer review.
- Perform a mapping to DO-178C objectives. As noted in Chapter 5, it is helpful to provide a mapping between the DO-178C objectives, the PSAC, and the other plans. If any of the DO-178C supplements are used, plans should map to those objectives as well.
- Fix any planning problems prior to the SOI audit. Any issues noted during the peer review should be resolved prior to the formal SOI audit.
- Put the plans under configuration control.
- Ensure that standards exist, have been reviewed against the DO-178C criteria, are applicable to the project, and are being used by the development team.
- Ensure that all team members are knowledgeable of and are following the plans, procedures, and standards.
- Consider the Job Aid questions during the development and review of the plans.
- Provide a response to all of the Job Aid questions prior to the formal SOI audit.

12.4.6.2 SOI 2 Entry Criteria, Expectations, and Preparation Recommendations

12.4.6.2.1 *SOI 2: When It Occurs*

Typically, the SOI 2 occurs after at least 50% of the code has been developed and reviewed. Sometimes a preliminary SOI 2 (an informal activity to reduce risk) may be performed earlier, but the formal SOI 2 usually requires a more mature product.

12.4.6.2.2 *SOI 2: What to Expect*

SOI 2 is typically performed on-site. Depending on the size and nature of the project, it might actually be performed in multiple phases. For example, I served as a consultant DER (and SOI team lead) on one project that was divided into 47 features (varying from 500 lines of code to 5000 lines of code each). Because the features were developed by different teams in various geographical locations and it was a high-risk project, the FAA requested that every feature be audited (all 47 of them). Therefore, the SOI 2 was divided into five phases in order to examine all features (some troublesome features were examined multiple times). On projects where the initial SOI 2 results in numerous systemic findings, SOI 2 audits will be performed until the issues are resolved. I compare it to baking a cake—you keep putting the toothpick in (performing SOI 2 reviews) until it is fully baked and the toothpick comes out clean (i.e., no more systemic issues are identified).

Following is a summary of what the SOI audit team normally does during a SOI 2 audit:

1. Close out SOI 1 audit report. Any issues not closed out from SOI 1 audit will generally be discussed and resolved at the beginning of SOI 2 audit. Most of the SOI 1 audit issues are expected to be resolved prior to the SOI 2 audit, since the SOI 1 audit closure is considered a prerequisite for submitting the PSAC to the certification authority. However, there may be some projects where the SOI 1 audit occurs late and it runs directly into the SOI 2 audit.

2. Have the applicant/developer walk through a top-down requirement thread (system requirements to software requirements to software design to source code) and a bottom-up thread (source code to software design to software requirements to system requirements). As noted earlier, the purpose of this activity is to familiarize the auditor with the applicant/developer's data and tracing mechanism.

3. Pick several requirement threads and perform top-down and bottom-up consistency checks. The SOI auditors will usually pick a variety of threads considering the following:

   a. *Functionality*: They will pick threads from different functional areas.

   b. *Development team*: They will sample data from the various development teams (to ensure that each team is following the defined processes and standards).

   c. *Complexity*: They will pick some easy threads and some complex threads. Some teams do a great job on the hard functions, but ease off on the easier ones (or vice versa).

   d. *Known problems*: If there are known problem areas (identified by lab testing, aircraft testing, or PRs), the auditors may sample data in that functional area to determine if there are any systemic issues.

   e. *Safety features*: They will often pick requirements that are most pertinent to safety to ensure they are being properly implemented.

4. Evaluate traceability completeness and accuracy. This occurs while performing the thread traces.

5. Look for inconsistencies among requirements, design, and code. Auditors will also evaluate if the lower level requirements and/or code fully implement the requirements that they trace up to.

6. Examine tool qualification data for any tools used in the development process that require qualification. Additionally, there may be an evaluation of all tools to ensure the correctness of the qualification determination (i.e., to confirm that all tools that need to be qualified are being qualified).

7. Evaluate the compliance to the plans and standards.

8. Examine review records and completed checklists (to ensure the reviews were thorough and appropriate checklists were completed).

9. Look at PRs and/or change requests and the change process.

10. Ensure that the identified development environment (in the SLECI or SCI) is being used, including compiler settings/options.

11. Evaluate the software configuration management (SCM) processes.

12. Witness the build process to ensure the written procedures are repeatable.

13. Examine SQA records and interview SQA personnel.

14. If some test cases exist, auditors may sample them to ensure that the testing effort is on the right path (this is an informal assessment during the SOI 2 audit but is helpful for early feedback and SOI 3 risk reduction).

12.4.6.2.3 *SOI 2: How to Prepare*

Following are some suggestions for how to prepare for a SOI 2 audit:

- Assign a point of contact to coordinate with the SOI audit team lead.
- Ensure that the previous SOI audit issues have been resolved. Provide the responses to the SOI audit team lead prior to the SOI 2 audit.
- Ensure that all data are available (depending on the audit team size, this may require several work stations). Some auditors may request a hard copy of some data, so be prepared to print it if needed.
- Perform a dry run SOI 2 using the Job Aid questions. Document the responses to the Job Aid questions for the audit team's consideration.
- Identify any known issues (including issues identified in the dry run) and the plan for their resolution.
- Perform some preliminary thread traces to ensure the data are ready. This may be part of the dry run SOI 2. The preliminary threads may later be used as examples during the presentation to the SOI audit team.

- Ensure that the development team has been following the plans and standards. Identify any deviations or deficiencies, along with planned resolution.
- Prepare a presentation on the system, processes used, status, and any known problems (with corrective actions being taken).
- Make sure review records are complete and available.
- Have tool qualification data available if qualifying any tools used for the development processes (e.g., a code generator or a coding standards compliance checker).
- Ensure that traceability data are accurate and available.
- Coordinate the audit agenda and ensure that all team members know what to expect. A meeting with the entire development team is valuable. If it is the team's first time to experience an audit, it's good to let them know how to respond. Encourage them to respond honestly and accurately and to only respond to direct inquiries.
- Have appropriate software developers available during the audit.
- Involve the SQA in the SOI 2 preparation and in the actual SOI 2 event, since SQA may be responsible for ensuring that corrections are made.

12.4.6.3 SOI 3 Entry Criteria, Expectations, and Preparation Recommendations

12.4.6.3.1 SOI 3: When It Occurs

The SOI 3 audit usually occurs after at least 50% of the test cases and procedures have been developed and reviewed. Additionally, during the SOI 3 audit, some sample data or a well-defined approach for the following analyses is needed (as appropriate for the software level): structural coverage, data and control coupling, worst-case execution time (WCET), stack usage, memory margin, interrupt, source-to-object code traceability, etc.

12.4.6.3.2 SOI 3: What to Expect

Like the SOI 2 audit, the SOI 3 audit is typically performed on-site and may occur in multiple phases depending on the size of the project and the challenges that are encountered. Following is a summary of what the SOI 3 audit team will typically do while on-site:

1. Examine open items from previous SOI audits and close them or determine additional required action.
2. Examine additional data that were developed since the SOI 2 audit (e.g., new requirements, design, and code) to ensure that the same processes were used or process issues were consistently resolved.

3. Choose some high-level requirements and examine the corresponding test cases and procedures to ensure they have the following characteristics:

   a. *Traceable*—ensure that bidirectional tracing between requirements and test cases, test cases and test procedures, and test procedures and test results exists and is accurate.

   b. *Complete*—ensure that the whole requirement is tested.

   c. *Robust*—ensure that the requirement has been exercised for robustness, if appropriate.

   d. *Appropriate*—ensure the test exercises the requirement properly, is effective, has pass/fail criteria, etc.

   e. *Repeatable*—ensure the test procedures are clear and can be run by someone who didn't write the procedures; also confirm that the same results will be obtained each time the test is run.

   f. *Passing*—ensure that the tests are producing expected results; typically, only informal results are examined at this point (formal results are examined in SOI 4).

4. Choose some low-level requirements and examine their corresponding test data to ensure they have the same characteristics noted for the high-level tests (traceable, complete, robust, appropriate, repeatable, passing).

5. Examine the test cases and procedures review records to ensure the reviews were thorough and appropriate checklists were completed.

6. Examine the structural coverage data to ensure coverage is being properly measured and analyzed. The structural coverage analysis will likely be in progress, but some data will be examined to ensure the approach addresses the DO-178C objectives.

7. Examine existing data for integration analyses, such as data and control coupling, timing, and memory. These analyses may still be in work, but the overall approach should be defined and some data drafted. The repeatability of analyses will be scrutinized.

8. Examine the appropriateness of integration during the testing to prove software/software and software/hardware integration required by DO-178C. If a lot of breakpoints are used during testing or tests are module-based rather than integrated, the integration approach will be closely examined.

9. Ensure that all requirements are tested. If analysis or code inspection is used, the approach will be examined to determine if it is sufficient to prove that the executable object code satisfies the identified requirements.

10. Evaluate tool qualification data for any tools used in the verification process that require qualification. Other tools may be examined as well to confirm that qualification is not needed for them.

11. Evaluate PRs to assess that they are filled out completely, including description, analysis, fix, and verification of changes to development and test data.

12. Evaluate changes implemented as a result of verification activities.

13. Witness test runs to ensure repeatability.

14. Determine if test data and verification data are under configuration control and change control.

15. Look at SCM records for requirements or test cases/procedures that are changed or added.

16. Look at SQA data associated with the test process and integration analyses, and interview SQA personnel.

17. Evaluate the verification environment for correctness (i.e., ensure that it is consistent with the environment identified in the SLECI or SCI).

12.4.6.3.3 *SOI 3: How to Prepare*

Following are some suggestions for how to prepare for SOI 3:

- Assign a point of contact to coordinate with the SOI audit team lead.
- Ensure items from previous SOIs have been addressed and be ready to discuss them with the auditors.
- Be prepared to present an overview of the test approach, test data, status, known issues, structural coverage approach, and other analyses.
- Ensure verification cases and procedures have been reviewed. If not all of them have been reviewed, clearly identify which ones have been reviewed, since the SOI audit normally focuses on those.
- Have verification cases and procedures, as well as review records, available for review. If a separate test plan exists, in addition to the verification plan, provide it to the auditors and present an overview of its contents.
- Provide verification results. The results may be from dry runs or even more informal runs at this point in the project. The auditors will mostly want to see how you intend to run and document the test results (to confirm organization, completeness, and repeatability).
- Be ready to run selected tests for the auditors to witness. Use the procedures to run the tests, rather than running them from memory.
- Have bidirectional trace data between requirements and verification cases, verification cases and verification procedures, and verification procedures and verification results (if they exist) available for review. If results don't yet exist, have a sample to show how the tracing to results will be performed.

- Be ready to show both normal and robustness test data for both high- and low-level requirements (except for level D, where low-level testing is not required).
- Have structural coverage, data and control coupling data, and other analysis data available. Be sure the technical expert for each analysis is available and ready to explain the approach.
- Ensure that the test environment identified in the SLECI or SCI is being used by the testers.
- Have tool qualification data ready for examination, if tools are being qualified.
- Review Job Aid questions ahead of time; identify any known issues; and have the answers to the Job Aid questions available for the auditor's consideration.
- Coordinate the audit agenda and ensure that all team members know what to expect. A meeting with the entire verification team is useful. If it is the team's first time to experience a SOI audit, let them know how to respond. Encourage them to respond honestly and accurately and to only respond to direct inquiries.
- Have appropriate testers available during the audit.
- Involve SQA in the SOI 3 preparation and in the actual SOI 3 audit, since SQA may be responsible for ensuring that corrections are made.

12.4.6.4 SOI 4 Entry Criteria, Expectations, and Preparation Recommendations

12.4.6.4.1 SOI 4: When It Occurs

SOI 4 occurs after issues from previous SOI audits have been resolved, and the verification results, SCI, and SAS have been reviewed and baselined. If the SOI audit is performed by a certification authority, they will typically require released data. If the SOI audit is performed by a designee, they may evaluate prereleased (but still baselined) data during the SOI audit and released data before closing the SOI.

12.4.6.4.2 SOI 4: What to Expect

SOI 4 is often performed remotely but some auditors prefer to finish it on-site, particularly if there were a lot of issues during the previous SOIs. The on-site versus desktop determination will depend on the nature of the project and the preferences of the auditor(s). Here is what the SOI 4 audit team will typically do:

1. Go over all previous SOI findings, actions, and observations to ensure that all are satisfactorily addressed, and close out the SOI 3 report (and any other SOI reports not previously closed).

2. Examine additional data that were developed since the last SOI audit to ensure that the same processes were used or process issues were consistently resolved (e.g., new test cases/procedures, test results, completed structural coverage data, and various analyses results).

3. Examine the software verification results.

4. Evaluate the analyses and justification for any code that was not covered by structural coverage analysis.

5. Examine the final SCI and SLECI for correctness and consistency with the released data.

6. Ensure that all PRs are properly closed or dispositioned.

7. Review the SAS and ensure the following:

   a. Any negotiations are documented (e.g., extraneous code or limitations on the system).

   b. Deviations from plans are documented.

   c. Open/deferred PRs are analyzed for safety, performance, operation, or regulatory compliance impact.

   d. The SAS agrees with what actually transpired during the project.

8. Examine the Tool Configuration Index and Tool Accomplishment Summary, if any TQL-1* to TQL-4 tools were used, or if a TQL-5 tool has a Tool Configuration Index and/or Tool Accomplishment Summary separate from the SAS. (See Chapter 13 for information on tool qualification.)

9. Examine SCM records.

10. Examine SQA records.

11. Examine conformity review results.

12.4.6.4.3 SOI 4: How to Prepare

Following are some suggestions for how to prepare for SOI 4:

- Ensure that issues from previous SOI(s) have been resolved and be ready to discuss them.

- Complete SCI, SLECI, software verification report, and SAS; review them; and address issues noted during the reviews.

- Ensure the software conformity review has been completed and any issues resolved. Have the conformity review records available.

- Be prepared to show any data that were not evaluated in previous SOIs (e.g., final verification results, structural coverage data, data and control coupling analysis data, WCET analysis, stack usage

---

* TQL is the tool qualification level assigned during the planning phase.

analysis, linker analysis, memory mapping analysis, load analysis, and source-to-object code analysis).

- Review Job Aid questions ahead of time; address any issues noted; have the answers to the Job Aid questions available for the auditor's consideration. Be sure to consider any Job Aid questions that were not answered during previous SOI audits as well (e.g., SOI 3 questions).

12.5 Software Maturity Prior to Certification Flight Tests

One question that invariably arises during a certification effort is: "How mature does the software need to be before it can be used for official flight testing?" The software needs to be proven before system-level and aircraft-level testing can be successfully performed. There are certification regulations and guidelines that essentially require that the system be in a certifiable state prior to official FAA flight testing.*

Ideally, all DO-178C compliance activities are completed prior to certification flight testing of the system(s) that uses the software. The certification authorities definitely prefer this state of maturity. However, since software is often one of the last items to mature, the ideal state is often not feasible. Therefore, certification authorities have identified software maturity criteria. At this time, the software maturity criteria have not yet been firmly established, and therefore tend to vary some from project to project. However, there are some general guidelines that are normally used as a starting point for negotiation. The following criteria are typically used to ensure the software is mature enough for certification flight testing [5]:

1. SOI audits 1–3 have been conducted and all significant findings (noncompliances) are closed. Significant review findings are ones that could impact safety, performance, operations, and/or compliance determination; therefore, they need to be addressed prior to flight testing of the software.
2. The software should be at a production (black label) equivalent state prior to certification flight testing of that software. This means that there is high confidence that the software in the flight test aircraft is the software that will actually be installed on the certified and

---

* Regulations include Title 14 of the Code of Federal Regulations Parts 21.33(b), 21.35(a), and XX.1301 (where XX may be 23, 25, 27, or 29). FAA Order 8110.4[ ] section 5-19, items d–f also provide guidelines.

production aircraft. Confidence that the software is production equivalent normally requires the following:

a. All system requirements allocated to software have been implemented in the software.

b. All software requirements-based tests (including, high-level, low-level, normal, and robustness tests) have been executed on the target.* Software test execution means: (1) test cases and procedures have been reviewed; (2) test environment, test cases and procedures, test scripts, and software under test are under configuration control; (3) tests have been executed and results are retained and controlled; and (4) PRs are created for any test failures.

c. Any significant software PRs that affect functionality and safety are fixed, verified, and incorporated into the software to be used for flight test.

d. The SCI is generated, complies with DO-178C section 11.16, and is provided with the software to be installed for flight test.

Any software that does not meet these criteria usually requires special coordination with the certification authority prior to certification flight testing. There tends to be a fair amount of negotiation on this front. Many times, the software is deemed to be mature, the flight tests are run, and then the software changes. In this scenario, some or all of the flight tests may need to be reexecuted based on the system- and software-level change impact analyses.

References

1. RTCA DO-178C, *Software Considerations in Airborne Systems and Equipment Certification* (Washington, DC: RTCA, Inc., December 2011).
2. Federal Aviation Administration, *Software Approval Guidelines*, Order 8110.49 (Change 1, September 2011).
3. European Aviation Safety Agency, *Software Aspects of Certification*, Certification Memorandum CM-SWCEH-002 (Issue 1, August 2011).
4. Federal Aviation Administration, *Conducting Software Reviews Prior to Certification*, Aircraft Certification Service (Rev. 1, January 2004).
5. W. Struck, Software maturity, presentation at *2009 Federal Aviation Administration National Software and Complex Electronic Hardware Conference* (San Jose, CA, August 2009).

---

* Formally executed software tests are preferred; however, there have been situations where dry run executions have been accepted, if the dry run is complete (all requirements exercised) and successful (no unacceptable failures).

Part IV

Tool Qualification and DO-178C Supplements

Part IV examines four technology-specific guidance documents that were developed by the RTCA Special Committee #205 (SC-205) and EUROCAE Working Group #71 (WG-71):*

- DO-330/ED-215, *Software Tool Qualification Considerations*
- DO-331/ED-218, *Model-Based Development and Verification Supplement to DO-178C and DO-278A*
- DO-332/ED-217, *Object-Oriented Technology and Related Techniques Supplement to DO-178C and DO-278A*
- DO-333/ED-216, *Formal Methods Supplement to DO-178C and DO-278A*

Chapter 4 provided a brief overview of each of these documents generated by SC-205/WG-71 and their overall relationship to DO-178C. Each document is over 100 pages long; therefore, a detailed examination is not possible. A chapter is dedicated to each of the four documents. Chapter 13 discusses DO-330 and tool qualification. Chapter 14 briefly explores model-based development and verification and DO-331. Chapter 15 covers DO-332 and object-oriented technology, as well as some related techniques. Chapter 16 surveys formal methods and DO-333.

---

* The EUROCAE ED- numbers are included for reference purposes. The EUROCAE ED-documents are exactly the same as the RTCA DO-, except the page size differs and the French translation is added. Throughout Part IV, only the RTCA document numbers will be referenced.

Because each document builds upon DO-178C, which was discussed in detail in Chapters 5 through 12, Part IV examines the relationship of DO-330, DO-331, DO-332, and DO-333 to DO-178C and identifies primary differences. Each chapter presents an overview of the identified technology, the key differences from DO-178C, and some of the challenges that will likely be encountered when applying the guidance. This overview is intended to give you a glimpse into the four documents and help you better understand how to apply them.

13

DO-330 and Software Tool Qualification

Acronyms

CNS/ATM	communication, navigation, surveillance, and air traffic management
COTS	commercial off-the-shelf
FAA	Federal Aviation Administration
FAQ	frequently asked question
MC/DC	modified condition/decision coverage
MISRA	Motor Industry Software Reliability Association
PSAC	Plan for Software Aspects of Certification
SAS	Software Accomplishment Summary
SC-205	Special Committee #205
SLECI	Software Life Cycle Environment Configuration Index
TOR	Tool Operational Requirement
TQL	tool qualification level
TQP	Tool Qualification Plan
WG-71	Working Group #71

13.1 Introduction

A *software tool* is "a computer program or a functional part thereof, used to help develop, transform, test, analyze, produce, or modify another program, its data, or its documentation" [1]. The use of tools in the software life cycle process, as well as other domains such as systems, programmable hardware, and aeronautical databases has exploded in recent years. I recently worked on a project that utilized around 50 software tools. It was a huge project with many specialized needs, so it is not typical. However, it

TABLE 13.1

Examples of Tool Usage in the Software Life Cycle

Development Tools	Verification Tools	Other Tools
• Requirements capture and management	• Debugging	• Project management
• Design	• Static analysis	• Configuration management
• Modeling	• Worst-case execution timing analysis	• Problem reporting
• Text editing	• Model verification	• Peer review management
• Compiling	• Coding standards conformity	
• Linking	• Trace verification	
• Automatic code generation	• Structural coverage analysis	
• Configuration file generation	• Automatic testing	
	• Emulation	
	• Simulation	
	• Automatic test generation	
	• Configuration file verification	
	• Formal methods	

does show how the use of tools has increased and will continue to increase. Table 13.1 identifies some common ways that tools are used in the software life cycle and divides them into the following categories: development, verification, and other.

Tools cannot replace the human brain; however, they can prevent errors and identify errors that humans might insert or fail to identify. Tools can help engineers do their job better and allow them to concentrate on the more challenging problems that require engineering skill and judgment.

Some engineers are resistant to using tools, while others go completely overboard and use tools for everything. There is a happy medium that must be reached when using tools to successfully develop software that performs its intended function.

In 2004, I was privileged to lead the *Software Tools Forum*. The forum was cosponsored by Embry-Riddle Aeronautical University and the Federal Aviation Administration (FAA). The purpose of the effort was to assess the state of tools in aviation and to identify issues that needed to be resolved to enable the industry to safely reap the benefits of tools. Approximately 150 participants from industry, government, and academia attended. The objectives of the forum were to (1) share information on software tools used in aviation projects, (2) discuss lessons learned to date regarding software tools used in aviation, (3) discuss the challenges of tools in safety-critical systems, and (4) consider next steps.

At the end of the forum, a brainstorming session was held to identify some of the most significant issues regarding effective use of tools in aviation. The brainstorming session identified six needs categories, which were prioritized by the industry. The first five categories are specific; the last one is a catchall for miscellaneous tool issues. The six categories are listed here in order of priority (items 1–3 were weighted much higher than items 4–6):

1. Development tool qualification criteria need to be modified.
2. Criteria for model-based development tools need to be established.
3. Criteria that enable tool qualification credit to be carried from one program to another need to be developed.
4. Different approaches to autocode generator usage and qualification need to be developed and documented.
5. Integration tools pose new challenges that need to be addressed.
6. A number of miscellaneous tool issues need to be addressed.

Soon after the tools event, the RTCA Special Committee #205 (SC-205) and EUROCAE Working Group #71 (WG-71) joint committee was formed to update DO-178C and provide necessary guidance in a number of technical areas. The recommendations from the *Software Tools Forum* and other sources were used as input to the committee. The SC-205/WG-71 tool qualification subgroup* took the issues seriously and worked to document guidance that would address them. The results were an update to DO-178C section 12.2 and the production of DO-330, entitled *Software Tool Qualification Considerations*. During the 6.5 years that it took to finish the SC-205/WG-71 effort, I also reviewed and approved dozens of software tools and their qualification data. The subject of software tool qualification is one near and dear to my heart.

This chapter provides an overview of DO-178C section 12.2 and DO-330 in order to explain when a tool needs to be qualified, what level of qualification is required, and how to qualify a tool. DO-330 is 128 pages long; therefore, only an overview is provided. The attention in this chapter is on the most critical aspects of DO-330 for tool developers and tool users. The differences between the DO-178B and the updated guidance are also discussed, since there are numerous tools in use today that were qualified using DO-178B criteria. In addition, some special topics related to tool qualification and some potential pitfalls to avoid when qualifying or using qualified tools are discussed.

---

* Co-led by yours truly.

13.2 Determining Tool Qualification Need
and Level (DO-178C Section 12.2)

Table 13.1 identifies common uses for tools in the software life cycle. Some of those tools may need to be qualified and others may not. This section examines DO-178C section 12.2 to answer the following questions:

- What is tool qualification?
- When is it required?
- To what level must a tool be qualified?
- How do DO-178B and DO-178C guidance on tool qualification differ?

Tool qualification is the process used to gain certification credit for a software tool whose output is not verified, when the tool eliminates, reduces, or automates processes required by DO-178C. Tool qualification is granted in conjunction with the approval of the software that uses the tool; it is not a stand-alone approval. The tool qualification process provides confidence in the tool functionality; the confidence is at least equivalent to the process(es) being eliminated, reduced, or automated [2]. The rigor required for the tool qualification effort "varies based upon the potential impact that a tool error could have on the system safety and upon the overall use of the tool in the software life cycle process. The higher the risk of a tool error adversely affecting system safety, the higher the rigor required for tool qualification" [2]. Figure 13.1 graphically shows the process to determine if a tool needs to be qualified and to what level.

DO-178B defined two types of tools: (1) software development tools and (2) software verification tools. Because these two categories were often tied to the life cycle phase rather than the tool impact, DO-178C does not use the terms *development tools* and *verification tools*. Instead, DO-178C identifies three criteria which focus on the potential impact the tool could have. Table 13.2 shows the comparison of the DO-178B and DO-178C tool categories.

The qualification process for *Criteria 1 tools* under DO-178C is very similar to *development tools* in DO-178B. Examples of *Criteria 1 tools* include autocode generators, configuration file generators, compilers, linkers, requirements management tools, design tools, and modeling tools. Likewise, the qualification process for *Criteria 3 tools* under DO-178C is about the same as *verification tools* in DO-178B. Examples of these tools include test case generators, automated test tools, structural coverage tools, data coupling analysis tools, and static code analyzers. The primary difference between DO-178B and DO-178C is that DO-178C introduces a third category of tools: the *Criteria 2 tool*. The need for this special category was primarily driven by considering tools in the future that may make more critical decisions than today's verification tools but do not have as much impact on the resultant software

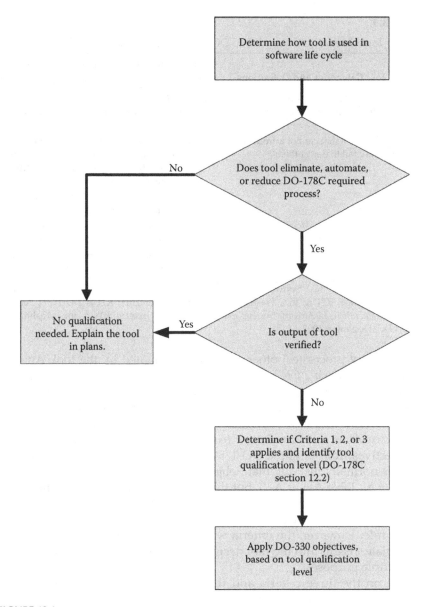

FIGURE 13.1
Determining tool qualification need and level.

as today's development tools. Some formal methods tools fall into this category. For example, a single proof tool may be used to automate some of the source code verification steps and to reduce the amount of testing needed. Either of these actions alone would make it a *Criteria 3 tool*, but the combination pushes it into the *Criteria 2 tool* realm. In this situation, the proof tool is

TABLE 13.2

DO-178B Tool Categories Compared to DO-178C Tool Criteria

DO-178B Tool Categories and Definitions	DO-178C Tool Qualification Criteria and Definitions
Development tool: Tool whose output is part of airborne software and thus can introduce errors	*Criteria 1*: A tool whose output is part of the resulting software and thus could insert an error
Verification tool: Tool that cannot introduce errors but may fail to detect them	*Criteria 2*: A tool that automates verification process(es) and thus could fail to detect an error, and whose output is used to justify the elimination or reduction of: • Verification process(es) other than that automated by the tool, or • Development process(es) that could have an impact on the airborne software *Criteria 3*: A tool that, within the scope of its intended use, could fail to detect an error

Sources: RTCA DO-178C, Software *Considerations in Airborne Systems and Equipment Certification*, RTCA, Inc., Washington, DC, December 2011; RTCA DO-178B, *Software Considerations in Airborne Systems and Equipment Certification*, RTCA, Inc., Washington, DC, December 1992.

used to verify process(es) other than that automated by the tool. Another example of a *Criteria 2 tool* is a static code analyzer which is used to replace source code review (verification step) and to reduce design mechanisms to detect overflow (development step). This tool performs some verification but it also reduces the software development process [1]. I tend to think of *Criteria 2 tools* as *super verification tools*. With a single tool, you can knock out objectives from multiple DO-178C Annex A tables. Traditional *verification tools* normally just automate objectives from a single DO-178C Annex A table (e.g., Table A-6 or A-7), but these *super verification tools* might satisfy objectives from multiple tables (e.g., A-5 through A-7). I do not expect that there will be many *Criteria 2 tools*; they really should be the exception and not the rule. However, the criteria was included in DO-330 to ensure that appropriate insight into the tool design is visible when a single tool carries out multiple purposes.

Based on the three criteria and the level of the software that the tool supports, a tool qualification level (TQL) is assigned. The TQL determines the amount of rigor required during the qualification process. There are a total of five TQLs; TQL-1 requires the most rigor and TQL-5 the least. Table 13.3 shows the TQL for each tool criteria and software level. Table 13.4 shows the correlation between the DO-178B software levels approach and the DO-178C TQLs. The intent of DO-178C and DO-330 is to make the tool qualification criteria clearer than what was included in DO-178B, not to change what is required. Therefore, in most cases, tools qualified under DO-178B will also be acceptable under DO-178C.

TABLE 13.3

Tool Qualification Level

Software Level	Criteria		
	1	2	3
A	TQL-1	TQL-4	TQL-5
B	TQL-2	TQL-4	TQL-5
C	TQL-3	TQL-5	TQL-5
D	TQL-4	TQL-5	TQL-5

Source: RTCA DO-178C, Software *Considerations in Airborne Systems and Equipment Certification*, RTCA, Inc., Washington, DC, December 2011. DO-178C Table 12-2 used with permission from RTCA, Inc.

TABLE 13.4

Correlation between DO-178B and DO-178C Levels

DO-178B Tool Qualification Type	DO-178B Software Level	DO-178C TQL
Development	A	TQL-1
Development	B	TQL-2
Development	C	TQL-3
Development	D	TQL-4
Verification	All	TQL-4[a] or TQL-5[b]

Source: RTCA DO-330, *Software Tool Qualification Considerations*, RTCA, Inc., Washington, DC, December 2011. DO-330 Table D-1 used with permission from RTCA, Inc.

[a] TQL-4 applies for Criteria 2 tools used on Level A and B software.

[b] TQL-5 applies for Criteria 2 tools used on Level C and D software, as well as Criteria 3 tools.

13.3 Qualifying a Tool (DO-330 Overview)

Once the TQL is established using DO-178C section 12.2, DO-330 provides guidance for qualifying the tool. This section explains why DO-330 is needed and the DO-330 tool qualification process.

13.3.1 Need for DO-330

Over the last few years, the number and types of tools have grown. Furthermore, the tools are often developed by third party vendors who may have little understanding of DO-178C. DO-330 was developed to provide tool-specific guidance for both tool developers and users. The document focuses on tools used in the software life cycle. However, the document may also be applied to other domains, such as programmable hardware, databases, and systems. How the document is applied to the domain will be dictated by the domain's guidance (e.g., DO-178C, DO-254 [3], DO-200A [4]).

The SC-205/WG-71 committee debated quite vigorously whether or not a separate tool qualification document was needed. There are advantages and disadvantages of generating such a document. Table 13.5 summarizes some of the pros and cons. These and others were considered during committee's deliberations. In the end, it was decided that the benefits of a stand-alone document outweighed the consequences and that a stand-alone document was the best long-term solution for the aviation industry as a whole.

While developing DO-330, the SC-205/WG-71 committee kept the following goals in mind:

- Maintain the DO-178B approach for traditional verification tools (as much as possible).
- Develop an approach that will support emerging tool technologies.
- Provide an approach that enables reuse of tool qualification credit on multiple projects.

TABLE 13.5

Pros and Cons of a Stand-Alone Tool Qualification Document

Pros for a Stand-Alone Document	Cons for a Stand-Alone Document
• For many projects there are actually more lines of code for the tools than there are for the airborne software; therefore, specific guidance is needed.	• There is considerable redundancy between the tool qualification document (DO-330) and DO-178C.
• Tools form a separate domain from the airborne software. A stand-alone document allows the tool-specific needs to be addressed head-on rather than in generalities. This prevents misinterpretations.	• The tool qualification guidance is difficult to apply when tools are developed using nontraditional technologies (such as model-based development, object-oriented technology, or formal methods) (i.e., the supplements explain how they apply to DO-178C and DO-278A but not DO-330).
• A stand-alone document provides guidance specifically for tool developers, who may have little to no understanding of DO-178C.	• Another document leads to challenges when maintaining the guidance; particularly when there is overlap between the document and DO-178C.
• Since other domains (such as programmable hardware, systems, aeronautical databases) also use tools, a stand-alone document can benefit more than just the software domain.	• An extra document may lead to additional work by the developers. Particularly for those who only qualify lower level tools (verification tools in DO-178B or TQL-5 in DO-178C).

- Identify general tool user and tool developer roles (in order to address integration of a tool into the development environment, to support reuse, and to help with integration of commercial off-the-shelf [COTS] tools).
- Develop an approach that is clear but flexible for tool developers and users.
- Develop an objective-based approach (to help with reuse and flexibility).
- Provide an approach that may be adopted by multiple domains.

13.3.2 DO-330 Tool Qualification Process

Once the need to qualify and the applicable TQL are determined (using DO-178C section 12.2), DO-330 provides the guidance and objectives for qualifying the tool. DO-330 is organized similar to DO-178C. The tool development goes through a life cycle, just like the airborne software. The tool life cycle processes include planning, requirements, design, verification, configuration management, quality assurance, and qualification liaison. Table 13.6 shows the DO-178C and DO-330 table of contents side by side, illustrating that DO-330 is organized very similar to DO-178C. DO-178C was used as the foundation for DO-330; then the content was modified to make it tool-specific.

Because DO-178C and DO-330 are similar and DO-178C was examined in depth in Chapters 5 through 12, this section emphasizes 18 primary differences between DO-178C and DO-330.

Difference 1: Different domains. As already mentioned, DO-178C applies to the airborne software domain, whereas DO-330 applies to the tools that may be used in the software life cycle. Additionally, DO-330 may be used to qualify tools used in other domains (such as communication, navigation, surveillance, and air traffic management [CNS/ATM] software, systems, and electronic hardware).

Difference 2: Introductory sections are different. DO-178C (section 2) includes an overview of the systems process and the software levels. DO-330 (sections 2 and 3) explains the purpose and characteristics of tool qualification. The information in DO-330 sections 2 and 3 is similar to what has already been discussed in this chapter.

Difference 3: DO-330 combines the life cycle process and planning sections. DO-178C has separate sections for software life cycle and planning (sections 3 and 4). However, DO-330 combines these subjects into a single section (section 4) since the life cycle plays a significant role in the planning process.

Difference 4: DO-330 adds the word tool to life cycle processes and data. In order to distinguish the tool development process from the airborne software

TABLE 13.6

Comparison of DO-178C and DO-330 Table of Contents

DO-178C Table of Contents [2]	DO-330 Table of Contents [1]
1. Introduction	1. Introduction
2. System aspects relating to software development	2. Purpose of tool qualification
	3. Characteristics of tool qualification
3. Software life cycle	4. Tool qualification planning process
4. Software planning process	5. Tool development life cycle and processes
5. Software development processes	6. Tool verification processes
6. Software verification processes	7. Tool configuration management process
7. Software configuration management process	8. Tool quality assurance process
	9. Tool qualification liaison process
8. Software quality assurance process	10. Tool qualification data
9. Certification liaison process	11. Additional considerations for tool qualification
10. Overview of certification process	
11. Software life cycle data	Annex A—Tool qualification objectives
12. Additional considerations	Annex B—Acronyms and glossary of terms
Annex A—Process objectives and outputs by software level	Appendix A—Membership list
Annex B—Acronyms and glossary of terms	Appendix B—Example of determination of applicable tool qualification levels
Appendix A—Background of DO-178 document	Appendix C—Frequently asked questions related to tool qualification for all domains
Appendix B—Committee membership	Appendix D—Frequently asked questions related to tool qualification for airborne software and CNS/ATM software domains

development processes, the term *tool* is added where appropriate. For example, *source code* in DO-178C is called *tool source code* in DO-330.

Difference 5: DO-330 enhances the information required in the PSAC and SAS when qualified tools are used. DO-178C sections 11.1 and 11.20 identify the expected contents of the Plan for Software Aspects of Certification (PSAC) and Software Accomplishment Summary (SAS). DO-330 sections 10.1.1 and 10.1.16 identify additional information that should be added in the PSAC and SAS when qualified tools are to be used in the software life cycle processes. Per DO-330 section 10.1.1, the following information should be added to the PSAC for each qualified tool that will be used [1]:

- Identify the tool and its intended use in the software life cycle process.
- Explain credit sought for the tool (i.e., explain what processes or objectives it eliminates, reduces, or automates).
- Justify the maturity of the technology automated by the tool (e.g., if the tool is used to perform formal proofs, one needs to ensure the maturity of the approach in general before automating it).
- Propose the TQL, along with rationale for why that TQL is adequate.

- Identify tool developer or source.

- Explain division of roles and responsibilities for the tool qualification. In particular, explain what the tool user and tool developer are doing and who is satisfying which DO-330 objectives.

- Explain the Tool Operational Requirements (TORs) development, tool integration, and tool operational validation and verification processes.

- Identify the intended tool operational environment and confirm that it is representative of the environment used during tool verification.

- Explain any additional considerations for the tool (e.g., reuse, COTS usage, service history).

- Reference the Tool Qualification Plan (TQP) where additional information can be found on the tool qualification details (for TQL-1–TQL-4). (For TQL-5, a TQP is not required.)

DO-330 section 10.1.16 also explains additional information that should be included in the SAS when utilizing qualified tools in the software life cycle. Table 13.7 summarizes the additional items for TQL-1–TQL-4 and TQL-5.

Difference 6: DO-330 distinguishes tool operation from tool development. DO-178B did not distinguish these and it led to some confusion and limited reuse. In order to promote tool reuse and clarity, DO-330 provides guidance for

TABLE 13.7

Additional Inclusions in the SAS When Using Qualified Tools

TQL-1–TQL-4	TQL-5
• Tool identification (the specific part number or other identifier).	• Tool identification (the specific part number or other identifier).
• Details of the credit sought for the tool (explanation of what processes and/or objectives it eliminates, reduces, or automates).	• Details of the credit sought for the tool (explanation of what processes and/or objectives it eliminates, reduces, or automates).
• Reference to the Tool Accomplishment Summary where additional information can be found on the tool qualification details.	• Listing of or reference to tool qualification data (including the version information).
• Statement confirming that the tool development, verification, and integral processes comply with the tool plans, including the tool user activities.	
• Summary of all open tool problem reports along with an analysis to ensure that the behavior of the tool complies with the TORs.	
• Summary of any variance in tool usage from what was explained in the PSAC, if applicable.	

Source: RTCA DO-330, *Software Tool Qualification Considerations*, RTCA, Inc., Washington, DC, December 2011.

tool development and verification (performed by a tool developer) and tool operation (performed by a tool user). However, it gives the flexibility for each project to assign roles and responsibilities appropriate for the situation. In general, DO-330 Annex A Tables T-0 and T-10 objectives primarily apply to the tool user/operator, whereas the DO-330 Annex A Tables T-1 through T-9 objectives mainly apply to the tool developer.

Difference 7: DO-330 guidance requires operational requirements in order to document the user's needs. In DO-178C the system requirements drive the high-level software requirements. However, for a tool, *system requirements* don't exist; instead, TORs drive the tool requirements. The TORs identify how the tool is used within the specific software life cycle process. Multiple users of a common tool could use it differently; therefore, the TORs are user specific. The TOR document includes a description of the following [1]:

- The context of the tool usage in the software life cycle, interfaces with other tools, and integration of the tool output files into the airborne software.
- The operational environment(s) where the tool will be used.
- The input to the tool, including file format, language description, etc.
- The format and content summary of the tool's output files.
- Functional tool requirements.
- Requirements for abnormal activation modes or inconsistency inputs that should be detected by the tool (needed for TQL-1–TQL-4, but not required for TQL-5).
- Performance requirements (the behavior of the tool output).
- User information that explains how to properly use the tool.
- Explanation of how to operate the tool (including selected options, parameters values, command line, etc.).

Difference 8: DO-330's requirements types differ. DO-178C identifies system requirements, high-level software requirements, and low-level software requirements. DO-330 identifies TORs (same basic level as DO-178C's system requirements), tool requirements (same intent as DO-178C's software high-level requirements), and low-level tool requirements (same level of granularity as DO-178C's low-level software requirements). In some situations three levels of requirements may not be needed. For example, the TORs and tool requirements might be a single set of requirements, or the tool requirements and the low-level tool requirements might be one level of requirements.* Even if levels of requirements are merged, all of the applicable objectives of

---

* The various levels of tool requirements should be closely coordinated with certification authority since there are divergent opinions on this topic.

DO-330 need to be satisfied. As with DO-178C, bidirectional traceability between the various levels of tool requirements is needed.

Difference 9: DO-330 requires validation of the TORs. In DO-178C, the system requirements that come into the software process are *validated* (using an ARP4754A-compliant process); therefore, DO-178C doesn't require requirements validation. However, TORs are not validated by the system process; therefore, they must be validated for correctness and completeness.

Difference 10: Per DO-330 derived requirements for a tool are evaluated for impact on tool functionality and TORs, rather than the safety assessment process. Because tools do not fly, their impact on the safety assessment process is indirect. Instead, the impact of derived requirements on the software life cycle, expected functionality, and TORs is evaluated. As with DO-178C, all derived requirements need to be justified. Users may not know the details of the tool design; therefore, the justification for derived requirements helps them properly assess the impact on their software life cycle. The justifications should be written with the user in mind.

Difference 11: DO-330 uses the concept of tool operational environment instead of target computer. DO-330 defines the tool operational environment as: "The environment and life cycle process context in which the tool is used. It includes workstation, operating system, and external dependencies, for example interfaces to other tools and manual process(es)" [1]. Tools are normally launched on desktop computers rather than embedded in the avionics. Therefore, the concept of a *target computer* (which is used in DO-178C) does not fit well in the qualification process. Since tools are often developed to be utilized in multiple operational environments, identification of each operational environment is important.

Difference 12: DO-330 gives more flexibility for structural coverage approaches and focuses on identifying unintended functionality. Structural coverage was probably the most controversial subject during the SC-205/WG-71 deliberations on DO-330. Modified condition/decision coverage (MC/DC) was originally not included in DO-330, since the tools themselves do not fly and historical data indicate that MC/DC has limited value for tools. Decision coverage and statement coverage were included. However, the MC/DC objective was added (in addition to decision coverage and statement coverage objectives) to obtain committee consensus. Per DO-330, statement coverage applies for TQL-3; statement and decision coverage apply for TQL-2; and statement coverage, decision coverage, and MC/DC apply for TQL-1. However, the wording in DO-330 emphasizes that the purpose of structural coverage is to identify unintended functionality; this opens the door for alternatives to MC/DC (as long as those alternatives meet the same intent as MC/DC). Another difference related to structural coverage is that DO-330 does not require source-to-object code traceability analysis for the TQL-1 tool code as is required for DO-178C level A software.

Difference 13: DO-330 addresses interfaces with external components. External components are "components of the tool software that are outside the control of the developer of the tool. Examples include primitive functions provided by the operating system or compiler run-time library, or functions provided by a COTS or open source software library" [1]. DO-330 includes objectives and activities to carry out the following [1]:

- Identify interfaces with external components in the design (DO-330 section 5.2.2.2.g).
- Verify the correctness and completeness of the interfaces (DO-330 section 6.1.3.3.e).
- Ensure correct integration of interfaces to external components (DO-330 section 6.1.3.5).
- Ensure that interfaces to external components are exercised through requirements-based testing (DO-330 section 6.1.4.3.2).

Difference 14: DO-330 requires a tool installation report. The tool installation report is generated during the operational integration phase to confirm the tool's integration into its operational environment. The installation report identifies (1) the configuration of the operational environment, (2) the version of the tool's executable object code and any supporting configuration files, (3) any external components, and (4) how to execute the tool [1].

Difference 15: DO-330 life cycle data are tool-specific. Most of the tool life cycle data required by DO-330 are similar to what is required by DO-178C for airborne software or DO-278A for CNS/ATM software. However, the titles are tool-specific and the recommended content is focused on the needs of the tool qualification. DO-330 section 10 describes the tool life cycle data and DO-330 Annex A objectives tables identify what data are required for each TQL. DO-330 section 10 divides the data into three categories, each identified in a different subsection. The three categories and the data for each category are shown in Table 13.8.

Difference 16: DO-330 objectives tables vary some from DO-178C objectives. The objectives tables in DO-330 Annex A are similar to the ones in DO-178C Annex A. However, there are some differences. One notable difference is that the DO-330 Annex A tables are numbered as T-x, rather than A-x to distinguish them from DO-178C. Another general difference is that DO-330 has 11 Annex A tables instead of 10 Annex A tables like DO-178C. The titles of the DO-330 Annex A tables are listed here, followed by a summary of the primary differences between the DO-330 and DO-178C Annex A tables:

- *Table T-0*: Tool operational processes
- *Table T-1*: Tool planning process
- *Table T-2*: Tool development processes

TABLE 13.8

Tool Qualification Data in DO-330

Data for the Tool Qualification Liaison Process and Other Integral Processes (DO-330 Section 10.1)	
10.1.1	PSAC (identifies tool-specific information to add to the airborne software's PSAC)
10.1.2	Tool qualification plan
10.1.3	Tool development plan
10.1.4	Tool verification plan
10.1.5	Tool configuration management plan
10.1.6	Tool quality assurance plan
10.1.7	Tool requirements standards
10.1.8	Tool design standards
10.1.9	Tool code standards
10.1.10	Tool life cycle environment configuration index
10.1.11	Tool configuration index
10.1.12	Tool problem reports
10.1.13	Tool configuration management records
10.1.14	Tool quality assurance records
10.1.15	Tool accomplishment summary
10.1.16	SAS (identifies tool-specific information to add to the airborne software's SAS)
10.1.17	Software Life Cycle Environment Configuration Index (SLECI) (identifies tool-specific information to add to the airborne software's SLECI)
Data Produced during the Tool Development and Verification Effort (DO-330 Section 10.2)	
10.2.1	Tool requirements
10.2.2	Tool design description
10.2.3	Tool source code
10.2.4	Tool executable object code
10.2.5	Tool verification cases and procedures
10.2.6	Tool verification results
10.2.7	Trace data
Data Produced to Support the Tool's Operational Approval (DO-330 Section 10.3)	
10.3.1	TORs
10.3.2	Tool installation report
10.3.3	Tool operational verification and validation cases and procedures
10.3.4	Tool operational verification and validation results

Source: RTCA DO-330, *Software Tool Qualification Considerations*, RTCA, Inc., Washington, DC, December 2011.

- *Table T-3*: Verification of outputs of tool requirements processes
- *Table T-4*: Verification of outputs of tool design process
- *Table T-5*: Verification of outputs of tool coding and integration processes
- *Table T-6*: Testing of outputs of integration process
- *Table T-7*: Verification of outputs of tool testing
- *Table T-8*: Tool configuration management process
- *Table T-9*: Tool quality assurance process
- *Table T-10*: Tool qualification liaison process

Table T-0 is a DO-330 tool-specific table that does not have a DO-178C equivalent. It includes seven objectives for four tool processes to address the tool operation (the user's perspective). The processes and objectives are summarized in Table 13.9.

DO-330 Tables T-1 through T-10 are similar to DO-178C Tables A-1 through A-9, with the following significant exceptions:

- DO-330 Table T-2 objective 8 states: "Tool is installed in the tool verification environment(s)" [1]. This objective is unique to tools, since the tool verification environment may differ from the operational environment. Additionally, if there are multiple verification environments, the tool will need to be integrated into each environment during the qualification effort.
- DO-330 Table T-3 objective 3 states: "Requirements for compatibility with the tool operational environment are defined" [1]. This objective

TABLE 13.9

Summary of DO-330 Table T-0 Processes and Objectives

Table T-0 Process	Table T-0 Objective(s)
Planning	1. "The tool qualification need is established."
Tool Operational Requirements development	2. "Tool Operational Requirements are defined."
Tool operational integration	3. "Tool Executable Object Code is installed in the tool operational environment."
Tool operational validation and verification	4. "Tool Operational Requirements are complete, accurate, verifiable, and consistent."
	5. "Tool operation complies with the Tool Operational Requirements."
	6. "Tool Operational Requirements are sufficient and correct."
	7. "Software life cycle process needs are met by the tool."

Source: RTCA DO-330, *Software Tool Qualification Considerations*, RTCA, Inc., Washington, DC, December 2011.

focuses on operational environment compatibility instead of target computer environment, as previously noted in Difference #11.

- DO-330 Table T-3 objective 4 states: "Tool requirements define the behavior of the tool in response to error conditions" [1]. This objective is unique for tools and focuses on ensuring the tools are robust and respond predictably to errors.

- DO-330 Table T-3 objective 5 states: "Tool Requirements define user instructions and error messages" [1]. Again, this is a tool-specific objective. It ensures that the tool has considered the user's perspective.

- DO-330 Table T-4 objective 11 states: "External component interface is correctly and completely defined" [1]. This tool-specific objective is intended to verify the external component interface that was previously discussed; see Difference #13.

- DO-330 Table T-7 objective 5 states: "Analysis of requirements-based testing of external components is achieved" [1]. There is no equivalent objective in DO-178C. As noted in Difference #13 earlier, this objective is intended to confirm that the requirements-based testing has exercised the tool's interfaces to external components.

- DO-330 Table T-10 (all objectives) focuses on tool *qualification* rather than on *certification*.

Difference 17: TQL-5 in DO-330 clarifies what was required for verification tools in DO-178B. DO-330 is intended to require the same amount of rigor for qualification of TQL-5 tools as was required for *verification tools* in DO-178B. However, the DO-330 criteria clarify the DO-178B expectations in order to ensure compliance and to support tool reuse. DO-330 frequently asked question (FAQ) D.6 summarizes the DO-178B *verification tool* to DO-330 TQL-5 differences this way:

> This document [DO-330] provides more accurate and complete guidance for tools at TQL-5 than DO-178B (and hence DO-278) did for verification tools. The intent is not to ask for more activities or more data (for example, the qualification does not require any data from the tool development process). However, it clarifies the content of the TOR, the compliance of the tool to the airborne (or CNS/ATM) software process needs, and the objectives of other integral processes applicable for TQL-5 [1].*

Some of the clarifications in DO-330 for TQL-5 that were not present for DO-178B's *verification tool* criteria are summarized as follows:

- DO-330 identifies specific objectives for TQL-5 (and all TQLs). This was not the case in DO-178B (i.e., there were no objectives for verification tool qualification in DO-178B).

---

* Brackets added for clarity.

- DO-330 provides additional clarification of what is expected in the TORs.

- DO-330 separates the TORs from the tool requirements.

- DO-330 adds an integration objective to ensure the qualification occurs within a specific operational environment.

- DO-330 includes objectives to ensure validation of the TORs.

Difference 18: DO-330 contains FAQs in appendices C and D. FAQs and discussion papers for DO-178C are included in DO-248C. However, tool-related FAQs are included in the DO-330 appendices. Appendix C includes domain independent FAQs. Appendix D provides FAQs specific to the DO-178C and DO-278A domains. Table 13.10 briefly describes each FAQ.

13.4 Special Tool Qualification Topics

There are a few special topics related to DO-330 that deserve some additional explanation.

13.4.1 FAA Order 8110.49

Chapter 9 of FAA Order 8110.49 includes some clarification of DO-178B regarding tool qualification. It is anticipated that this section of the Order will be deleted (or significantly modified) since DO-178C and DO-330 provide more comprehensive guidance than DO-178B did for tool qualification. However, there may be some additional guidance added to explain transition from DO-178B to DO-178C tool qualification criteria. Additionally, guidance for how to use DO-331, DO-332, and DO-333 supplements with DO-330 may be needed.

13.4.2 Tool Determinism

DO-178B section 12.2 included the following statement: "Only deterministic tools may be qualified, that is, tools which produce the same output for the same input data when operating in the same environment" [5].

Order 8110.49 clarifies what is meant by deterministic tools. Section 9-6.d of the Order explains that determinism is often interpreted too restrictively for tools. The Order states:

> A restrictive interpretation [of determinism] is that the same apparent input necessarily leads to exactly the same output. However, a more accurate interpretation of determinism for tools is that the ability to determine correctness of the output from the tool is established. If it can be shown that all possible variations of the output from some given input

TABLE 13.10

Summary of DO-330 FAQs (in Appendices C and D)

FAQ #	FAQ Title	Description of FAQ
C.1	What does "protection" mean for tools and what are some means to achieve it?	Explains how the word *protection* is used throughout DO-330. The word *protection* is used instead of *partitioning* to separate tool functions when differing levels of qualification are applied to a multifunction tool or a collection of tools.
C.2	What are external components and how does one assess their correctness?	Explains the concept of external components, which was discussed earlier in this chapter (Difference #13 in Section 13.3.2).
C.3	How can one maximize reusability of tool qualification data?	Explains how to package the tool qualification data in order to support reusability.
D.1	Why are the terms "verification tool" and "development tool" not used to describe tools that may be qualified?	Explains how the terminology in DO-330 differs from DO-178B, as was discussed earlier in this chapter (Section 13.2).
D.2	Can TQL be reduced?	Explains that it might be feasible to reduce a tool's TQL, if something like architectural mitigation, monitoring, and/or independent evaluations are performed. Such an approach would need to be thoroughly justified and closely coordinated with the certification authority.
D.3	When do target computer emulators or simulators need to be qualified?	As the title suggests, this FAQ explains when emulators and simulators need to be qualified and offers some suggestions for what to consider during the qualification.
D.4	What credit can be granted for tools previously qualified using DO-178B/DO-278?	Provides some evaluation criteria to consider when using tools qualified under DO-178B/DO-278.
D.5	What is the rationale for tool qualification criteria definition?	Explains the rationale for Criteria 1–3 in DO-178C. A summary of this information was provided earlier in this chapter (Section 13.2).
D.6	What "verification tool" qualification improvements were made in DO-178C/DO-278A?	Explains the variances from DO-330 TQL-5 and DO-178B verification tool. Most of the differences are related to DO-330 Table T-0 which contains the majority of the objectives for TQL-5.
D.7	How might one use a qualified tool to verify the outputs of an unqualified tool?	Explains how to verify the output of an unqualified tool using a lower level qualified tool (e.g., TQL-5). This FAQ explains how a tool, rather than a human, might be used to verify the output on an unqualified tool. As with the other FAQs, this FAQ is based on experience from actual projects.

(continued)

TABLE 13.10 (continued)

Summary of DO-330 FAQs (in Appendices C and D)

FAQ #	FAQ Title	Description of FAQ
D.8	How might one use a qualified autocode generator?	This is the lengthiest FAQ. It explains how an autocode generator may be qualified to automate, replace, or eliminate source code verification, executable object code verification, and/or verification of output of software testing. This paper was somewhat controversial, so when opting to use the concepts, please review the paper very carefully and coordinate closely with the certification authority.
D.9	Is qualification of a model simulator needed?	This FAQ is closely related to the model-based development approach and considers when qualification may be needed for model simulation.

Source: RTCA DO-330, *Software Tool Qualification Considerations*, RTCA, Inc., Washington, DC, December 2011.

are correct under any appropriate verification of that output, then the tool should be considered deterministic for the purposes of tool qualification. This results in a bounded problem. This interpretation of determinism should apply to all tools whose output may vary beyond the control of the user, but where that variation does not adversely affect the intended use (e.g., the functionality) of the output and the case for the correctness of the output is presented. However, this interpretation of determinism does not apply to tools that have an effect on the final executable image embedded into the airborne system. The generation of the final executable image should meet the restrictive interpretation of determinism [6].*

DO-330 has the same intent as DO-178B and Order 8110.49; however, the words *deterministic* and *determinism* are avoided because they carry specific meaning in software engineering that may be overly restrictive for tools which do not fly. Instead, the following statement was made to explain the expectations:

> During the qualification effort, the output of all qualified tool functions should be shown to be correct using the objectives of this document [DO-330]. For a tool whose output may vary within expectations, it should be shown that the variation does not adversely affect the intended use of the output and that the correctness of the output can be established. However, for a tool whose output is part of the software and thus could insert an error, it should be demonstrated that qualified tool functions produce the same output for the same input data when operating in the same environment [1].*

---

* Brackets added for clarification.

13.4.3 Additional Tool Qualification Considerations

DO-330 section 11 includes guidance for several *additional considerations* that may be encountered when developing or using qualified tools. The additional considerations are briefly summarized here.

Additional Consideration 1: Multifunction tools: The name is self-descriptive: a multifunction tool is a single tool that performs multiple functions. The multiple functions may be within a single executable file, multiple executable files (which allows disabling of certain functions), or some other arrangement that allows selection or disabling of some tool functionality [1]. The DO-330 section 11.1 guidance explains that multifunction tool(s) should be explained in the plans and that a TQL needs to be established for each function. If there is a mixture of TQLs, the functions either need to be developed to the highest TQL or separated to ensure that lower TQL functions don't impact higher TQL functions (i.e., protection). If functions will not be used, they need to be disabled and have assurance that the disabling mechanism is sufficient to protect the enabled functions from the disabled functions. If unused functions cannot be disabled, they need to be developed to the appropriate TQL (usually the highest TQL of the tool). For tools that have a role in development *and* verification processes, the independence of tool development functions and verification functions must be considered (e.g., they might be developed by independent teams).

Additional Consideration 2: Previously qualified tools: DO-330 section 11.2 provides guidance for reusing tools that were previously qualified. It considers reuse in the following scenarios [1]:

- The tool and its operational environment are unchanged.
- The tool is unchanged but will be installed in a different operational environment.
- The tool is changed but will be installed in the same operational environment.
- The tool is changed and will be installed in a different operational environment.

The potential for reuse is significantly improved when a tool is designed and packaged to be reusable. DO-330 FAQ C.3 provides some suggestions for how to develop and package a tool for maximum reuse. Reusability does usually require more planning, robustness, and testing, but it can save significant resources down the road.

Additional Consideration 3: Qualifying COTS tools: DO-330 section 11.3 provides guidance to successfully qualify a COTS tool. COTS tools at TQL-5 are relatively easy to qualify because no insight into the tool development is needed. However, TQL-1–TQL-4 tools require insight into the tool development itself. Essentially, COTS tools still need to comply with the DO-330 objectives.

DO-330 section 11.3 explains what is typically expected of the tool developer and the tool user when qualifying a COTS tool. It is important to remember that tools are qualified in conjunction with the DO-178C software approval; that is, tool qualification is not a stand-alone approval. DO-330 attempts to make tool qualification as stand-alone as possible, but the actual qualification is only received in the context of a certification project. For COTS tools, DO-330 section 11.3 explains what objectives the tool developer is typically responsible for, as well as some suggestions for how to package the life cycle data so that it can be more easily rolled into the tool user's project. Suggestions include the development of a developer-TOR, developer-TQP, developer-Tool Configuration Index, and developer-Tool Accomplishment Summary, which the tool user can evaluate and either use directly or reference from their own tool qualification data. The concept is that the tool developer anticipates the user's needs and proactively develops the tool qualification data to meet those needs.

Additional Consideration 4: Tool service history: DO-330 section 11.4 explains the process if one decides to use service history for a tool. Service history may be feasible for a tool with considerable in-service experience but that has some missing life cycle data. Service history may also be used to raise the TQL of a tool, supplement the tool qualification data, or increase confidence in TOR compliance. As with service history for airborne software (see Chapter 24), making a service history case is quite challenging and will likely be met with skepticism from the certification authority.

Additional Consideration 5: Alternative methods for tool qualification: DO-330 section 11.5 explains that an applicant may propose alternate methods to those described in DO-330. Any alternative must be explained in the plans, thoroughly justified, assessed for impact on the resulting software's life cycle processes and life cycle data, and coordinated with the certification authority.

13.4.4 Tool Qualification Pitfalls

Having assessed dozens of tools over the years, I've observed some common pitfalls. Each project differs and some of these challenges are more prevalent than others. However, they are provided here for your awareness. To be forewarned is to be forearmed.

Pitfall 1: Missing user instructions. Tools are designed to be used, typically by a team that did not develop the tool. However, user instructions are often missing or are inadequate. This makes it uncertain if the tool will be used as intended.

Pitfall 2: Tool version not specified. Sometimes, there are multiple versions of a tool. It must be clearly identified what version of the tool is being used. This is typically documented in the SLECI or Software Configuration Index. It should also be confirmed that the team is using the identified version.

Pitfall 3: Configuration data not included in tool qualification package. Many tools require some configuration data. For example, a tool that verifies conformance to Motor Industry Software Reliability Association (MISRA) C standard may only be activated to check a subset of the rules. The specific rules are enabled or disabled in a configuration file. The configuration file needs to be under configuration management and identified in the qualification data.

Pitfall 4: Need for tool qualification not accurately assessed. Occasionally, when reviewing a team's data, I come across a tool that was not listed in the PSAC because the team didn't think it needed to be qualified; however, it turns out that qualification is needed. Late discoveries of this sort lead to unplanned work. As explained in Chapter 5, to avoid this pitfall it is recommended that all tools be listed in the PSAC, along with a justification for why they do or do not need to be qualified. This allows the project and the certification authority to reach agreement early on, rather than late in the project.

Pitfall 5: TORs or tool requirements not verifiable. Some of the worst requirements I've encountered were tool requirements. Since TORs must be verified for all TQLs and tool requirements are verified for TQL-1–TQL-4, the requirements must be verifiable. The TORs and tool requirements should have the same quality attributes discussed in Chapters 2 and 6.

Pitfall 6: Incomplete COTS tool qualification data. Several COTS tool vendors offer tool qualification packages. Some of them are great, and some of them are not so great. Caution should be used when investing in a tool qualification package—consider getting some kind of guarantee along with the tool data.

Pitfall 7: Overreliance on tools without adequate understanding of tool functionality. Some software teams, particularly inexperienced teams, rely on the tools without really understanding why they are used, what they do, or how they fit into the overall software life cycle. Tools can be terrific additions to the engineering process. However, they must be understood. As one of my friends often says, "A fool with a tool is still a fool." Users need to understand what the tool does for them.

Pitfall 8: PSAC and SAS do not explain the tool qualification approach. Oftentimes, the PSAC states that a tool needs to be qualified, but does not go beyond that. For TQL-1–TQL-4 tools, the PSAC should summarize the tool's use and reference its TQP. For TQL-5 tools, the qualification approach should be explained in the PSAC or in a TQP. Likewise, the SAS frequently provides inadequate details about the tool and its qualification data. See Section 13.3.2, Difference #5, of this chapter for a summary of what tool information should go in the PSAC and SAS.

Pitfall 9: Team spends more time on the tool development than on the software that uses it. Many engineers enjoy developing tools; sometimes they get carried away and focus so much on the tool that they lose sight of the purpose for the tool.

Pitfall 10: The role of the tool(s) in the overall life cycle not explained in the plans. Oftentimes, the description of the development, verification, configuration management, and quality assurance processes fail to mention the tools that are used in those processes. The plan lists the tools in a tool section but does not explain how the tools are used in the life cycle phases (i.e., when the processes themselves are discussed). For example, a requirements management tool may be explained in a tool section but not mentioned in the requirements capture and verification sections of the plans. This makes it difficult to ensure that the tools are being properly utilized in the life cycle.

Pitfall 11: Qualification credit not clearly identified. When tools replace, automate, or eliminate DO-178C (and/or the supplements) objectives, it should be clear what objectives they impact. This should be specified in the PSAC, as well as the software development plan, software verification plan, and/or software configuration management plan where the tool's use is explained. Additionally, for TQL-1–TQL-4 tools the TQP may explain this.

Pitfall 12: Tool operational environment not clearly identified. As mentioned earlier (Section 13.3.2, Differences #11 and #14), it is important to ensure that the operational environment assumed during the qualification is representative of the environment used in operation. This is commonly missed in the SLECI and plans.

Pitfall 13: Tool code not archived. Since tools are part of the software development and verification environment, they should be archived along with other software life cycle data. For some tools that require special hardware, the hardware may also need to be archived.

13.4.5 DO-330 and DO-178C Supplements

The model-based development (DO-331), object-oriented technology (DO-332), and formal methods (DO-333) supplements apply DO-330 the same as DO-178C does. Basically, tool qualification is the same regardless of the technology that utilizes the tool. However, if a tool is implemented using model-based development, object-oriented technology, or formal methods, the technology supplements would likely need to be applied to the tool development and verification.

13.4.6 Using DO-330 for Other Domains

DO-330 was developed to be usable by multiple domains. If another domain (such as aeronautical databases or programmable hardware) chooses to use DO-330, they will need to explain how to determine the TQL for their domain, explain how to adapt software terminology in DO-330 for their domain, and clarify any domain-specific needs. DO-330 Appendix B provides an example of how the CNS/ATM domain implemented the DO-330 tool qualification document. Other domains could use a similar approach, or they could create their own approach.

References

1. RTCA DO-330, *Software Tool Qualification Considerations* (Washington, DC: RTCA, Inc., December 2011).
2. RTCA DO-178C, *Software Considerations in Airborne Systems and Equipment Certification* (Washington, DC: RTCA, Inc., December 2011).
3. RTCA/DO-254, *Design Assurance Guidance for Airborne Electronic Hardware* (Washington, DC: RTCA, Inc., April 2000).
4. RTCA/DO-200A, *Standards for Processing Aeronautical Data* (Washington, DC: RTCA, Inc., September 1998).
5. RTCA/DO-178B, *Software Considerations in Airborne Systems and Equipment Certification* (Washington, DC: RTCA, Inc., December 1992).
6. Federal Aviation Administration, *Software Approval Guidelines*, Order 8110.49 (Change 1, September 2011).

14

DO-331 and Model-Based Development and Verification

Acronyms

EASA	European Aviation Safety Agency
EUROCAE	European Organization for Civil Aviation Equipment
FAA	Federal Aviation Administration
HLR	high-level requirement
LLR	low-level requirement
SC-205	Special Committee #205
SCADE	standard for the development of critical embedded display software
WG-71	Working Group #71

14.1 Introduction

The DO-331 glossary defines a *model* as follows:

> An abstract representation of a given set of aspects of a system that is used for analysis, verification, simulation, code generation, or any combination thereof. A model should be unambiguous, regardless of its level of abstraction.
>
> Note 1: If the representation is a diagram that is ambiguous in its interpretation, this is not considered to be a model.
>
> Note 2: The 'given set of aspects of a system' may contain all aspects of the system or only a subset [1].

Models may enter the requirements hierarchy as system requirements, software requirements, and/or software design. Systems and software engineers have used models for years to graphically represent controls. However, the use of qualified modeling tools that automatically generate code, and

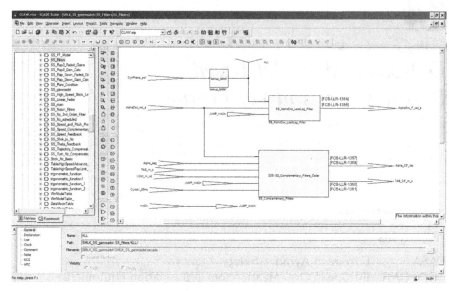

FIGURE 14.1
Example of model in Simulink®. (Courtesy of Embraer, São Paulo, Brazil.)

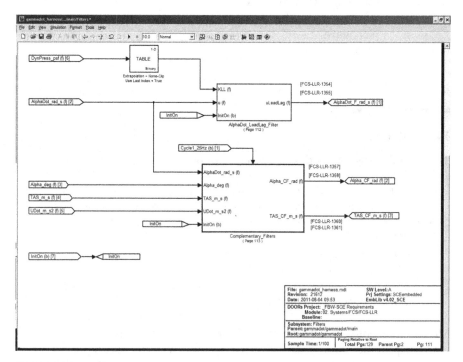

FIGURE 14.2
Example of model in SCADE. (Courtesy of Embraer, São Paulo, Brazil.)

in some cases even automatically generate test vectors, is a relatively recent paradigm shift for the aviation software community. The Airbus A380 used model-based development extensively for the following systems: flight controls, autopilot, flight warning, cockpit display, fuel management, landing gear, braking, steering, anti-ice, and electrical load management [2]. Other aircraft manufacturers are also modeling requirements for control systems. Figures 14.1 and 14.2 provide an example of a simple model. The same model is shown in Simulink® (Figure 14.1) and SCADE (standard for the development of critical embedded display software) (Figure 14.2).

With such a shift there is potential for incredible benefits, but the benefits are not without some risks and growing pains. This chapter examines both the advantages and the disadvantages of model-based development from a certification and safety perspective. Additionally, a brief overview of DO-331, *Model-Based Development and Verification Supplement to DO-178C and DO-278A*, is provided.

14.2 Potential Benefits of Model-Based Development and Verification

There are a number of potential benefits to model-based development and verification. The ability to reap these benefits depends on the implementation details and may not all be realized.

Potential Benefit 1: V life cycle to Y life cycle. One of the main motivators of model-based development is the ability to go from the traditional *V life cycle* to a *Y life cycle*, with the potential to reduce development time, cost, and possibly even human error. By putting focus on the system requirements and utilizing automation, the life cycle may shorten. Some estimates show a 20% reduction when using an unqualified code generator and up to 50% when using a qualified code generator [3]. Figures 14.3 and 14.4 illustrate the progression from V to Y life cycle.

There are a number of ways to introduce the models into the product life cycle. They may be generated at the system level, the software requirements level, and/or the software design level. Most companies attempt to have a platform independent level model that abstracts the model from the platform, making it more portable. The platform-specific details may be represented in a lower level model.

Potential Benefit 2: More focus on requirements. As Chapter 6 noted, requirements errors are often the root cause of software-related in-service problems. Additionally, the later a requirement error is found, the more expensive it is to fix. When a model represents the requirements, it has the potential to present a clearer view of the system or software functionality during requirements

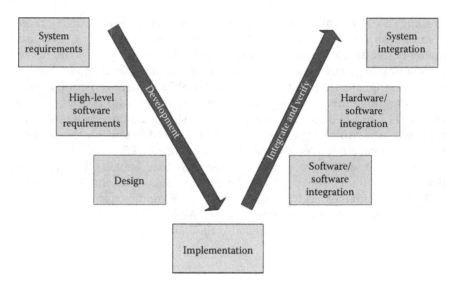

FIGURE 14.3
Traditional V life cycle.

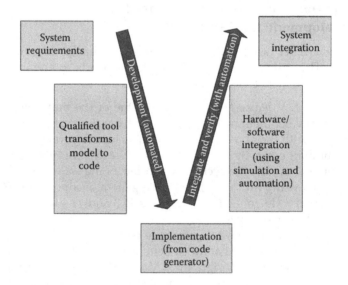

FIGURE 14.4
Y Life cycle using model-based development and verification.

capture than textual requirements. When automation is used to implement the model (such as qualified autocode generators), it allows development teams to concentrate on the graphical representation of the requirements, rather than focusing on implementation details. If properly executed, this can promote higher quality requirements and earlier detection of requirements errors.

Potential Benefit 3: Earlier verification. When qualified model-to-code tools are used, the focus shifts from the code to the model (i.e., less time is spent on code implementation details and more on the model itself). Use of model simulation and/or automatic test case generators promotes early verification, enabling earlier maturity of requirements and detection of errors. Figures 14.3 and 14.4 illustrate the differences between traditional and model-based development. Since verification often consumes about 50%–70% of the software resources, detecting errors earlier and reducing the manual verification effort can be a benefit to the project budget and schedule [4].

Potential Benefit 4: Reduction of unnecessary redundancy and inconsistencies. Traditional software development uses three levels of requirements: (1) system requirements allocated to software, (2) software high-level requirements (HLRs), and (3) software low-level requirements (LLRs). Depending on the requirements capture approach, there can be significant redundancy between these levels, which may lead to inconsistencies when changes are made. The use of qualified modeling tools has the potential to reduce these redundancies and inconsistencies. However, when there are multiple levels of models, there can also be a fair amount of redundancy and potential inconsistencies. Hence, like the other benefits noted, it really depends on the approach taken.

Potential Benefit 5: Improved requirements understandability. Some tools produce models that are easier to understand than textual requirements. Therefore, they may promote greater accuracy, consistency, and completeness. Models can be more intuitive than textual requirements, allowing easier detection of errors and gaps in planned behavior [2]. However, not all modeling tools generate easy-to-understand models and some of the models generated are quite confusing. Modeling tools should be selected with care. A good modeling standard is also essential to ensure that all developers are using the model notations, symbols, and tools properly.

Potential Benefit 6: Improved interaction with customers. As discussed in Chapter 6, customer input is crucial to capturing the right requirements. Due to the more intuitive nature of models, they can be useful to gain input from customers and systems engineers. This can increase early coordination between customers, systems, and software teams, which improves the overall quality and accuracy of the requirements.

Potential Benefit 7: Growing tool support. The capability and quality of model-based development and verification tools is growing. There are still not many qualified code generators available, but there are some. There are also tools available to verify the output of the code generators. When confidence in one tool cannot fully be realized, it might be possible to use two different tool paths to boost the confidence. For some of the more critical systems (e.g., flight controls), this might be a recommended approach even if you do have confidence in the tools, as it helps to reduce the concern of a common design error.

Potential Benefit 8: Evolving formal methods support. Interestingly, there appears to be a convergence of modeling technology, tool support, and formal methods which can significantly increase the benefits of model-based development and verification. When a modeling tool suite, including the code generator, is formally verified for correctness, the tool can be relied upon to perform its intended functionality. Without formal verification, there is always the uneasy feeling that the modeling tool or underlying code generator might insert an error given some corner-case scenario. However, formal methods can provide confidence in the model-to-code tool suite by verifying the entire input and output space. See Chapter 16 for more information on formal methods.

14.3 Potential Risks of Model-Based Development and Verification

As with any technology, there are a number of potential challenges when using model-based development and verification. Several of them are noted here.

Potential Risk 1: Loss of multiple requirements reviews. When models are implemented using automatic code generation, some of the traditional scrutiny of the requirements may be lost, particularly if the model is introduced at the systems level. In the traditional software development, there are reviews of the systems requirements, software requirements, design, and code. Additionally, as each level of requirements is analyzed to generate the lower level, issues are identified and resolved. For example, someone implementing the code might find an inconsistency with the requirements. Mats Heimdahl refers to this as *collateral validation*. He notes: "Experienced professionals designing, developing code, or defining test-cases provide informal validation of the software system" [4]. As these engineers examine the data, they notice errors and can ensure appropriate corrective action is taken. However, when the model is pushed to a higher level and implementation is handled by tools, this *collateral validation* is reduced or lost altogether [4].

Potential Risk 2: Expansion of systems engineering role. In the aviation world, at least at this point in history, software engineers have been more committed to process than systems engineers. DO-178B, and now DO-178C, has enforced this. Systems engineers, however, are really just now beginning to implement development assurance practices. Therefore, if the model is controlled by systems engineering, there is concern about the quality of the model to meet the DO-178C and DO-330 expectations. For example, it may not be common for systems engineers to ensure compliance to standards, perform documented reviews of their artifacts, or test all requirements. Systems engineers may also be unfamiliar with the concepts of requirements coverage, robustness testing, model coverage analysis, etc. To overcome this issue, it is advisable to supplement the systems engineering team with experienced software engineers. Both systems and software engineers benefit from such an arrangement.

Potential Risk 3: Traceability difficulties. Even when requirements are graphically presented as models, they still need to trace up to the higher level requirements and down to the lower level requirements or code. The traces to the higher level and to the lower level must be bidirectional. Some modeling tools do not support traceability well. The trace between textual requirements and model-based requirements can be particularly challenging.

Potential Risk 4: Difficulties with test completeness. Depending on the modeling approach and the test team's experience, it may be difficult to ensure that the requirements presented in the form of a model are fully verified. Complex models can be particularly challenging to completely verify. Detailed testing guidelines are needed to ensure that the requirements-based tests fully verify the model. Additionally, model complexity also needs to be controlled in order to ensure the testability of the models.

Potential Risk 5: Simulation credit controversy. One of the drivers for model-based development and verification is the desire to utilize simulation tools to identify errors as early as possible. However, when companies try to take formal verification credit for the simulation (which is not executed on the target computer), it can become controversial. DO-331 attempts to clarify what is expected and allowed for simulation, but prior to DO-331, it has been a case-by-case debate with the certification authorities. Even with DO-331, the credit requested for simulation will require close coordination with the certification authorities.*

Potential Risk 6: Models often mix what and how. Because models frequently include details about the implementation, it can become difficult to discern *what* the model does. Basically, the design and requirements get mixed. As noted in Chapter 6, this is often a problem in traditional development as well; however, models particularly lend themselves to this dilemma. To prevent this situation, DO-331 requires a level of requirements above the model. Hopefully, such guidance will encourage developers to identify what the model is supposed to do. Unfortunately, I've seen some rather *weak* requirements above the model. For example, I once saw a 98-page model that had one requirement above it. The requirement essentially said: "You shall have a model." It's rather difficult to test that!

Potential Risk 7: Separating specification models from design models challenges. As will be discussed shortly, DO-331 provides guidance for *specification models* and *design models*. The guidance states that the models cannot be combined.

---

* As experience with DO-330 is gained, this is an area where additional guidance will likely be developed by the certification authorities.

In reality, it may prove challenging to distinguish between specification and design models, particularly for already existing models. Most existing models combine *what* (requirements) and *how* (design) to some degree. Additionally, most modeling techniques encourage the inclusion of design detail (e.g., equations and algorithms). Separating specification models and design models will require a significant paradigm shift for most organizations.

Potential Risk 8: Model validation challenges. Specification models, particularly at the system level, need to be validated for correctness and completeness. Without this, the implementation of the model is of little value. Oftentimes, processes for model validation are limited. DO-331 attempts to address this by requiring model coverage analysis (which is discussed later in this chapter) in addition to traceability to the requirements above the model.

Potential Risk 9: Blurring the systems and software roles. Depending on where the model is introduced, the systems team may be more involved in lower level details than they are accustomed, and the software team may be more involved in higher level details than is traditional. If handled correctly, the closer connection between systems engineers and software engineers can be beneficial. However, without proper consideration there may be unnecessary overlap or large gaps in the process.

Potential Risk 10: Challenges combining traditional and model-based development. Most model-based development projects have some traditional (nonmodel) requirements and design in addition to the models. Likewise, there is typically some manual code required, even when a code generator is used. The combination of traditional and model-based development must be carefully defined to ensure consistency and completeness. Likewise, both DO-178C and DO-331 will need to be used (DO-178C for traditional and DO-331 for models); this requires some extra thought, coordination, and planning. The plans should clearly define the development, verification, and integration of both traditional and model-based techniques.

Potential Risk 11: Inconsistent model interpretation. If the modeling notation is not well defined and carefully implemented, it can lead to inconsistent interpretations of the model. Thorough modeling standards and adherence to them should address this issue.

Potential Risk 12: Model maintenance. Without adequate documentation, the models may not be maintainable. Not too long ago, I was involved in a flight control project. The engineer who created the model left the company. Other key engineers also left the company. The team left to finish the certification effort had no idea what the model did or why it did it. All that existed was the model itself; there were no higher level requirements, assumptions, justification, or rationale. It was unclear why certain decisions were made and why some model elements were included. It was a nightmare to validate the correctness and completeness of the model and to verify its implementation.

The moral of the story is: maintainable models require thorough documentation. Just as it's important to put comments in the code (see Chapter 8), it is crucial to explain and justify modeling decisions. Section 14.4 (Difference #3) explains this concept further.

Potential Risk 13: Automatic test generation controversy. Some modeling tools not only automatically generate code, but also automatically generate tests against the model. This is beneficial to informally verify and build confidence in the model. However, when a project attempts to use the automatically generated tests for certification credit, it causes several concerns. If the model is wrong, the error may not be caught. In traditional software testing, the human test authors often find issues with the requirements and weaknesses in the system (see Chapter 9). Machine generated tests may not be as effective. Model coverage analysis may help alleviate some of this concern, if implemented properly. However, in addition to model coverage analysis, manually generated tests to consider integration, safety requirements, key functionality, challenging functions, and robustness may be needed to supplement the automatically generated tests. When using automatically generated tests one must also ensure adequate independence; for higher software levels, the test generator and the code generator must be independently developed.

Potential Risk 14: Tool instability. Because of the reliance of model-based development and verification on tools, any instability in the tools can be problematic. Immature tools may change frequently, which can lead to the need to regenerate and/or re-verify models. I recently heard of an issue found in a modeling tool that has been used by multiple aircraft manufacturers and their suppliers. Each tool user had to evaluate the impact of the newly discovered issue on their product. The ripple effect of a tool problem can be quite significant in the safety-critical domain. It is recommended that if you are using qualified tools to support model-based development and verification, you ensure that they are mature and rigorous enough for the task at hand. An independent assessment of the tool and its supporting data may be beneficial to ferret out its suitability and maturity.

Potential Risk 15: Modeling limitations. Not every system is suitable to modeling. Model-based development and verification will not be the right approach for every project or organization. Despite some claims, it is not a silver bullet.

14.4 Overview of DO-331

DO-331, entitled *Model-Based Development and Verification Supplement to DO-178C and DO-278A*, was the most challenging supplement developed by RTCA Special Committee #205 (SC-205) and European Organization for Civil Aviation Equipment (EUROCAE) Working Group #71 (WG-71). It was also the last document approved by the committee and may be the least mature. There are a variety of approaches to modeling and opinions on how to best implement models in a safety-critical domain. The DO-331 guidance attempts to provide flexibility for model implementation, while at the same time ensuring that the software generated using the model performs its intended function and only its intended function. If used properly, the supplement helps to address several of the concerns mentioned earlier in this chapter.

Like the other supplements, DO-331 uses the DO-178C outline and modifies, replaces, or adds objectives, activities, and guidance to DO-178C, as needed.

The DO-331 definition of model is included earlier in Section 14.1. DO-331 recognizes that a model may come into the software life cycle as system requirements, software requirements, and/or software design and that there may be multiple levels of model abstraction. Regardless of the level that a model enters the software life cycle, there must be requirements above and external to the model to explain details and constraints to enable the model-based development and verification activities.

DO-331 section MB.1.6.2 defines two types of models: (1) specification model and (2) design model. The description of each is provided in the following [1]:*

> A *Specification Model* represents high-level requirements that provide an abstract representation of functional, performance, interface, or safety characteristics of software components. The Specification Model should express these characteristics unambiguously to support an understanding of the software functionality. It should only contain detail that contributes to this understanding and does not prescribe a specific software implementation or architecture except for exceptional cases of justified design constraints. Specification Models do not define software design details such as internal data structures, internal data flow, or internal control flow. Therefore, a Specification Model may express high-level requirements but neither low-level requirements nor software architecture.
>
> A *Design Model* prescribes software component internal data structures, data flow, and/or control flow. A Design Model includes low-level requirements and/or architecture. In particular, when a model expresses software design data, regardless of other content, it should be classified as a Design Model. This includes models used to produce code.

There are two important things to note about these model types. First, a model cannot be classified as both specification model and design model. Second, since DO-330 requires that there must be requirements above the model, there will always be at least two levels of requirements when using models [1].

DO-331 retains most of the guidance of DO-178C, but adds some model-specific information. In general, most of the guidance in DO-178C related to

---

* Italics added for clarification.

HLRs applies to specification models and the guidance for software design applies to design models. The most significant differences between DO-330 and DO-178C are summarized in the following. These are the areas where DO-331 modifies or clarifies (i.e., supplements) DO-178C for application to model-based development and verification.

Difference 1: Model planning. During the planning phase, the plans should explain the use of modeling and how it fits into the software life cycle. Plans should explain what software life cycle data are represented by each model, what model standards will be used, and the intended verification approach. If simulation will be used for credit, the plans should explicitly detail the approach and what credit is sought. The planning phase should also define the model simulation environment, including methods, tools, procedures, and operating environment if simulation is used to formally support verification. Depending how the simulator is used, it may need to be qualified; the rationale for why qualification is or is not needed should be included in the plans. Chapter 13 explains the tool qualification criteria and approach.

If there are both *traditional* textual requirements and design elements used in addition to the models, it should be explained in the plans. In this situation, the plans need to explain how both the DO-178C and DO-331 objectives will be satisfied. There are various ways to accomplish this. One way is to include both DO-178C and DO-331 objectives in the PSAC (perhaps as an appendix) to show how both sets of objectives will be satisfied.

Difference 2: Model standards. Each type of model used must have standards to explain the modeling techniques, constraints, instructions, etc. DO-331 section MB.11.23 explains that there should be a model standard for each type of model and each standard includes the following kinds of information, as a minimum [1]:

- Explanation and justification for the methods and tools to be used.
- Identification of the modeling language to be utilized and any language syntax, semantics, features, and limitations.
- Style guidelines and complexity restrictions that allow the modeling approach and its implementation to be unambiguous, deterministic, and compliant with DO-331 objectives. Complexity restrictions are particularly important. Typically, the depth of the model (nesting levels), number of architectural layers, and number of elements per diagram need to be restricted.
- Constraints to ensure proper use of the modeling tools and supporting libraries.
- Methods to identify requirements, establish bidirectional traceability between requirements layers, identify derived requirements, document justification for all derived requirements, and identify any model element that is not a software requirement or design (such as comments).

The model standards are the guide for the developers to enable them to produce high quality and compliant models. Clear guidelines and realistic examples are beneficial.

Difference 3: Supporting model elements. DO-331 explains that model elements that do not contribute to or realize the requirements or design implementation must be clearly identified. DO-331 adds three objectives to address this. All three objectives are in DO-331 Annex MB.A Table MB.A-2 and are listed in the following [1]:

- *Objective MB8*: "Specification Model elements that do not contribute to implementation or realization of any high-level requirement are identified."
- *Objective MB9*: "Design Model elements that do not contribute to implementation or realization of any software architecture are identified."
- *Objective MB10*: "Design Model elements that do not contribute to implementation or realization of any low-level requirement are identified."

Difference 4: Model element libraries. Libraries are used extensively in most modeling tools. For example, the symbols used to graphically illustrate a model come from a symbol library. Each library element that can be used in a model must be assured to the proper software level per DO-178C. Basically, the library elements need plans, development standards, requirements, design, code, verification cases and procedures, etc., just like any other safety-critical software. If there are elements in the library that do not have the appropriate level of assurance, they should not be used. It is preferable to remove them from the library altogether to avoid inadvertent use. However, if that is not feasible, there should be explicit standards to prohibit unassured element use and reviews to ensure the standards are followed. Additionally, the modeling standards need to provide guidelines to properly use the library elements.

Difference 5: Model coverage analysis for design models. The DO-331 glossary defines model coverage analysis as:

> An analysis that determines which requirements expressed by the Design Model were not exercised by verification based on the requirements from which the Design Model was developed. The purpose of this analysis is to support the detection of unintended function in the Design Model, where coverage of the requirements from which the model was developed has been achieved by the verification cases [1].

Model coverage analysis is not the same as structural coverage analysis. DO-331 section MB.6.7 explains the recommended model coverage analysis activities and criteria, as well as resolution activities if any coverage issues are noted. Interestingly, DO-331 section MB.6.7 is referenced as an activity and not an objective in the DO-331 Annex MB.A Table MB.A-4 (i.e., the *MB.6.7* reference is in the *Activity* column—not the *Objective* column). Objective 4 states: "Test coverage of low-level requirements is achieved." Model coverage analysis is only needed for design models. Its inclusion in DO-331 was somewhat controversial, which is why it was included as an activity and not an objective. Even though it is only listed as an activity in DO-331 Table MB.A-4 and not an objective, most certification authorities will expect it to be performed. They will likely allow alternate approaches to be used but will still need evidence that the design model has been fully verified to the same level of rigor that model coverage analysis provides.

Difference 6: Model simulation. The DO-331 glossary defines *model simulation* and *model simulator* as follows [1]:

> *Model simulation*: The activity of exercising the behavior of a model using a model simulator.
>
> *Model simulator*: A device, computer program or system that enables the execution of a model to demonstrate its behavior in support of verification and/or validation.
>
> *Note*: The model simulator may or may not be executing code that is representative of the target code.

DO-331 section MB.6.8 provides specific guidance on model simulation. Model simulation may be used to satisfy some of the DO-331 verification objectives. Table 14.1 summarizes DO-331 objectives that may or may not be satisfied using model simulation. As a model may represent HLRs, LLRs, or software architecture, the objectives in Table 14.1 apply to the applicable representation of the model.

If simulation cases and procedures are used for formal verification credit, the simulation cases and procedures need to be verified for correctness and the simulation results need to be reviewed and any results discrepancies explained. DO-331 adds new objectives to Annex MB.A Tables MB.A-3 (objectives MB8–MB10), MB.A-4 (objectives MB14–MB16), and MB.A-7 (objectives MB10–MB12) to address the verification of the simulations cases, procedures, and results. The three objectives are the same but are repeated in the three Annex tables since they apply to different phases of the software life cycle [1]:

- "Simulation cases are correct." (Table MB.A-3 objective MB8, Table MB.A-4 objective MB14, and Table MB.A-7 objective MB10)

- "Simulation procedures are correct." (Table MB.A-3 objective MB9, Table MB.A-4 objective MB15, and Table MB.A-7 objective MB11)

- "Simulation results are correct and discrepancies explained." (Table MB.A-3 objective MB10, Table MB.A-4 objective MB16, and Table MB.A-7 objective MB12)

TABLE 14.1

Summary of Objectives Related to Model Simulation

DO-331 Objective	Reference	Verification Activity	Model Simulation Allowed?
Specification Model Verification			
Table MB.A-3, Obj 1	DO-331: MB.6.3.1.a	Compliance to system requirements for HLRs	Yes
Table MB.A-3, Obj 2	DO-331: MB.6.3.1.b	Accuracy and consistency of HLRs	Yes
Table MB.A-3, Obj 3	DO-331: MB.6.3.1.c	Compatibility of HLRs with the target	No
Table MB.A-3, Obj 4	DO-331: MB.6.3.1.d	Verifiability of HLRs	Yes
Table MB.A-3, Obj 5	DO-331: MB.6.3.1.e	Conformance of HLRs to standards	No
Table MB.A-3, Obj 6	DO-331: MB.6.3.1.f	Bidirectional traceability between HLRs and system requirements	No
Table MB.A-3, Obj 7	DO-331: MB.6.3.1.g	Algorithm aspects of HLRs	Yes
Design Model Verification			
Table MB.A-4, Obj 1	DO-331: MB.6.3.2.a	Compliance to software HLRs for LLRs	Yes
Table MB.A-4, Obj 2	DO-331: MB.6.3.2.b	Accuracy and consistency of LLRs	Yes
Table MB.A-4, Obj 3, 10	DO-331: MB.6.3.2.c, MB.6.3.3.c	Compatibility of with the target computer (LLRs and software architecture)	No
Table MB.A-4, Obj 4, 11	DO-331: MB.6.3.2.d, MB.6.3.3.d	Verifiability of LLRs and software architecture	Yes
Table MB.A-4, Obj 5, 12	DO-331: MB.6.3.2.e, MB.6.3.3.e	Conformance of LLRs to standards	No
Table MB.A-4, Obj 6	DO-331: MB.6.3.2.f	Bidirectional traceability between LLRs and HLRs	No
Table MB.A-4, Obj 7	DO-331: MB.6.3.2.g	Algorithm aspects of LLRs	Yes
Table MB.A-4, Obj 8	DO-331: MB.6.3.3.a	Compatibility to software HLRs for software architecture	Yes
Table MB.A-4, Obj 9	DO-331: MB.6.3.3.b	Consistency of software architecture	Yes
Table MB.A-4, Obj 13	DO-331: MB.6.3.3.f	Partitioning integrity is confirmed	No

TABLE 14.1 (continued)

Summary of Objectives Related to Model Simulation

DO-331 Objective	Reference	Verification Activity	Model Simulation Allowed?
		Testing and Test Coverage	
Table MB.A-6, Obj 1	DO-178C: 6.4.a	Executable object code complies with the HLRs	Partial[a]
Table MB.A-6, Obj 2	DO-178C: 6.4.b	Executable object code is robust with the HLRs	Partial[a]
Table MB.A-6, Obj 3	DO-178C: 6.4.c	Executable object code complies with the LLRs	No
Table MB.A-6, Obj 4	DO-178C: 6.4.d	Executable object code is robust with the LLRs	No
Table MB.A-6, Obj 5	DO-178C: 6.4.e	Executable object code is compatible with the target computer	No
Table MB.A-7, Obj 1	DO-178C: 6.4.5.b	Test procedures are correct	No
Table MB.A-7, Obj 2	DO-178C: 6.4.5.c	Test results are correct and discrepancies explained	No
Table MB.A-7, Obj 3	DO-178C: 6.4.4.a	Test coverage of HLRs is achieved	Partial[a]
Table MB.A-7, Obj 4	DO-178C: 6.4.4.b	Test coverage of LLRs is achieved	No
Table MB.A-7, Objs 5–7	DO-178C: 6.4.4.c	Test coverage of software structure to the appropriate coverage criteria is achieved	Partial[a]
Table MB.A-7, Obj 8	DO-178C: 6.4.4.d	Test coverage of software structure, both data coupling and control coupling is achieved	Partial[a]
Table MB.A-7, Obj 9	DO-178C: 6.4.4.c	Verification of additional code that cannot be traced to source code is achieved	No

Source: RTCA DO-331, *Model-Based Development and Verification Supplement to DO-178C and DO-278A*, RTCA, Inc., Washington, DC, December 2011.

[a] Some testing on target still required.

14.5 Certification Authorities Recognition of DO-331

Prior to the publication of DO-331, the international certification authorities have used project-specific issue papers (or equivalent) to identify model-based development and verification issues that need to be addressed.* It is anticipated that these issue papers (or equivalent) will no longer be needed when DO-331 is used as the means of compliance. As experience with DO-331 is gained, there may be some additional certification authority guidelines in the future to clarify model-based development and verification issues.

Of all the DO-178C supplements, DO-331 will likely be the most difficult to apply, especially when trying to apply the guidance to existing models. Some specific challenges are as follows:

- Developing a complete set of requirements above the model, particularly when the model is outside the software life cycle.
- Clarifying the scope of DO-331 in the systems domain.
- Separating specification model and design model.
- Performing model coverage analysis.
- Performing bidirectional tracing between the model elements and the higher and lower levels.
- Integrating the new guidance into existing processes.
- Using model simulation for certification credit.
- Automatically generating tests against the models.

References

1. RTCA DO-331, *Model-Based Development and Verification Supplement to DO-178C and DO-278A* (Washington, DC: RTCA, Inc., December 2011).
2. S. P. Miller, Proving the shalls: Requirements, proofs, and model-based development, *Federal Aviation Administration 2005 Software & Complex Electronic Hardware Conference* (Norfolk, VA, 2005).
3. B. Dion, Efficient development of airborne software using model-based development, *2004 FAA Software Tools Forum* (Daytona, FL, May 2004).
4. M. Heimdahl, Safety and software intensive systems: Challenges old and new, *Future of Software Engineering Conference* (Minneapolis, MN, 2007).

---

* For example, the European Aviation Safety Agency's (EASA) certification memo CM-SWCEH-002 ("Software Aspects of Certification") dated August 11, 2011, includes a section on model-based development (section 23). This certification memo is called out in EASA certification review items which are the EASA equivalent of Federal Aviation Administration (FAA) issue papers.

15

DO-332 and Object-Oriented Technology and Related Techniques

Acronyms

AVSI	Aerospace Vehicle Systems Institute
CAST	Certification Authorities Software Team
CRI	certification review item
DC/CC	data coupling and control coupling
EASA	European Aviation Safety Agency
FAA	Federal Aviation Administration
FAQ	frequently asked question
IEEE	Institute of Electrical and Electronics Engineers
LSP	Liskov substitution principle
NASA	National Aeronautics and Space Administration
OOT	object-oriented technology
OOTiA	object-oriented technology in aviation
OOT&RT	object-oriented technology and related techniques

15.1 Introduction to Object-Oriented Technology

Object-oriented technology (OOT) has been around since the late 1960s with the introduction of the Simula programming language [1]. OOT is a software development paradigm for analysis, design, modeling, and programming that centers around *objects*. The Institute of Electrical and Electronics Engineers (IEEE) refers to OOT as "a software development technique in which a system or component is expressed in terms of objects and connections between those objects" [2]. An object is like a *black box* at the software level; each object is capable of receiving messages, processing data, and sending messages to other objects. An object contains both code (methods) and data (structures). The user does not require insight into the internal details of the object in order to use the object—hence, the comparison to a black box. An object can model real-world entities, such as a sensor or hardware controller, as separate software components with defined behaviors. A major concept in OOT is that of a *class*. A class defines attributes, methods, relationships, and semantics that share a common structure and behavior representative of a real-world entity.

DO-332 section OO.1.6.1.1 provides the following description of objects and classes [3]:

> The feature that distinguishes OOT is the use of classes to define objects along with the ability to create new classes via subclassing. In procedural programming, a program's behavior is defined by functions, and the state of a running program is defined by the contents of the data variables. In object-oriented programming, functions and data that are closely related are tightly coupled to form a coherent abstraction known as a class...
>
> A class is a blueprint from which multiple concrete realizations can be created. These concrete realizations are known as objects...
>
> The subprograms defined for a class are referred to as its methods, or member functions. Methods that operate on or use the data contained within an object instance are referred to as instance methods. This is in contrast to class methods that are associated only with a class, and do not require an associated instance to be invoked.

Throughout DO-332, the term *subprogram* is used to refer generically to all forms of methods (instance methods and class methods), functions, or procedures. DO-332 also explains:

> Data variables associated with a class are referred to as attributes, fields, or data members. An attribute may be classified as an instance attribute, in which case a separate copy of the attribute exists for each object, or it may be a class attribute, in which case a single, shared copy of the attribute exists for all objects of the class [3].

15.2 Use of OOT in Aviation

OOT is used widely in non-safety-critical software development (e.g., in web-based and desktop applications) and is taught almost exclusively in universities. OOT has also been used in safety-critical medical and automotive systems. OOT is appealing to the aviation industry for several reasons, including strong tool support, availability of programmers, perceived cost savings, and potential for reusability.

However, OOT has not been widely used in aviation to date. The Federal Aviation Administration (FAA) and the aviation industry have been researching and investigating the use of OOT in safety-critical systems and developing guidelines for its safe implementation for over 10 years. There are several technical challenges related to OOT (mostly related to the programming languages) that have delayed its widespread use in real-time systems and in aviation. FAA issue papers and European Aviation Safety Agency (EASA) certification review items (CRIs) have been issued for projects using OOT to ensure the issues were addressed.

15.3 OOT in Aviation Handbook

In 1999, the FAA and the industry began actively exploring the use of OOT in aviation. Several projects were considering using the approach, but little research existed to evaluate its suitability for safety-critical and real-time applications. An Aerospace Vehicle Systems Institute (AVSI)* team supported by Boeing, Honeywell, Goodrich, and Rockwell Collins collaborated on a project titled "Certification Issues for Embedded Object-Oriented Software," the goal of which was to mitigate the risk that individual projects face when certifying airborne systems with object-oriented software. The AVSI project proposed a number of guidelines for producing object-oriented software that complies with DO-178B [4]. The AVSI work became input into an FAA and National Aeronautics and Space Administration (NASA) effort called object-oriented technology in aviation (OOTiA). NASA's Langley Research Center and the FAA hosted two workshops to gather input from the industry and to collaborate on the development of a four-volume document entitled *Handbook for Object-Oriented Technology in Aviation*, which was completed in October 2004 [5].

The issues identified in the *OOTiA Handbook* became the foundation for evaluating OOT in aviation projects. Applicants and software developers were requested to ensure that their OOT approach addressed the issues identified in volume 2 of the *OOTiA Handbook*. Volume 3 of the *OOTiA Handbook* suggested some ways to address the issues raised in the following 10 areas: (1) single inheritance and dynamic dispatch, (2) multiple inheritance, (3) templates, (4) inlining, (5) type conversion, (6) overloading and method resolution, (7) dead and deactivated code and reuse, (8) object-oriented tools, (9) traceability, and (10) structural coverage.

---

* AVSI is an aerospace research consortium whose goals are to reduce the costs and maintain the safety of complex subsystems in aircraft.

15.4 FAA-Sponsored Research on OOT and Structural Coverage

In the 2002–2007 timeframe, the FAA sponsored a three-phase research effort, which considered structural coverage issues that may occur when using OOT.*

The first phase surveyed the issues and resulted in a report entitled *Issues concerning the structural coverage of object-oriented software* [6].

The second phase investigated issues and proposed acceptance criteria for the confirmation of data coupling and control coupling (DC/CC) within OOT in commercial aviation. As noted in Chapter 9, the intent of the confirmation of DC/CC is to provide an objective assessment (measurement) of the completeness of the requirements-based tests for the integrated components. The second phase resulted in a report entitled *Object-oriented technology verification phase 2 report—Data coupling and control coupling* [7]. This report provided recommendations for the coverage of intercomponent dependencies as a measurable adequacy criterion to satisfy DO-178B (and now DO-178C) Table A-7 objective 8. In addition to the report, the phase 2 effort produced a handbook entitled *Object-Oriented Technology Verification Phase 2 Handbook—Data Coupling and Control Coupling* [8]. The handbook provides guidelines into issues and acceptance criteria for the confirmation of DC/CC within OOT in commercial aviation.

The third phase of the effort investigated issues and acceptance criteria for the use of structural coverage analysis at the source code versus object code or executable object code levels within OOT in order to satisfy DO-178B (and now DO-178C) Table A-7 objectives 5–8. Like the second phase, the third phase resulted in a report and a handbook:

- *Object-Oriented Technology Verification Phase 3 Report—Structural Coverage at the Source-Code and Object-Code Levels* [9]
- *Object-Oriented Technology Verification Phase 3 Handbook—Structural Coverage at the Source-Code and Object-Code Levels* [10]

15.5 DO-332 Overview

DO-332 is entitled *Object-Oriented Technology and Related Techniques Supplement to DO-178C and DO-278A*. It provides guidance on the use of OOT and techniques that are closely related to OOT. DO-332 is also referred to as the *OOT&RT Supplement*. The *OOTiA Handbook*, FAA research reports, FAA issue

---

* The investigation was led by John Joseph Chilenski of the Boeing Company. Other investigators from Boeing included John L. Kurtz, Thomas C. Timberlake, and John M. Masalskis. The reports and handbooks are available on the FAA website (www.faa.gov).

papers, EASA CRIs, and Certification Authorities Software Team (CAST) position papers were all inputs to DO-332. The supplement modifies and adds to DO-178C objectives, activities, explanatory text, and software life cycle data when OOT and the related techniques are used in the software life cycle. Since OOT terminology and the overall understanding of OOT vary among the aviation industry, the supplement provides an overview of OOT and the related techniques. The related techniques include parametric polymorphism, overloading, type conversion, exception management, dynamic memory management, and virtualization. Most, but not all, object-oriented languages employ these techniques. Additionally, some of these techniques are applicable even beyond OOT. The primary differences between DO-332 and DO-178C are summarized here.

15.5.1 Planning

DO-332 adds three activities to the DO-178C planning process. First, if virtualization is used, it should be explained in the plans. Second, if components are reused (which is one of the goals of OOT), the reuse should be described in the plans, including the "maintenance of type consistency, requirements mapping, and exception management strategy between the components and the using system" [3]. Third, the plans and standards should explain how the DO-332 Annex OO.D vulnerabilities will be addressed (these are discussed in Section 15.5.4).

15.5.2 Development

DO-332 adds some additional OOT-specific development guidance beyond what is included in DO-178C. In particular DO-332 section OO.5 adds guidance on the following topics: class hierarchy, type consistency, memory management, exception management, and deactivated functionality when applying reuse [3].

DO-332 section OO.5.5 also adds clarification on traceability for OOT. Since object-oriented design is implemented using methods,

> traceability is from requirements to the methods and attributes that implement the requirements. Classes are an artifact of the architecture for organizing the requirements. Due to subclassing, a requirement, which traces to a method implemented in a class, should also trace to the method in its subclasses when the method is overridden in a subclass. This is in addition to tracing requirements that are specific to the subclass [3].

15.5.3 Verification

DO-332 adds or modifies four activities to verify (1) consistency of class hierarchy with the requirements, (2) local type consistency, (3) consistency of memory management with the architecture and requirements, and (4) consistency of exception management with the architecture and requirements.

DO-332 also adds an activity to normal-range testing to "ensure that class constructors properly initialize the state of their objects and that the initial state is consistent with the requirements for the class" [3].

Two object-oriented technology and related techniques (OOT&RT)-specific verification objectives were added to DO-332:

- *DO-332 Table OO.A-7 objective OO10*: "Verify local type consistency" [3]. Activities for this objective are included in DO-332, section OO.6.7.

- *DO-332 Table OO.A-7 objective OO11*: "Verify the use of dynamic memory management is robust" [3]. Activities for this objective are included in DO-332 section OO.6.8.

15.5.4 Vulnerabilities

A unique aspect of DO-332 is the vulnerabilities concept. DO-332 Annex OO.D includes a list of vulnerabilities that may be present in an OOT system or a system using the related techniques. The vulnerabilities are classified under two categories: (1) key features and (2) general issues. Annex OO.D is an important part of the DO-332 document and provides valuable guidance for addressing some of the more challenging issues related to OOT&RT.

The vulnerabilities for the *key features* are identified in DO-332 section OO.D.1, along with a summary of supporting guidance and activities. Each of the following key features is discussed: inheritance, parametric polymorphism, overloading, type conversion, exception management, dynamic memory management, and virtualization.

The vulnerabilities for *general issues* of OOT&RT are discussed in DO-332 section OO.D.2. These issues are not limited to OOT (which is why the supplement includes the term *related techniques* in the title) and may be present in a non-OOT system. DO-332 section OO.D.2 describes how traceability, structural coverage, component usage, and resource analysis may be more complicated for OOT&RT and includes additional guidance to consider when using OOT&RT.

The vulnerabilities in DO-332 Annex OO.D point back to the guidance in sections OO.4–OO.12 of the supplement that provide the objectives and activities to ensure that the vulnerabilities are addressed.

15.5.5 Type Safety

One unique aspect of DO-332 is identification of *type safety* as a means to mitigate several vulnerabilities in OOT systems and to control the level and depth of testing needed to fully verify the OOT system. Type safety is concerned with behavior between classes and subclasses. DO-332 states: "For a type to be a proper subtype of some other type, it should exhibit the same behavior as each of its supertypes. In other words, any subtype of a given type should be usable wherever the given type is required" [3]. DO-332 relies on the Liskov Substitution Principle (LSP) to specify and ensure type safety. LSP defines what constitutes a proper subclass (type safe), which limits how a subclass may behave. DO-332 explains:

> The principle is, "Let q(x) be a property provable about objects x of type T. Then q(y) should be true for objects y of type S where S is a subtype of T." ... Each subprogram redefined in the subclass should meet the following requirements of the same subprogram in any of its superclasses:
>
> - Preconditions may not be strengthened,
> - Postconditions may not be weakened, and
> - Invariants may not be weakened [3].

In terms of the requirements-based verification objectives defined by DO-178C, compliance to LSP means that any subtype (or subclass) of a given type (parent class) should fulfill all the requirements of the given type (parent class). The application of formal methods or testing can be used to demonstrate LSP [3].

15.5.6 Related Techniques

As mentioned earlier, DO-332 provides specific guidance on dynamic memory management and virtualization. These topics are related to OOT but are not specific to OOT. The guidance for dynamic memory management provides a means to specify, design, and evaluate a memory management system for predictable use. The guidance on virtualization is important for interpreters and hypervisor technology.

15.5.7 Frequently Asked Questions

DO-332 includes 39 frequently asked questions (FAQs) about OOT&RT and the guidance provided in the supplement. The FAQs help to clarify the intent of the supplement and some of the more challenging technical topics related to OOT. The FAQs are divided into five categories: (1) general questions, (2) requirements considerations, (3) design considerations, (4) programming language considerations, and (5) verification considerations.

15.6 OOT Recommendations

When considering the use of OOT in safety-critical applications, the following recommendations are offered:

Recommendation 1: Review DO-332 to ensure that the technical challenges related to OOT are understood. It is advisable to do this before making the definite decision to use OOT.

Recommendation 2: If new to OOT, start with a small, less critical project (e.g., level C or D software or a tool) and work up to larger, more complex, and/ or more critical software. This provides the opportunity to work out some of the process issues, as well as technical challenges.

Recommendation 3: Follow the guidance of DO-332. The certification authorities and aviation industry have spent over 10 years investigating OOT, identifying the issues, and exploring sound approaches to addressing the issues. DO-332 is the culmination of the efforts.

Recommendation 4: Provide a mapping to the DO-332 guidance to ensure that the guidance and the identified vulnerabilities are fully addressed. As noted earlier, the vulnerabilities are a critical part of the OOT&RT supplement.

Recommendation 5: Expect some challenges. Implementing a new technology for the first time can be a minefield of unexpected challenges. Instead of *leading edge*, it is sometimes called *bleeding edge*. Hopefully, the effort invested by the certification authorities and industry over the last several years has identified the bulk of the issues, but there will undoubtedly be some project-specific surprises.

Recommendation 6: Coordinate closely with the certification authorities. Because OOT is still relatively new in the aviation industry, the certification authorities tend to oversee it closely. Hopefully, as positive experience is gained, there will be less need for such intense oversight. Until that time, the certification authorities will want to know the details of an OOT project.

Recommendation 7: Consider if the guidance applies to any non-OOT projects. As noted earlier, some of the guidance of DO-332 may apply to non-OOT projects. In particular, new technologies that use dynamic memory management or virtualization will likely need to apply the guidance of DO-332.

15.7 Conclusion

With the technical progress over the last several years, as well as the development of clear guidance for OOT in DO-332, the door for OOT in aviation has now been opened. It is likely that OOT will become more commonplace in aviation software.

References

1. Federal Aviation Administration, *Handbook for Object-Oriented Technology in Aviation (OOTiA)*, Vol. 1 (October 2004).
2. ANSI/IEEE Standard, *Glossary of Software Engineering Terminology* (1983).
3. RTCA DO-332, *Object-Oriented Technology and Related Techniques Supplement to DO-178C and DO-278A* (Washington, DC: RTCA, Inc., December 2011).
4. Aerospace Vehicle Systems Institute, *Guide to the Certification of Systems with Embedded Object-Oriented Software* (2001).
5. Federal Aviation Administration, *Handbook for Object-Oriented Technology in Aviation (OOTiA)*, Vols. 1–4 (October 2004).
6. J. J. Chilenski, T. C. Timberlake, and J. M. Masalskis, *Issues Concerning the Structural Coverage of Object-Oriented Software*, DOT/FAA/AR-02/113 (Washington, DC: Office of Aviation Research, November 2002).
7. J. J. Chilenski and J. L. Kurtz, *Object-Oriented Technology Verification Phase 2 Report—Data Coupling and Control Coupling*, DOT/FAA/AR-07/52 (Washington, DC: Office of Aviation Research, August 2007).
8. J. J. Chilenski and J.L. Kurtz, *Object-Oriented Technology Verification Phase 2 Handbook—Data Coupling and Control Coupling*, DOT/FAA/AR-07/19 (Washington, DC: Office of Aviation Research, August 2007).
9. J. J. Chilenski and J. L. Kurtz, *Object-Oriented Technology Verification Phase 3 Report—Structural Coverage at the Source-Code and Object-Code Levels*, DOT/FAA/AR-07/20 (Washington, DC: Office of Aviation Research, August 2007).
10. J. J. Chilenski and J. L. Kurtz, *Object-Oriented Technology Verification Phase 3 Handbook—Structural Coverage at the Source-Code and Object-Code Levels*, DOT/FAA/AR-07/17 (Washington, DC: Office of Aviation Research, June 2007).

Recommended Readings*

1. M. Weisfeld, *The Object-Oriented Thought Process* (Indianapolis, IN: SAMS Publishing, 2000): This book provides a simple introduction to OOT fundamentals. It is a good resource for those transitioning from the functional approach to OOT.
2. G. Booch, *Object-Oriented Analysis and Design*, 2nd edn. (Reading, MA: Addison-Wesley, 1994): This book provides a practical introduction to OOT concepts, methods, and applications.
3. B. Webster, *Pitfalls of Object-Oriented Development* (New York: M&T Books, 1995): Although somewhat dated, this book provides a sound overview of the potential problems in OOT development.

---

* These references are also identified in the *OOTiA Handbook*, volume 1. Even though several years have passed since the handbook's publication, these are still the references I turn to when I need to investigate an OOT issue.

4. E. Gamma, R. Helm, R. Johnson, and J. Vlissides, *Design Patterns: Elements of Reusable Object-Oriented Software* (Reading, MA: Addison-Wesley, 1995): Patterns are widely used by the OOT community to address analysis and design problems. This book provides a guide for effective development and use of patterns.
5. B. Meyer, *Object-Oriented Software Construction*, 2nd edn. (Upper Saddle River, NJ: Prentice Hall, 1997): This large book provides good fundamental information for OOT developers.
6. R. V. Binder, *Testing Object-Oriented Systems: Models, Patterns, and Tools* (Reading, MA: Addison-Wesley, 2000): This book addresses OOT testing, which is one of the more difficult aspects of OOT.

16

DO-333 and Formal Methods

Acronyms

CFD   computational fluid dynamics
DARPA   Defense Advanced Research Projects Agency
FAA   Federal Aviation Administration
IEC   International Electrical Technical Commission
ISO   International Standards Organization
NASA   National Aeronautics and Space Administration

16.1 Introduction to Formal Methods

Engineering as a whole relies heavily on mathematical models to make informed judgments about designs. However, software development has traditionally been less formal—maybe even *ad hoc* at times, with testing at the end of the project to confirm its goodness. The goal of formal methods is to bring the same mathematical basis used for other engineering disciplines to the digital world, to both software and programmable hardware. Formal methods apply logic and discrete mathematics to model "the behavior of a system and to formally verify that the system design and implementation satisfy functional and safety properties" [1].

Although I am not a formal methodist, I have been exposed to formal methods for over 15 years and have a great respect for both the technical discipline of formal methods and its incredible potential to improve the quality and safety of digital systems.

Formal methods have been advocated by academia for years. Some of the brightest, most educated, and most passionate people I know are formal methodists. They have been persistent in their research and pursuing their convictions. It is exciting to see that their hard work is starting to connect with applied engineering.

To date, formal methods have had more application in the electronic hardware world—primarily because the hardware tools are more standardized and stable, allowing formal methods to be rolled into the day-to-day activities of the engineers. Software tools and technology (such as requirements and design methodology, languages, target dependencies) are still changing rapidly. Software engineering hasn't reached the level of maturity of electronic hardware and other engineering disciplines.

Daniel Jackson et al. summarize the state of formal methods in software engineering well. They explain that traditional software development relies on human inspection and testing in order to validate and verify. Formal methods utilize testing as well, but they also use notations and languages for rigorous analysis. Formal methods also use tools to reason about the properties of requirements, designs, and code. "Practitioners have been skeptical about the practicality of formal methods. Increasingly, however, there is evidence that formal methods can yield systems of very high dependability in a cost-effective manner..." [2].

Some standards (such as ISO/IEC* 15408, *Common Criteria for Information Technology Security Evaluation*, and the United Kingdom's Defence Standard 00-55, *Requirements for Safety Related Software in Defence Systems*) require at least some application of formal methods for the most critical levels.† DO-178B identified formal methods as an acceptable alternative method. However, for the civil aviation world, formal methods have scarcely been used. Some avionics and tool vendors are beginning to use formal methods, and their use is expected to continue to grow. It seems that the practical experience and the supporting formal methods tools have finally reached a level of maturity to be used by the aviation industry.

My first introduction to formal methods was really quite humorous. I was asked to present an overview of DO-178B and the Federal Aviation Administration's (FAA) software-related policy and guidance at a Defense Advanced Research Projects Agency (DARPA) meeting in the Washington, DC, area. I had presented the topic many times, so I simply replaced the title page and added some background information to my PowerPoint presentation, took the metro to Crystal City, and arrived to give the presentation. I was first on the agenda; therefore, not long after I arrived, I was introduced and began giving the presentation. Probably 5 minutes into the 60-minute presentation, I started seeing unusual looks on the faces of the audience. There were about 50 people and I had never met any of them before. Some looked confused. Many of them looked disgusted or flat out mad. Some of the looks just cannot be described. I knew I had hit a nerve of some sort,

---

* ISO/IEC stands for International Standards Organization/International Electrical Technical Commission.
† Although Defence Standard 00-56 (*Safety Management Requirements for Defence Systems*), which replaced Defence Standard 00-55, does not require it. Defence Standard 00-56 does, however, recognize the use of formal proof and analysis as acceptable for providing evidence of safety.

but I had no idea what it could be, so I just kept on talking. They didn't laugh at any of my jokes, so I just quit trying to add humor. When I finally got to the question and answer time, I learned "the rest of the story." I was speaking to a group of formal methodists who had no experience with the real world of civil aviation. I, on the other hand, had certified several aircraft but had no experience with formal methods. And, so the two worlds collided. I got away relatively unscathed and was happy to get back to my safe office that afternoon. I've learned more about formal methods since that time and have great hopes for their role in the future of safety-critical software.

16.2 What Are Formal Methods?

A 2002 document published by National Aeronautics and Space Administration (NASA) entitled *NASA Langley's research and technology-transfer program in formal methods* describes formal methods as follows:

> Formal methods refers to the use of techniques from logic and discrete mathematics in the specification, design, and construction of computer systems (both hardware and software) and relies on a discipline that requires the explicit enumeration of all assumptions and reasoning steps. Each reasoning step must be an instance of a relatively small number of allowed rules of inference. In essence, system verification is reduced to a calculation that can be checked by a machine. In principle, these techniques can produce error-free design; however, this requires a complete verification from the requirements down to the implementation, which is rarely done in practice. Thus, formal methods is the applied mathematics of computer systems engineering. It serves a similar role in computer design as Computational Fluid Dynamics (CFD) plays in aeronautical design, providing a means of calculating and hence predicting what the behavior of a digital system will be prior to its implementation.
>
> The tremendous potential of formal methods has been recognized by theoreticians for a long time, but the formal techniques have remained the province of a few academicians, with only a few exceptions… It is important to realize that formal methods is not an all-or-nothing approach [1].

As John Rushby states: "Formal methods allow properties of a computer system to be predicted from a mathematical model of the system by a process akin to calculation" [3].

DO-333, entitled *Formal Methods Supplement to DO-178C and DO-278A*, was recently published by RTCA. The DO-333 glossary defines *formal methods* as follows: "Descriptive notations and analytical methods used to construct, develop and reason about mathematical models of system behavior. A formal method is a formal analysis carried out on a formal model" [4]. Basically, a formal method is a formal model plus a formal analysis. A *formal model* is "an abstract representation of a given set of aspects of a system that is used for analysis, simulation, code generation, or any combination thereof. A formal model is a model defined using a formal notation" [4].* A *formal notation* is "a notation having a precise, unambiguous, mathematically defined syntax and semantics" [4]. A *formal analysis* is: "The use of mathematical reasoning to guarantee that properties are always satisfied by a formal model" [4].

The formal model is typically used during the development phase of a project to establish different properties. DO-333 section FM.1.6.1 provides some examples of formal models, including graphical models, textual models, and abstract models. Such models use mathematically defined syntax and semantics [4].

Currently, formal models are typically only used to model some of the software behavior, since higher level requirements input to the model "may include properties that cannot be verified with a formal method. Models can also be insufficiently detailed to allow meaningful analysis of some properties and yet be perfectly adequate for others" [4].

Formal models can be used to describe certain properties with a high degree of assurance, thus supporting safety. Many times the models only assure certain properties. Therefore, it is important to identify the limits of the model. Properties outside those limits are addressed by other models or by traditional DO-178C approach [4].

Formal models can benefit the software development process; however, the most beneficial aspect of formal methods is the formal analysis of the formal models. Formal analysis is used to prove or guarantee that software complies with the requirements. In order to prove or guarantee compliance to requirements, a set of software properties is defined (either created or embedded in the formal analysis tool). When a set of software properties fully define a set of requirements, the formal analysis can be used to prove that the set of software properties hold true [4].

An analysis method may only be considered formal if its determination of property is *sound*. DO-333 explains this as follows:

> Sound analysis means that the method never asserts a property to be true when it is not true. The converse case, the assertion that a property is false when it may be true, colloquially "the raising of false alarms", is a usability issue but not a soundness issue. Furthermore, it is acceptable for a method to return "don't know" or not to return an answer when trying to establish whether a property holds, in which case additional verification is necessary [4].

---

* DO-333 essential defines formal methods as follows: *formal methods = formal model + formal analysis*. Formal methods are not normally described this way; this approach was used in DO-333 to explain formal methods using the DO-178C life cycle processes.

DO-333 identifies three typical categories of formal analysis [4]:

1. *Deductive methods* which use mathematical arguments to establish each property of a formal model. Proofs are normally constructed using a theorem proving tool to show that the software properties are sound.

2. *Model checking* which "explores all possible behaviors of a formal model to determine whether a specified property is satisfied."

3. *Abstract interpretation* which constructs conservative representations of a programming language's semantics for "sound determination of dynamic properties of infinite-state programs... It can be viewed as a partial execution of a computer program which determines specific effects of the program (e.g., control structure, flow of information, stack size, number of clock cycles) without actually performing all the calculations."

These three categories of formal analysis share a couple of attributes. First, they rely on formal models. "Compliance between artifacts can never be demonstrated between an informal model and a formal model, using formal analysis. For example, using formal methods to demonstrate compliance between specification and code requires that both be formal models. Second, all of the formal analyses are generally implemented using a tool" [4]. Therefore, tools used for formal analysis may need to be qualified. See Chapter 13 for information on tool qualification.

16.3 Potential Benefits of Formal Methods

Formal methods offer a number of potential benefits, as discussed in this section. As the experience and technology matures, the benefits will likely increase.

Potential Benefit 1: Improved requirements definition. Because of the formal nature of formal models, they can ensure that requirements are complete and well thought out. Traditional requirements are often incomplete, inaccurate, ambiguous, and inconsistent. Rushby writes: "Many of the problems in software and hardware design are due to imprecision, ambiguity, incompleteness, misunderstanding, and just plain mistakes in the statement of top-level requirements, in the description of intermediate designs, or in the specifications of components and interfaces. Some of these problems can be attributed to the difficulty of describing large and complex artifacts in natural language" [3]. Using a formal model helps to identify such weaknesses earlier.

Potential Benefit 2: Reduced errors. Formal methods help to reduce reliance on human intuition and judgment by using a more rigorous and mathematically based approach. Because of the rigor required to define a formal model, it can reduce the number of errors in the requirements and/or design.

Potential Benefit 3: More errors detected. Formal analysis can find errors that might go totally undetected using traditional verification means. It can uncover errors, misunderstandings, and subtle unexpected properties that might go unnoticed using the traditional review and test approach.

Potential Benefit 4: Increased confidence in safety properties. Using formal methods to model and verify the safety characteristics of a system can boost confidence in safety and quality.

Potential Benefit 5: Increased confidence for highly complex systems. Formal methods offer the ability to more thoroughly analyze input and output space. For highly complex functions, formal methods can verify functionality more thoroughly than traditional testing that only exercises a portion of the input and output space.

Potential Benefit 6: Improved quality in error-prone areas. While it may be impractical to apply formal methods to all of the software design, it can effectively be applied in areas that are error-prone. These tend to also be the most complex areas.

Potential Benefit 7: More effective tool implementation. Commercial tools (e.g., automatic code generators) are being developed for use by multiple avionics manufacturers. When formal methods are used to build the tools, it can improve the robustness of the tool and decrease the amount of additional verification required by the tool user. For example, an automatic code generation tool developed using formal methods may reduce or eliminate the need for structural coverage analysis on the tool's output, since formal methods ensure the tool will not generate inaccurate, untraceable, incomplete, or extraneous code.

Potential Benefit 8: Reduced cost. While the application of formal methods might require more rigor and resources initially, it has the potential to find errors earlier and with more certainty. This can reduce the overall cost of the software development, since finding errors late in the process is quite costly.

Potential Benefit 9: Maintainable software. Since formal methods have the potential to produce cleaner requirements and architecture, they can make the system easier to maintain. It's worth noting that even with formal methods, caution must be exercised when reusing or modifying existing software.

16.4 Challenges of Formal Methods

There are a number of challenges that projects face when implementing formal methods.

Challenge 1: Lack of education leads to a high fear factor. Most software engineers and their managers are still uneducated and even afraid of formal methods.

The symbology looks strange and many engineers may not have the mathematical background to apply the tools. This challenge can be overcome by training and user-friendly tools.

Challenge 2: Formal methods are not applicable to all problems. Formal methods have their limits. It is important to use them where they are most effective and where they have the most potential (e.g., complex or problematic functions).

Challenge 3: It can be difficult to mix formal and not-so-formal methods. Since most projects apply formal methods to a subset of the overall requirements definition and verification activities, the formal methods must be integrated with a more traditional process. There may be situations where formal methods are used with other not-so-traditional techniques, such as object-oriented technology or model-based development. To address the challenge, it's important to keep the big picture in mind and to remember the objectives that drive the activities.

Challenge 4: Limited number of formal methods experts available. There is not an abundance of formal methods practitioners out there, so this can be a barrier to successful use of the techniques. As noted in Challenge #1, training and user-friendly tools can help alleviate this challenge as well.

Challenge 5: Formal methods can be resource intensive. Formal methods require specialized skills and tools. They can also be quite challenging to apply. For this reason it is best to use them in the most critical, complex, and error-prone areas, where they will have the most return on investment.

Challenge 6: Too much confidence in the verification ability of formal methods. There can be a misconception that formal methods can guarantee correctness. They do offer greater confidence in the software's compliance to the requirements; however, as Bowen and Hinchey explain: "If the organization has not built the right system (validation), no amount of building the system right (verification) can overcome that error" [5].

Challenge 7: False assumptions exist about testing. Some contend that formal methods mean no testing is needed. Formal methods can help to reduce the likelihood of certain errors or help to detect them, but they must be accompanied with appropriate testing [5]. DO-333 emphasizes the need to test the software, even when applying formal methods. In particular, target-based testing is needed to verify the hardware/software integration.

Challenge 8: Industry and certification guidance are not yet standardized. If three people were asked to explain *formal methods,* they would probably provide at least four answers. The completion of DO-333 is a major step toward standardization; however, there are still many project-by-project issues to be resolved.

Challenge 9: Tool support is still incomplete. The formal methods tools have improved significantly over the last few years, but there is still considerable work needed to make the tools practical for the aviation community. The tools need to fit well into the software life cycle and be usable by domain experts. If the tools are not practical for those who have the domain expertise (e.g., a flight controls engineer), they will not be effectively used, and the aforementioned potential benefits will not be realized.

16.5 DO-333 Overview

16.5.1 Purpose of DO-333

DO-333 supplements DO-178C by modifying, deleting, and adding objectives specific to formal methods. The DO-333 supplement is based on the following key principles [6]:

- A formal method is the application of a formal analysis to a formal model.
- A formal model must be in a notation with mathematically defined syntax and semantics.
- Formal methods may be used at different verification steps in the software life cycle, for all or part of a step and for all or part of the system being developed.
- A formal method must never produce a result which may not be true (that is, the formal analysis must be sound).
- It is possible to apply the results of formal analysis of Source Code to the corresponding object code by understanding the compilation, link, and load processes in sufficient detail.
- Test is always required to ensure compatibility of the software with target hardware and to fully verify the understanding of the relationship between source and object code.

16.5.2 DO-333 and DO-178C Compared

16.5.2.1 Planning and Development

DO-333 clarifies DO-178C's planning and development processes when formal methods are used. DO-333 explains that the plans and standards need to be developed to describe and address the formal methods approach. If a formal model is used in development without formal analysis, DO-333 does not need to be applied; DO-178C itself is adequate to address this scenario. That is, if a formal model or multiple layers of models are used in the requirements, design, and/or coding phases, the DO-178C objectives apply. The supplement provides some model-specific clarification for development, but it is basically the same process as identified in DO-178C.

16.5.2.2 Configuration Management, Quality Assurance, and Certification Liaison

Configuration management, quality assurance, and certification liaison processes in DO-178C are unchanged in DO-333.

16.5.2.3 Verification

The major differences between DO-178C and DO-333 are in the area of verification. Since formal analysis has the ability to prove or disprove the correctness of a formal model, some conventional DO-178C review, analysis, and test objectives may be replaced by formal analysis. DO-333 modifies objectives and activities for review and analysis of high-level requirements, low-level requirements, software architecture, and source code to be specific to formal methods. Formal analysis may be used to satisfy objectives related to compliance of input to output, accuracy, consistency, compatibility with the target computer, verifiability, conformance to standards, traceability, algorithm accuracy, and correctness of requirements formalization [4].

Two primary challenges exist for formal methods verification: (1) executable object code verification (DO-178C Table A-6 and DO-333 Table FM.A-6) and (2) verification of verification (DO-178C Table A-7 and DO-333 Table FM.A-7).

First, let's consider *executable object code verification*. At this point in time, it is not possible to replace executable object code testing with formal analysis. Formal analysis may be used to supplement the testing activities and to verify compliance with the requirements; however, because of the target dependencies, models are not yet adequate to replace testing. Therefore, the objectives for verifying executable object code are the same whether formal methods or traditional software development methods are used. That is, DO-333 Table FM.A-6 has the same objectives as DO-178C Table A-6. However, an additional objective was added to DO-333 to address the situation when formal analysis is used to verify properties of the executable object code. DO-333 Table FM.A-7 Objective FM9 states: "Verification of property preservation between source and object code" [4].

> By verifying the correctness of the translation of source to object code, formal analysis performed at the source code level against high or low level requirements can be used to infer correctness of the object code against high or low level requirements. This is similar to the way that coverage metrics gained from source code can be used to establish the adequacy of tests to verify the target system [7].

Doing such an analysis will likely prove difficult. However, as technical advances in emulation and portability continue, it may become more feasible.

The second area of challenge is the *verification of verification*. DO-178C Table A-7 has nine objectives. Table 16.1 shows the correlation between the DO-178C Table A-7 objectives and the DO-333 Table FM.A-7 objectives that replace the DO-178C objectives. The table shows that objectives 1–4 and 9 (from DO-178C) have an equivalent in DO-333. However, the four structural coverage objectives of DO-178C (objectives 5–8) are replaced by one objective in DO-333. Because structural coverage involves the execution of tests and measurement of code coverage, an alternative is proposed when using formal methods. The alternative still needs to satisfy the intent of structural coverage (i.e., to detect shortcomings in requirements-based tests, to identify inadequacies in requirements, and to identify extraneous (including dead) or deactivated code). The supplement proposes the four following activities to satisfy the structural coverage objectives [4,7]:

- *Complete coverage of each requirement:* When assumptions are made for the formal analysis, all assumptions must be verified to ensure complete coverage of each requirement.

- *Completeness of the set of requirements:* For formally modeled requirements, it needs to be demonstrated that the set of requirements is

TABLE 16.1

Comparison of DO-178C and DO-333 Verification of Verification Objectives

DO-178C Table A-7 Objective	DO-333 Table FM.A-7 Objective
1. Test procedures are correct.	FM1. Formal analysis cases and procedures are correct.
2. Test results are correct and discrepancies explained.	FM2. Formal analysis results are correct and discrepancies explained.
3. Test coverage of HLRs is achieved.	FM3. Coverage of HLRs is achieved.
4. Test coverage of LLRs is achieved.	FM4. Coverage of LLRs is achieved.
5. Test coverage of software structure (modified condition/decision coverage) is achieved.	FM5—FM8. Verification coverage of software structure is achieved. (A single objective that replaces the four structural coverage objectives in DO-178C.)
6. Test coverage of software structure (decision coverage) is achieved.	
7. Test coverage of software structure (statement coverage) is achieved.	
8. Test coverage of software structure (data coupling and control coupling) is achieved.	
9. A verification of additional code, that cannot be traced to source code, is achieved.	FM9. Verification of property preservation between source and object code.
N/A	FM10. Formal method is correctly defined, justified, and appropriate.

Sources: RTCA DO-333, *Formal Methods Supplement to DO-178C and DO-278A*, RTCA, Inc., Washington, DC, December 2011; RTCA DO-178C, *Software Considerations in Airborne Systems and Equipment Certification*, RTCA, Inc., Washington, DC, December 2011.

complete with respect to the intended functions. It must be verified that for all input conditions, the required output has been specified, and for all outputs, the required input conditions have been specified. If requirements are incomplete, the requirements need to be updated. If completeness of the requirements cannot be demonstrated, structural coverage is needed.

- *Detection of unintended data flow relationships:* The intent is to confirm that that data flow in the source code complies with the requirements and that there are no unintended dependencies between code inputs and outputs. Unintended dependencies need to be resolved (e.g., requirements are added or erroneous code is removed).

- *Detection of extraneous code and deactivated code:* The goal is the same as for DO-178C with respect to extraneous code (which includes dead code) and deactivated code. Chapter 17 explains extraneous, dead, and deactivated code.

The application of DO-333 Table FM.A-7 may be challenging. The selected approach needs to ensure that all noncovered code is identified. Additionally, the DO-333 guidance doesn't seem to address control flow analysis. Those who implement formal methods will need to detect unintended control flow, as well as unintended data flow. When using formal methods, it is important to coordinate closely with the certification authorities regarding the approach selected to satisfy the DO-333 Table FM.A-7 objectives.

Since formal methods will typically only be used for part of the software and not all of it, the selected approach needs to clearly identify where DO-333 is used and where DO-178C and/or another supplement applies.

16.6 Other Resources

This chapter has merely provided an introduction to formal methods. There are numerous books, reports, and articles on formal methods. There are people who have committed many years of their lives to this subject. They've been quite industrious at generating reports to educate. If you choose to use formal methods in a project, you'll want to examine the topic much deeper and probably hire some specialists to assist and educate your team. The "References" section includes some of the resources I've found most useful. Each reference points to other beneficial resources.

References

1. R. W. Butler, V. A. Carreno, B. L. Di Vito, K. J. Hayhurst, C. M. Holloway, J. M. Maddalon, P. S. Miner, C. Munoz, A. Geser, and H. Gottliebsen, NASA Langley's research and technology-transfer program in formal methods (Hampton, VA: NASA Langley Research Center, May 2002).
2. D. Jackson, M. Thomas, and L. I. Millett, *Software for Dependable Systems: Sufficient Evidence?* (Washington, DC: Committee on Certifiably Dependable Software Systems, National Research Council, 2007).
3. J. Rushby, Formal methods and the certification of critical systems, Technical Report CSL-93-7 (Menlo Park, CA: SRI International, December 1993).
4. RTCA DO-333, *Formal Methods Supplement to DO-178C and DO-278A* (Washington, DC: RTCA, Inc., December 2011).
5. J. P. Bowen and M. G. Hinchey, Ten commandments of formal methods… Ten years later, *IEEE Computer* January, 40–48, 2006.
6. RTCA DO-248C, *Supporting Information for DO-178C and DO-278A* (Washington, DC: RTCA, Inc., December 2011).
7. D. Brown, H. Delseny, K. Hayhurst, and V. Wiels, Guidance for using formal methods in a certification context, *ERTS Toulouse Conference* (Toulouse, France, May 2010).
8. RTCA DO-178C, *Software Considerations in Airborne Systems and Equipment Certification* (Washington, DC: RTCA, Inc., December 2011).

Part V

Special Topics

Part III provided an overview of DO-178C and examined the processes required for compliance, including planning, development, verification, configuration management, quality assurance, and certification liaison processes. Part IV presented the tool qualification guidance (DO-330) and the DO-178C technology supplements on model-based development (DO-331), object-oriented technology (DO-332), and formal methods (DO-333). Part V explores technologies and specialized topics that are relevant to many safety-critical software development efforts. The topics examined are as follows:

- Noncovered code (dead, extraneous, and deactivated code) (Chapter 17)
- Field-loadable software (Chapter 18)
- User-modifiable software (Chapter 19)
- Real-time operating systems (Chapter 20)
- Software partitioning (Chapter 21)
- Configuration data (Chapter 22)
- Aeronautical databases (Chapter 23)
- Software reuse (including previously developed software and commercial off-the-shelf software) (Chapter 24)
- Reverse engineering (Chapter 25)
- Outsourcing and offshoring (Chapter 26)

There are several other subjects that I had hoped to cover in this part, including integrated modular avionics, aircraft electronic hardware, security, and electronic flight bags. However, because of time and space limitations, they are not addressed. Some of these topics are, however, slated to be addressed in the new edition of CRC Press's *Digital Avionics Handbook*.

17

Noncovered Code (Dead, Extraneous, and Deactivated Code)

Acronyms

PSAC   Plan Software Aspects of Certification
SAS   Software Accomplishment Summary

17.1 Introduction

When structural coverage is performed to ensure that the requirements-based tests fully exercise the code structure (as discussed in Chapter 9), there may be some code that is not *covered* (i.e., not exercised by the tests). If the lack of coverage is due to missing requirements or tests, the requirements and tests are updated and re-executed as appropriate. There may be some code that cannot be exercised in the test environment but still meets the requirements (e.g., code that functions during initialization or defensive code). Such code needs to be verified through an alternate approach, such as analysis or code inspection. However, even after this, there may still be some noncovered code that is classified as extraneous, dead, or deactivated. This chapter discusses these classes of code and provides recommendations for how to handle them.

17.2 Extraneous and Dead Code

The term *extraneous code* was added in the update from DO-178B to DO-178C. There was considerable debate about what to call this class of code that is (1) not exercised during requirements-based testing and (2) not traceable to the functional requirements. Several alternate terms were considered, such as *untraceable code, unreached code,* and *noncovered code.* The term *extraneous code* won the contest.

DO-178C defines extraneous code as: "Code (or data) that is not traceable to any system or software requirement. An example of extraneous code is legacy code that was incorrectly retained although its requirements and test cases were removed. Another example of extraneous code is dead code" [1].

The definition of extraneous code identifies *dead code* as a subset of extraneous code. DO-178C explains that dead code is "executable object code (or data) which exists as a result of a software development error but cannot be executed (code) or used (data) in any operational configuration of the target computer environment. It is not traceable to a system or software requirement…" [1]. The following illustrates a simple example of dead code; as you can see, the line "calculate airspeed" cannot be reached because of the way the code is written.

```
if (UP and OVER) and not OUT
  calculate wheelspeed
  if OUT
    calculate airspeed
  end if
end if
```

DO-178C requires that extraneous code (including dead code) be removed. In addition to the removal, an analysis is performed to determine (1) the effects of the code removal and (2) the need for re-verification prior to approval of the code in the airborne system [1]. This guidance has been a hard pill to swallow for many projects. However, it should be noted that DO-178C also provides some common exceptions to the *rip-it-out guidance*, including the following [1]:

- *Embedded identifiers* in the code are not considered extraneous or dead code. Examples of embedded identifiers include checksums or part numbers.

- *Defensive programming structures* to improve robustness are not considered extraneous or dead code. However, such defensive programming practices need to be clearly identified in the coding standards. Additionally, the defensive programming structure must be shown to support the implementation of the requirements and design. As an example, a commonly required defensive coding practice is to have an *else* for every *if* statement.

- *Deactivated code* is not extraneous or dead code (deactivated code will be discussed later in this chapter).

- *Source or object code that does not exist in the executable object code* (e.g., due to compiler or linker settings) is not considered extraneous or dead code. However, an analysis is required to show that the code does not exist in the executable object code. Additionally, procedures must be in place to ensure that such code will not be inadvertently inserted into the executable object code in the future (e.g., build procedures that explicitly state the settings and provide the reason for their existence).

When structural coverage analysis first identifies noncovered code, the data should be examined and analyzed to determine if it is a requirements problem, a test case issue, or a code issue. Don't automatically assume that noncovered code is extraneous or dead code. In my experience, most of the initially discovered noncovered code is due to missing requirements, missing test cases, or logic errors. Once it has been determined that the noncovered code is truly extraneous or dead, then time must be spent analyzing the impact of the code and determining the best solution. Some code may appear to be *dead*, but when it is removed, chaos occurs. One company told me that they called some of their code *zombie code* because it appeared to be dead, but every now and then it came to life (not an acceptable scenario in safety-critical systems).

17.2.1 Avoiding Late Discoveries of Extraneous and Dead Code

Many developers have learned the hard way that rapid removal of dead or extraneous code late in the project's life cycle can be problematic. Removal of dead or extraneous code after formal software testing (i.e., the testing for certification credit) has been performed must be treated with extreme caution. If code is erroneously determined to be extraneous or dead and is removed, it may be difficult to determine the impact without substantial re-verification (possibly even a re-execution of all tests). Therefore, it is important to make every effort to find coding issues as early as possible. Here are some suggestions to avoid late discoveries of dead or extraneous code:

- Implement high-quality code reviews that specifically focus on the code's compliance with and traceability to the requirements.
- Keep trace data current. Each time the requirements or code change the trace data should be modified.
- Run tests and analyze structural coverage as early as possible (as mentioned in Chapter 9, execution of tests prior to reviewing the test cases and procedures is typical). As the requirements or code changes, tests should be rerun and coverage reanalyzed. These early test runs and coverage analysis help to identify issues prior to the final test run that will be used for certification credit (i.e., the *formal* test run or the *for-score* test run).

- Schedule time for addressing issues discovered during testing. As indicated in Chapter 9, many projects implement success-based scheduling and assume all tests will execute without issue, but this sets the project up for failure (or at least significant delays).
- Define processes and procedures for addressing noncovered code and include examples.

17.2.2 Evaluating Extraneous or Dead Code

Even after implementing these suggestions, there may still be some extraneous code (which includes dead code) discovered late in the program. If this occurs, Figure 17.1 illustrates the steps that normally take place. Each box of the flowchart is described in the following:

1. Thoroughly analyze the noncovered code to confirm that it is indeed extraneous or dead.
2. If the noncovered code is not extraneous, dead, or deactivated, make the appropriate changes to requirements or tests. If it is deactivated, ensure that it follows the guidance for deactivated code (discussed in Section 17.3).
3. If the noncovered code is extraneous or dead, analyze the removability of the code considering the architecture. Some code is easy to remove and some is quite challenging. The feasibility determination may also need to consider the schedule. Extraneous or dead code found late in the program may have schedule constraints, as well as technical constraints.
4. Evaluate the effects of activating the extraneous or dead code if it is not removed. Consider what would happen if the extraneous or dead code was inadvertently activated (evaluate safety and functional effects).
5. If the analysis ensures with a high level of confidence that inadvertent activation of the extraneous or dead code is benign from a safety perspective (e.g., unused variables), it may be possible to obtain a special agreement from the certification authority. This agreement typically requires the removal of the code during the first post-certification modification. This approach is subjective and must be thoroughly analyzed and justified, and presented to the certification authority as soon as possible. The criticality of the software and the nature of the extraneous or dead code will affect the details of the proposed strategy.
6. Seek certification authority's agreement on the proposed strategy as soon as possible.

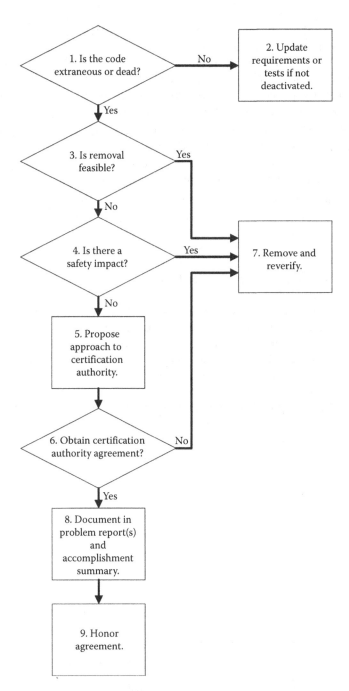

FIGURE 17.1
Steps for evaluating extraneous or dead code.

7. Remove the extraneous or dead code and perform re-verification, including rerun of the appropriate tests, if one of the following is true: (1) removal is feasible (from a technical and schedule perspective), (2) the effects of activation are not benign (i.e., safety could be impacted), or (3) the certification authority does not agree to approve the software with the extraneous or dead code remaining.

8. If the certification authority agrees to allow the extraneous or dead code in for the initial approval, document the terms of the agreement and identify the specific problem report(s) that addresses the extraneous or dead code in the Software Accomplishment Summary (SAS).

9. If an agreement is reached, ensure that steps are in place to honor the agreement (i.e., the schedule for removal is implemented). The team that modifies the software (normally not the same team that developed the software) must be made aware of the agreement when they begin the software upgrade, so they can ensure the changes are tracked, implemented, and verified.

It should be noted that extraneous or dead code discussions and evaluations normally occur at the source code level because most projects perform their structural coverage measurement on the source code. However, some companies perform structural coverage on the object or machine code. If this is the case, the same issues regarding extraneous or dead code still need to be addressed—the code just looks different.

As noted in Chapter 9, for level A software there is an additional objective required when performing structural coverage on the source code. The source to object code traceability is analyzed to ensure that the compiler did not generate extraneous or dead code. For level A software, any extraneous or dead code generated by the compiler must be analyzed and addressed as described earlier.

17.3 Deactivated Code

Deactivated code is often used to create flexible yet fully compliant software. Deactivated code is related to extraneous or dead code because it may show up as noncovered code during structural coverage analysis. However, unlike extraneous or dead code, deactivated code is planned and implemented by design.

The DO-178C glossary defines *deactivated code* as follows:

> Executable Object Code (or data) that is traceable to a requirement and, by design, is either (a) not intended to be executed (code) or used (data), for example, a part of a previously developed software component such as unused legacy code, unused library functions, or future growth code; or

(b) is only executed (code) or used (data) in certain configurations of the target computer environment, for example, code that is enabled by a hardware pin selection or software programmed options. The following examples are often mistakenly categorized as deactivated code but should be identified as required for implementation of the design/ requirements: defensive programming structures inserted for robustness, including compiler-inserted object code for range and array index checks, error or exception handling routines, bounds and reasonableness checking, queuing controls, and time stamps [1].

Some examples of deactivated code usage include the following:

- Debug code used during testing or troubleshooting in the lab. The debug code is deactivated when installed in the aircraft.
- Maintenance features used on the ground by an authorized technician. The maintenance features are deactivated during aircraft operation.
- Option-selectable software, which is used instead of hardware pins to select a preapproved software configuration based on the specific operator's needs. The software not selected is deactivated.
- A subset of an operating system that was developed to satisfy DO-178C objectives and to meet the needs of multiple users. The features not used are deactivated.
- A flight management system designed to take input from multiple sensors that can be selected depending on the aircraft configuration. The sensor code not used is deactivated.
- An engine controller developed for one-engine or two-engine configuration. Either the one-engine or two-engine code is deactivated.
- Selected functions from a library designed to meet the needs of multiple software teams. The unused library functions are deactivated.
- Compiler options (e.g., #ifdef) added to use the same code in multiple places (e.g., same code for multiple lanes or for multiple platforms). During compilation, such code is removed from the executable object code; hence, it is deactivated.

The list goes on and on. Deactivated code allows configurability of the software to meet the needs of multiple users. As you can see from the definition of deactivated code and the examples, deactivated code falls into two basic categories: (1) code that will never be used in flight, and (2) code that may be used at some time, depending on the configuration of the system. Table 17.1 summarizes the two categories and some things to consider for each category.

I once encountered a company that tried to classify their dead code as deactivated code at the end of the project. However, there's a big

TABLE 17.1

Overview of Deactivated Code Categories

Category	Description	Considerations
Category one	Code that will NEVER be activated during aircraft operations. For example, debug code.	• The deactivated code must be disabled during aircraft operation. • The functionality of the deactivation mechanism must be proven to work (it needs requirements and tests). • The deactivation approach must support the software level of the activated software. • The deactivated code itself is generally treated like level E software from a certification perspective. However, it is still recommended that there be requirements and some level of testing on the deactivated code, in order to ensure it meets its intent.
Category two	Code that may be used in some configurations but not others.	• The deactivated code must be tied to requirements like all other airborne software. • The deactivated code must be assured to the appropriate software level since it is intended to be used in the future. That is, the deactivated software must meet the applicable DO-178C objectives for the software level. • The deactivation approach to ensure proper deactivation and configuration must be included in the requirements and design. • The deactivation approach must meet the appropriate development assurance level and will need to be tested.

difference between the two: deactivated code is planned and designed; dead code is not.

Figure 17.2 summarizes the processes to consider for deactivated code, as well as references to the related DO-178C sections. The planning, development, and verification processes are described in the following subsections.

17.3.1 Planning

One of the primary differences between extraneous or dead code and deactivated code is that deactivated code is planned. I've never seen anyone plan to add dead code (they plan to deal with it if it arises, but they don't plan to implement it). Deactivated code should be explained in the Plan Software Aspects of Certification (PSAC), as well as in the Software Development Plan and Software Verification Plan. The deactivated code plan is normally included as an *additional consideration* in the PSAC. The PSAC should explain the types of deactivated code planned, the category of the deactivated code

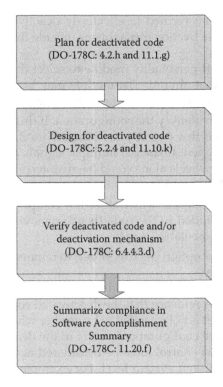

FIGURE 17.2
Summary of deactivated code process.

(*category one* or *category two*), the deactivation mechanism(s), and the development and verification approach for the deactivated code and the deactivation means. The Software Development Plan explains more details on the development approach for the deactivated code. It is important to give the developers direction on how to handle the deactivated code during requirements, design, and coding phases. The Software Verification Plan provides details on the verification approach for both the deactivated code itself (particularly for *category two* code) and for the deactivation mechanism(s).

17.3.2 Development

As the DO-178C definition of deactivated code implies, deactivated code is driven by the requirements and is documented in the design. DO-178C section 5.2.4 provides guidance (called activities) for designing deactivated code [1].

First, the deactivation mechanism needs to be designed and implemented so that the deactivated code has no adverse impact on the active code. There are a variety of ways to deactivate the code. Regardless of the mechanism used to deactivate the code, it must be considered in the safety assessment process and it must be verified to effectively deactivate the desired code. Some examples of deactivation mechanisms include the following:

- Using an aircraft personality module to select the appropriate configuration for that aircraft. The module may be a physical module with switches or it may be a configuration file.
- Selecting pins to identify the configuration. If this is the approach, it is recommended that two or more pins are used, since one pin can easily short or open. Likewise, multiple signals are generally used (e.g., weight-on-wheels alone could be misinterpreted). The possible failures should be considered by the safety team.
- Using a smart compiler which does not compile library functions that are not included in the code (hence removing the deactivated code from the executable image).
- Using a compiler option (such as #ifdef) to remove the code from the executable image.

Second, there must be evidence that the deactivated code is not active in the environments where it is not intended to be used or where it is not approved. This requires effective configuration control of the deactivated code and its use. The configuration control must be considered at the software level, as well as at the aircraft and system level. For example, if using a library of functions during the development, there must be a process to ensure that only approved library functions are used. If care isn't taken, future developers may just assume that the entire library is approved. One way to handle this is to only release the approved functions, so the unapproved functions are unavailable.

Third, the deactivated code itself needs to comply with applicable DO-178C objectives. The applicable objectives depend on the category of the code. For *category one* deactivated code, level E is usually acceptable because this software will not be used during flight operations. However, if the deactivated code will be used to make important maintenance decisions or to perform vital maintenance activities, it may need to be developed to a higher level. *Category two* deactivated code should to be treated like active code during development, since it is intended to be used in some configurations.

17.3.3 Verification

Both the deactivated code and the deactivation mechanism need to be verified to the appropriate DO-178C software level. If the compiler is not a qualified tool and *smart compiling* or compiler options were chosen as the deactivation mechanism, it is important to analyze the compiler output to ensure that the deactivated code truly is removed from the executable image.

This brings me to a somewhat common and controversial issue related to deactivated code. Sometimes, it can be challenging to classify the deactivated code as *category one* or *category two*. The project goal may be to have *category two* deactivated code in order to satisfy the needs of multiple customers. However, the actual implementation may end up different (due to schedule or changes in project direction). This may lead to some *category two* code being downgraded to *category one*. This must be carefully evaluated and discussed with the certification authority. If this approach is taken, there needs to be concrete limitations to prevent users and future developers from thinking that the deactivated code is approved for future use.

Reference

1. RTCA DO-178C, *Software Considerations in Airborne Systems and Equipment Certification* (Washington, DC: RTCA, Inc., December 2011).

18

Field-Loadable Software

Acronyms

CD	compact disk
DAL	development assurance level
DVD	digital video disk
FAA	Federal Aviation Administration
FADEC	full authority digital engine control
FLS	field-loadable software
LAN	local area network
LRM	line replaceable module
LRU	line replaceable unit
PMAT	portable multi-purpose access terminal
PSAC	Plan for Software Aspects of Certification

18.1 Introduction

On my very first day on the job at the Federal Aviation Administration (FAA) Headquarters, as the national software program manager, I was introduced to the issues surrounding field-loadable software (FLS). I sometimes wonder if it was some kind of mean initiation by my coworkers. They were probably standing on the other side of the cubicle laughing their heads off...

It started out as a typical first day. Having just moved from Kansas to the Washington, DC, area, I figured out how to ride the Virginia Railway Express train; I also found the right floor of the FAA Headquarters building, got my badge, filled out the plethora of paperwork, met everyone in the avionics branch, and was ushered to my cubicle. As I was messing with my phone to set up voicemail, it rang. I figured it was the administrative assistant apprising me of something we forgot to cover. Instead it was a lawyer—a very loud and angry woman from one of the alphabet soup aviation organizations in Washington, DC. She began chewing me out for the FAA's lack of responsiveness on the need for guidance on part manufacturer approval for FLS, etc. etc. etc. I kept reminding myself: "We're not in Kansas anymore, Toto."

18.2 What Is Field-Loadable Software?

DO-178C describes FLS as "software that can be loaded without removing the system or equipment from its installation" [1]. It is loaded into the aircraft or engine equipment using some kind of data port without the need to open the unit. FAA also treats software loaded through a data port at qualified labs (e.g., repair stations or service centers) as FLS [2]. Essentially, FLS is contrasted with factory-loadable software, where the line replaceable unit (LRU) or line replaceable module (LRM) may require the unit seal to be broken in order to load the software (e.g., flashing).

The media used to load FLS changes with time. In the early 1990s, 3.5-inch diskettes were used. For example, the original Boeing 777 carried around binders full of floppy disks. The technology has moved on to compact disks (CDs), digital video disks (DVDs), thumb drives, mass storage devices, and even local area networks (LANs).

As long as the appropriate safety measures are taken and the software is approved by the certification authorities, nearly any kind of equipment can be field-loadable—all the way from flight controls and engine controls to navigational and communication systems to flight management systems and traffic collision avoidance systems. Establishing processes and developing systems to be field-loadable are not trivial tasks. There are numerous regulations (e.g., part marking and approved repair station regulations) to consider at multiple levels (e.g., the equipment level, the aircraft level, the flight operations level). However, FLS is very common in today's aviation industry.

It should be noted that while aeronautical databases (e.g., navigational or terrain databases) are field-loadable, they are not treated the same as FLS. In general, aeronautical databases are covered by DO-200A guidance, whereas DO-178C guidance applies to FLS. Chapter 23 provides specific information on aeronautical databases.

18.3 Benefits of Field-Loadable Software

FLS allows more efficient maintenance processes that meet the needs of the worldwide aircraft fleets. Rather than shipping LRUs or LRMs, pulling equipment out of the aircraft to upgrade, and maintaining a large inventory of expensive equipment, software can be distributed and loaded in the aircraft. This allows less downtime for the aircraft, helping meet the needs of the airlines and the flying public.

18.4 Challenges of Field-Loadable Software

As with nearly everything in life, the benefits do not come without struggle. There are several challenges to address when implementing processes that impact lives across the world. There are also multiple stakeholders involved in managing the challenges, including software and equipment developers, aircraft or engine manufacturers, airlines, and certification authorities. Some of the challenges include the following:

- Designing the system to be safe, which involves protecting from unauthorized changes, performing integrity checks, etc.
- Managing configuration at multiple levels (software, equipment, and aircraft or engine) to ensure (1) the approved software is installed in approved equipment and (2) the approved equipment is installed in the certified aircraft or engine.
- Applying equipment-oriented regulations to software (e.g., part marking regulations). Most of the regulations are written for the aircraft or engines and their installed systems. Because FLS becomes a stand-alone *part*, such regulations require interpretation to apply them at the software level.
- Ensuring that all loaded software is approved by the certification authority and that no unapproved parts are installed on the aircraft or engine.
- Ensuring that there are no security issues (viruses, hackers, etc.), particularly when software is transferred or loaded using a network.
- Verifying that the software is completely loaded without corruption.
- Disabling loading capability during flight.
- Managing compatibility with other equipment, when software is updated. Updates to software could impact equipment that interfaces with the software, so equipment that interacts with the software must also be evaluated.

18.5 Developing and Loading Field-Loadable Software

Fortunately, after two decades of experience with FLS, many of the challenges have been overcome. FLS is routinely used now. This section considers four areas to be considered when developing FLS: (1) developing systems to be field-loadable, (2) developing FLS, (3) loading FLS, and (4) modifying FLS.

18.5.1 Developing the System to Be Field-Loadable

Designing FLS is not just about the software. The system itself must be designed to safely load the software. DO-178C section 2.5.5 points out several items to consider when developing a system to safely load software in the field, including the following [1]:

- *Protection from inadvertent loading* (e.g., loading during flight). This may be through hardware, software, or a combination of hardware and software mechanisms. The safety assessment must analyze the impact of the detection mechanism's failure. Unless justified in the safety assessment, the detection mechanism is normally designed to the highest development assurance level (DAL) of the software being loaded.

- *Detection of failed, partial, or corrupted loads.* Per DO-178C section 2.5.5.b, if a safety or default mode is entered when detecting a partial, corrupted, or inappropriate FLS load, "then each partitioned component of the system should have safety-related requirements specified for recovery to and operation in this mode" [1].

- *Determination of inappropriate (incorrect) software loads,* for example, inadvertently loading the wrong files(s) or not loading the software at all.

- *Integrity of the display mechanism* used to verify the aircraft configuration (i.e., to confirm the right software is loaded). The loss or corruption of the ability to display the software part identification also needs to be considered. This is only considered if the on-aircraft display system is used to verify software load.

Addressing these details requires system-level requirements and design and the involvement of the system safety personnel. It is advisable to involve safety personnel early. I recently discovered a flaw in one company's system-level field loading strategy while reviewing the Plan for Software Aspects of Certifiction (PSAC). The PSAC is prepared after the systems architecture is established and initial safety assessment is performed. This discovery led to a system-level redesign late in the project that could have been prevented if safety had been involved earlier.

Another system-level consideration is the part marking approach for FLS. Some manufactures update the hardware part number whenever FLS is loaded. Others manage the hardware and software part numbers separately. Either approach is acceptable; it just needs to be established as part of the system design.

18.5.2 Developing the Field-Loadable Software

Like any other software, FLS is developed to comply with DO-178C and/or applicable supplements. Additionally, the FLS must be designed to support the field loading. This typically means implementing an integrity check (e.g., a cyclic redundancy check), designing the software to address the issues noted earlier, and developing a field loading application (oftentimes, a separate application). Many applicants design their systems to comply with ARINC 615, which provides both "general and specific design guidance for the development of software data loading equipment for all types of aircraft" [3]. Additionally, ARINC 644 may be applied for the maintenance terminal. ARINC 644 provides guidance for development of a *portable multi-purpose access terminal* used by airlines [4].

Planning is an important part of any DO-178C effort. This is especially true of FLS. The plans describe the development, verification, configuration management, quality assurance, and certification liaison aspects of the FLS. The PSAC explains the plans for FLS, ensures that the issues noted earlier are addressed, explains the utilized integrity check and its sufficiency for the given software level (i.e., accuracy), and describes the configuration management details for the FLS development and verification. The Software Development Plan builds upon the PSAC contents by providing details for the FLS development team; it includes an explanation of the protective mechanisms, integrity checks, and any special standards that apply (e.g., ARINC 615). The Software Verification Plan includes details to ensure that the FLS, the integrity checks, and the protective mechanisms are verified and tested. The Software Configuration Management Plan discusses the configuration management process for the FLS. During the development phase, configuration management will probably be the same as non-field-loadable software, but there will need to be some explanation of who is responsible for configuration management after development (this is normally explained in the PSAC as well as the Software Configuration Management Plan).

Once the plans are developed, they should be followed, just as they are for any other airborne software.

18.5.3 Loading the Field-Loadable Software

After the FLS is developed and approved by the certification authority, it can be loaded. Most applicants use an integrity check (e.g., a cyclic redundancy check) to ensure the software is not corrupted during the loading process. The adequacy of the integrity check should be confirmed during the development. The accuracy is typically determined by the integrity check's algorithm and the size of the file(s) being protected. For large files, it may be necessary to use higher bit integrity checks (e.g., 32 bits vs. 8 bits or 16 bits) or divide the data into smaller packages and use additional integrity checks. The probability of error should be consistent with the level of the software being loaded.

If a reliable integrity check is not used, the equipment used to load the FLS may require some kind of qualification (e.g., the data loader software may need to be developed using DO-178C or DO-330 and the software may only be loaded using the approved data loader). This approach is rarely used anymore.

Per FAA Order 8110.49, the FLS loading process should also ensure the following [2]:

- The software being loaded has been approved by the certification authority.

- The loading procedure has been approved and is followed. (As noted in Chapter 10, loading procedures are normally referenced or included in the Software Configuration Index. If they are not the responsibility of the software team, they may be included in system-level or aircraft-level documentation. If this is the case, the Software Configuration Index should explain it.)

- The software is approved for the hardware it is being loaded on (i.e., it is an approved hardware/software configuration).

- The software is approved for the aircraft or engine it is being loaded on (i.e., approved aircraft or engine configuration). This includes ensuring that any redundant parts on the aircraft or engine are appropriately configured; for example, if there are two full authority digital engine control (FADECs), both may need to be loaded when upgraded software is loaded, unless the aircraft or engine configuration allows one FADEC to have the new software version and the other FADEC the older software version.

- The software is completely loaded without error. This is typically done by confirming that the cyclic redundancy check matches what is provided in the load instructions.

- The software part number (including version number, if applicable) is confirmed after the load.

- The change in configuration is documented in the aircraft or engine configuration records and any other applicable maintenance records.

18.5.4 Modifying the Field-Loadable Software

One of the motivators for FLS is that the software can be modified and loaded without having to return the equipment to the manufacturer.

When FLS is modified, it should go through a change impact analysis process, like any other airborne software. All changed and impacted data and code are re-verified. (See Chapter 10 for information on change impact analysis.) The software must be re-verified for the target hardware and aircraft or engine configuration intended for installation. And, the modified software must be approved by the certification authority for the equipment and aircraft or engine configuration prior to installation.

18.6 Summary

FLS is commonplace in today's aviation industry. However, the system and software design, loading considerations, and configuration management must still be carefully managed, particularly as newcomers come to the market and as loading approaches advance.

References

1. RTCA DO-178C, *Software Considerations in Airborne Systems and Equipment Certification* (Washington, DC: RTCA, Inc., December 2011).
2. Federal Aviation Administration, *Software Approval Guidelines*, Order 8110.49 (Washington, DC: Federal Aviation Administration, Change 1, September 2011).
3. Aeronautical Radio, Inc., Software data loader using ethernet interface, ARINC REPORT 615A-3 (Annapolis, MD: Airlines Electronic Engineering Committee, June 2007).
4. Aeronautical Radio, Inc., Portable multi-purpose access terminal (PMAT), ARINC REPORT 644A (Annapolis, MD: Airlines Electronic Engineering Committee, August 1996).

19

User-Modifiable Software

Acronyms

ACARS	Aircraft Communications Addressing and Reporting System
ACMS	Airplane Condition Monitoring System
AMI	Airline-Modifiable Information
EASA	European Aviation Safety Agency
FAA	Federal Aviation Administration
PSAC	Plan for Software Aspects of Certification
UMS	user-modifiable software

19.1 Introduction

This chapter defines user-modifiable software (UMS) and provides several examples of such software. The chapter then explains certification guidance to consider when designing an airborne system that contains UMS. The last section focuses on the user (e.g., an airline) who will modify and maintain the software after system certification.

19.2 What Is User-Modifiable Software?

UMS is "software intended for modification without review by the certification authority, the airframe manufacturer, or the equipment vendor, if within the modification constraints established during the original certification project" [1]. ARINC 667-1, entitled *Guidance for management of field loadable software*, describes UMS as: "software intended for modification by the aircraft operator without review by the certification authority, the TC [type certificate] or STC [supplemental type certificate] holder, or the equipment manufacturer. Modifications may include modifications to data or executable code, or both" [2].* There are various forms of UMS, including executable source code, aircraft-specific parameter settings, or databases [2].

The *user* is typically the aircraft operator or an airline. However, there may be situations where the user is the equipment customer (e.g., the aircraft manufacturer who installs a Technical Standard Order authorized unit, possibly as a commercial off-the-shelf unit, into their aircraft).

UMS is normally designed and initially approved as part of an airborne system that has both nonmodifiable software (e.g., flight controls, engine monitor, or navigation software) and modifiable software (e.g., airline modifiable checklists). The nonmodifiable software is not designed or intended to be changed by the user and has been the primary subject of this book. The UMS, on the other hand, *is* designed to be changed by the user utilizing approved procedures. The nonmodifiable software must be protected from any changes by the user. As will be discussed later in this chapter, there are several options for such protection.

During the initial certification, the system is designed to allow user modification and to protect the nonmodifiable software from user changes. Additionally, the UMS procedures are approved with the initial approval of the system. Since the UMS is outside the control of the equipment and aircraft manufacturers, it is typically treated as level E software during the initial certification. Some manufactures may treat it as level D, but that is for quality purposes rather than for certification.

Both DO-178C and Federal Aviation Administration (FAA) Order 8110.49 Chapter 7 address UMS. These documents concentrate on airborne software that is approved as part of type certification. They emphasize the need to design the system to be modifiable and to protect the nonmodifiable software from the UMS. Both documents also explain that UMS will not require involvement from the certification authority once it is approved as UMS. However, approval from the operational authority may still be required. Type certification and operational approval of the aircraft are typically granted by different divisions of the certification authority (e.g., in the United States, the FAA's Aircraft Certification Offices grant the aircraft type certificate and the Flight Standards Offices grant aircraft operational approval). Therefore, from the user perspective, there may be two classes of UMS: (1) UMS that requires operational approval, and (2) UMS that does not require operational approval. The first category is classified as *airline certifiable software* or *user certifiable software* by some entities, since the user or airline is responsible for obtaining the operational authority's approval [2].†

---

* Brackets added for clarification.
† It should be noted that some sources also use the term *airline certifiable software* to refer to software that also requires certification authority approval (e.g., an STC by the airlines); therefore, this term should be used cautiously.

19.3 Examples of UMS

The type and volume of UMS varies from aircraft to aircraft and airline to airline. A recent student of mine, who works for a major international airline, explained that one of the aircraft types that he manages has around 130 modifiable components. The following provides just a few examples of UMS.

Example 1: Aircraft Communications Addressing and Reporting System (ACARS). ACARS provides the ability to send messages between aircraft and ground stations. Users can optimize the system for the specific airline's needs. Some of the functions that ACARS may perform are abnormal flight condition identification, repair and maintenance plan, e-mail messaging between crew and air traffic control, weather reports, and engine reports.

Example 2: Airplane Condition Monitoring System (ACMS). The ACMS allows the user to collect data necessary to make decisions about maintenance. Users can program the ACMS to log specific data for future analysis. The ACMS receives data from the airborne system, but it does not send data to the nonmodifiable software.

Example 3: Airline-specific checklists. Some airlines have programmable checklists for their specific operations. Such checklists may require operational approval by the regulatory authority but are not approved as part of the type certification.

Example 4: Modifiable circuit breaker panel. A modifiable circuit breaker panel allows nonrequired equipment (e.g., coffee pot, flight entertainment, or stereo) to be added after type certification. UMS is used to set the circuit breaker limits based on the user input.

Example 5: Cabin placard and lighting system. The placards and lighting system varies from airline to airline. Therefore, some airlines program these using UMS. This UMS requires operational approval but is not approved as part of the aircraft type certificate.

Example 6: Airline-Modifiable Information (AMI). The AMI is a data file that allows users to specify preferences for cabin management data, recording, report formatting, and services provided to the various passenger seating zones (e.g., first class, business class, and economy class).

19.4 Designing the System for UMS

Systems with UMS must be designed to be modifiable. DO-178C (sections 2.5.2 and 5.2.3), FAA Order 8110.49 (chapter 7), and the European Aviation Safety Agency (EASA) Certification Memorandum CM-SWCEH-002 (chapter 9) provide insight into how to design a system to accommodate UMS [1,3,4]. This section briefly summarizes that process using _developer recommendations_ that are applicable during the initial certification of the system. The next section (Section 19.5) then provides _user recommendations_ for those who will modify the software after certification.

Developer Recommendation 1: Consider the UMS aspects during the type certification and system development. The intended presence of UMS should be planned (in the Plan for Software Aspects of Certification [PSAC], Software Development Plan, and possibly the Software Verification Plan) and coordinated with the certification authority. UMS is normally explained as an _additional consideration_ in the PSAC. Requirements and design should be in place to accommodate the UMS and to protect the nonmodifiable software. Such requirements must also be verified.

The UMS itself is not approved as part of the type certificate. As noted in DO-178C (section 7.2.2.b): "User-modifiable software is not included in the software product baseline, except for its associated protection and boundary components. Therefore, modifications may be made to user-modifiable software without affecting the configuration identification of the software product baseline" [1]. Basically, the system approved by the certification authority will be able to accommodate UMS, but will not include the UMS itself.

Developer Recommendation 2: Ensure that the UMS cannot adversely affect safety-related data. The UMS itself and changes to the UMS must not impact any of the following safety-related data: safety margins, operational capability of the aircraft or engine, flight crew workload, nonmodifiable components, protective mechanisms, software boundaries (such as preverified ranges of data) [3]. These should be considered as part of the system safety assessment process. If any of the safety-related data can be impacted by the UMS or a change to the UMS, the software should not be classified as UMS.

Developer Recommendation 3: Involve appropriate stakeholders. Since UMS crosses the airworthiness and operational boundaries (i.e., aircraft and system design and user operation), it is important to involve the appropriate stakeholders, including the certification authority, operational approval authority, aircraft manufacturer, systems developer, software developer, safety personnel, and users. Coordination with these entities early and throughout the development is essential to success.

Developer Recommendation 4: Put measures in place to protect nonmodifiable software. DO-178C section 5.2.3.a emphasizes that "the non-modifiable component should be protected from the modifiable component" [1].* The protection should be robust during operation of the UMS and when modifying the UMS. Protection may be implemented via hardware (e.g., separate

---

* DO-248C frequently asked question #7 also discusses this concept [5].

processors), software (e.g., a software partition), tools (e.g., a user interface that limits what can be entered), or a combination of the hardware, software, and tools [1]. One example of protection is a monitor function embedded in the core software to prevent any changes to nonmodifiable software. The monitor and the nonmodifiable software are inaccessible to the user [2]. The development assurance level of the protective mechanism should be the same as the most severe failure condition in the system that the UMS could impact [3]. For example, if the airborne (nonmodifiable) software is level A, and software partitioning is used to protect the nonmodifiable software, the software partition needs to be developed to level A. (Chapter 21 provides information on partitioning.)

Developer Recommendation 5: Evaluate tools used to enforce protection of the nonmodifiable software. Most systems that accommodate UMS provide a tool or toolset for users to modify the UMS. The impact of the tool on nonmodifiable software needs to be considered. During the initial development of the system, the tool use, control, design, functionality (modifications to UMS), and maintenance should be considered. The tool must be robustly designed to prevent incorrect entries by the user. Order 8110.49 section 7.6.b even goes as far as to state: "Software forming a component of the tool and used in the protective function should be developed to the software level of the most severe failure condition of the system, as determined by a system safety assessment" [3].

There are cases when the tool may need to be qualified. Order 8110.49 section 7.6.c encourages tool qualification "and approval of procedures to use and maintain the tool" [3]. Unfortunately, the guidance on when the tool requires qualification is somewhat unclear. DO-178C section 12.2 provides guidelines for when tools used in DO-178C processes should be qualified and to what level. However, for UMS, this is not without ambiguity (since tools to enable UMS do not normally reduce, automate, or eliminate a DO-178C process). I generally recommend that if the protection mechanism relies on the tool alone, the tool should be qualified; conversely, if the tool's impact is mitigated by system architecture (e.g., the embedded monitor), qualification may not be needed. Even if the tool does not require qualification, it should still be specified, verified, and configuration controlled. The determination of whether or not to qualify the tool must be coordinated with the certification authority. (See Chapter 13 for more information on tool qualification.)

Developer Recommendation 6: Consider displayed data. FAA Order 8110.49 (section 7.3) explains that any data displayed to the flight crew that were driven by UMS should be clearly noted to the crew as advisory or should have operational approval. Basically, it is important that the pilots understand the integrity of the data they observe in order to ensure that they use it properly. This type of data needs to be closely coordinated with flight test and human factors personnel during the initial certification of the aircraft.

Developer Recommendation 7: Prevent inadvertent and unauthorized changes. The UMS should only be modified by authorized personnel on the ground using the approved procedure. During the initial certification, it must be ensured that the approved procedure is the only way to change the UMS [3].

Developer Recommendation 8: Inform users of all UMS software. The type certificate holder should inform the aircraft user of all UMS on the aircraft. Additionally, approved procedures for changing the UMS should be clearly documented and provided to the users. Such procedures should be repeatable and under configuration control.

Developer Recommendation 9: Identify constraints and procedures for the users. The initial system development should clearly document constraints and procedures to ensure that users have adequate data to make modifications to UMS without impacting aircraft safety. DO-178C section 11.16.j indicates that the Software Configuration Index should identify "procedures, methods, and tools for making modifications to the user-modifiable software, if any" [1]. Such procedures need to be provided to the users. Interestingly, airline representatives indicate this is an area they frequently find lacking. Oftentimes, airlines receive unclear or incomplete documentation from the aircraft or systems manufactures. This causes airline employees to make assumptions which may or may not be accurate. This is one of the reasons why 11.16.j was added to DO-178C.

Additionally, any changes that impact the UMS should be clearly communicated to the users (e.g., changes to the system which accommodates the UMS, updates to the approved procedures, or modifications to the tool used to change the UMS). Airline operators indicate that they rarely receive updated procedures when the system, procedures, or tool are updated.

19.5 Modifying and Maintaining UMS

Once the software is approved under the type certificate as UMS, it is the user's responsibility to manage the UMS. This includes management of the environment, the changes to the UMS, and the overall aircraft configuration. The available guidance for modifying and maintaining UMS is very sparse; therefore, this section provides some recommendations for users who are required to manage UMS. Since the types of UMS vary greatly, these recommendations are general. The specific details must be worked out among the project-specific stakeholders.

User Recommendation 1: Ensure agreements are in place. Users should ensure that agreements are in place (e.g., technical service agreement) with the aircraft and/or equipment manufacturer(s) to obtain all of the necessary data, tools, and support.

User Recommendation 2: Manage the configuration of all UMS on the aircraft. It is important to know what version of software (UMS, as well as nonmodifiable software) is installed on each aircraft. Maintenance records should identify when specific software versions or part numbers are installed.

User Recommendation 3: Manage the environment of each UMS component. Each UMS component will have an environment to change and modify the UMS. The tools that comprise the environment are typically provided by the equipment manufacturer or aircraft manufacturer (examples included software editor, compiler, part number generator, and media set creation tools) [2]. The supporting tools should be under configuration management. If updated tools are provided by the equipment or aircraft manufacturer, the impact of these tools should be assessed prior to their use. Understanding this impact may require coordination with the equipment or aircraft manufacturer.

User Recommendation 4: Modify the software using constraints and procedures provided. As noted in Developer Recommendation #9, the equipment or aircraft manufacturer should identify any constraints and procedures to be followed by the user. These constraints and procedures are referenced by the type certification data (i.e., is approved as part of the type certificate). Therefore, the user is expected to follow the procedures. If the procedures are missing or unclear, the user should coordinate with the supplier. As noted in Order 8110.49 section 7.8, failure to follow the approved procedures could have serious consequences, including potential withdrawal of the aircraft type certificate [3].

User Recommendation 5: Make modifications to UMS using an organized and repeatable process. The modifications to UMS should be planned and driven by documented requirements. The configuration management process for the UMS modification should include change control (to ensure changes are authorized), change tracking (to ensure the changes are documented), configuration identification (update to software version or part number, as well as the supporting documentation), release, and archival. The modification should be verified (preferably using some testing) to meet the requirements and monitored by quality assurance. While the UMS is typically not required to comply with DO-178C, consider using a DO-178C-like process (such as DO-178C level D). Such a process helps to ensure the functionality, configuration management, and quality assurance of the UMS. Whatever process is used, it should be auditable.

User Recommendation 6: If needed, obtain operational approval of the UMS. As previously noted, some UMS may require approval by the operational authority, even though it may not need type certification authority review. Such approval should be obtained prior to using the modified UMS for flight operation.

User Recommendation 7: Ensure proper installation of the modified UMS on the aircraft. Once the appropriate approval is granted (e.g., operational approval), it should be ensured that the approved version of the UMS is actually the version installed on the aircraft. As noted in User Recommendation #2, maintenance records should be kept up to date.

References

1. RTCA DO-178C, *Software Considerations in Airborne Systems and Equipment Certification* (Washington, DC: RTCA, Inc., December 2011).
2. Aeronautical Radio, Inc., Guidance for the management of field loadable software, ARINC Report 667-1 (Annapolis, MD: Airlines Electronic Engineering Committee, November 2010).
3. Federal Aviation Administration, *Software Approval Guidelines*, Order 8110.49 (Washington, DC: Federal Aviation Administration, Change 1, September 2011).
4. European Aviation Safety Agency, Software aspects of certification, Certification Memorandum CM-SWCEH-002 (Issue 1, August 2011).
5. RTCA DO-248C, *Supporting Information for DO-178C and DO-278A* (Washington, DC: RTCA, Inc., December 2011).

20

Real-Time Operating Systems

Acronyms

AC	Advisory Circular
AFDX	ARINC 664 avionics full-duplexed switched ethernet
APEX	APplication EXecutive
API	application program interface
BSP	board support package
CAN	controller area network
CEO	chief executive officer
COTS	commercial off-the-shelf
CPU	central processing unit
EAL	evaluation assurance levels
EPROM	erasable programmable read-only memory
FAA	Federal Aviation Administration
IEEE	Institute of Electrical and Electronic Engineers
IMA	integrated modular avionics
I/O	input/output
ISR	interrupt service routine
MMU	memory management unit
MOS	module operating system
POS	partition operating system
POSIX	portable operating system interface
RAM	random access memory
RSC	reusable software component
RTOS	real-time operating system
SAP	support access port
SVA	software vulnerability analysis
UNIX	uniplexed information and computing system
VMM	virtual machine monitor

20.1 Introduction

Since the arrival of the new millennium, the real-time operating system (RTOS) has become a common component in aviation systems. This chapter explains what an RTOS is, why it is used, how it fits into the typical avionics system, desired RTOS functionality, issues to be addressed when using an RTOS, and some future RTOS-related challenges.

20.2 What Is an RTOS?

H.M. Deitel writes:

> Operating systems are primarily resource managers; the main resource they manage is computer hardware in the form of processors, storage, input/output (I/O) devices, communication devices, and data. Operating systems perform many functions such as implementing the user interface, sharing hardware among users, allowing users to share data among themselves, preventing users from interfering with one another, scheduling resources among users, facilitating I/O, recovering from errors, accounting for resource usage, facilitating parallel operations, organizing data for secure and rapid access, and handling network communications [1].

In general, an operating system is software that manages the hardware resources of a computer, providing controlled access for one or more applications running on the computer. A general-purpose RTOS performs these operations but is also specially designed to run applications with very precise timing. Safety-critical RTOSs are a subset of the general-purpose RTOS and tend to have the following characteristics: deterministic (predictable), responsive (in a guaranteed timeframe), controllable (by the software developer and integrator), reliable, and fail-safe [2]. These characteristics will be further discussed when considering the desirable features of a safety-critical RTOS.

The following terms are used in this chapter and in RTOS literature:*

- *Application*: Software which consists of tasks or processes that perform a specified function on the aircraft. An application may contain one or more partitions [3].

- *Application Program Interface (API)*: A formal set of software calls and routines that can be referenced by an application program in order to access supporting system or network services.

---

* These definitions are based primarily on ARINC 653 Part 1-3, entitled *Avionics Application Software Standard Interface: Part 1—Required Services* [3].

- *Partition*: "A program, including instruction code and data, that is loadable into a single address space in a core module" [3]. The RTOS has control over each partition's use of computer resources (processing time, memory, and other resources) in order to isolate each partition from all others that share the core processing hardware.
- *Interpartition*: Communication between partitions.
- *Robust Partitioning*: "A mechanism for assuring the intended isolation of independent aircraft operational functions residing in shared computing resources in all circumstances, including hardware and programming errors. The objective of robust partitioning is to provide the same level of functional isolation as a federated implementation (i.e., applications individually residing on separate computing elements). This means robust partitioning must support the cooperative coexistence of applications on a core processor, while assuring unauthorized, or unintended interference is prevented" [3].

20.3 Why Use an RTOS?

The RTOS is the heart of many modern avionics systems. An RTOS can impact software dependability, productivity, and maintainability [4]. Using an API, the RTOS provides a clearly defined and controlled interface between the application and the underlying hardware. Additionally, an RTOS narrows the possible interactions, making it easier to verify correctness of the applications. Without an RTOS, the programmer needs detailed knowledge of the underlying hardware and its functionality.

In the past, the software in embedded systems was written as a monolithic set of code and was tightly coupled to the underlying hardware. This was necessary to ensure that the performance requirements were met. Unfortunately, this type of design made the software difficult to maintain, reuse, and port to different hardware.

By abstracting and encapsulating the hardware interface, an RTOS largely eliminates the need for the application developer to write hardware-specific code. The RTOS improves portability and

> screens the complexities of the computer from the programmer, leaving him to concentrate on the job at hand. Detailed knowledge of interrupts, timers, analogue-to-digital converters, etc. is no longer needed. As a result the computer can be treated as a *virtual* machine, providing facilities for safe, correct, efficient, and timely operation. In other words, it makes life easy (or at least easier) [4].

Another driver of increased RTOS usage has been the microprocessor's increased capability. With the speed and capacity of today's processors, it is now possible to run multiple applications on a single processor. Additionally, since weight reduction saves cost over the life of an aircraft, there is a great desire to minimize hardware on the aircraft. In order to maximize the benefits of the processor, minimize weight on the aircraft, and improve maintenance, integrated modular avionics (IMA) systems are replacing many of the traditional federated systems. An IMA allows multiple applications to run on a single processor.

To enable IMA, robust partitioning between the applications is needed. A robustly partitioned RTOS ensures the following [6]:

- The execution of an application does not interfere with the execution of any other application.
- Dedicated computer resources allocated to applications do not conflict or lead to memory, schedule, or interrupt clashes.
- Shared computer resources are allocated to applications in a way that maintains the integrity of the resources and the separation of the applications.
- Resources are allocated to each application independently of the presence or absence of other applications.
- Standardized interfaces to applications are provided.
- Software applications and the hardware resources needed to host them are independent.

IMA implementation hinges on the robustly partitioned RTOS and its environment. Chapter 21 discusses partitioning in more detail.

The competitiveness of the aviation market has also made the use of commercial off-the-shelf (COTS) RTOSs appealing. Use of a COTS RTOS largely eliminates the need for system developers to develop their own RTOS and maintain operating system development as a core competency. A DO-178C (or DO-178B) compliant COTS operating system is quite expensive, but can be more cost effective than building an RTOS from scratch, maturing it to support safety needs, developing it to meet DO-178C, and maintaining it. Some of the common issues with COTS RTOSs are discussed later in this chapter.

20.4 RTOS Kernel and Its Supporting Software

Figure 20.1 illustrates the typical place of the RTOS in an avionics system. The applications software may be in one or multiple partitions. Both partitioned and nonpartitioned RTOSs are used in safety-critical systems. The partitioned RTOSs are becoming more common in aviation projects. Since the RTOS kernel and its supporting software are part of the embedded

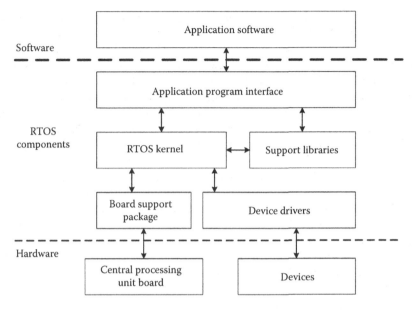

FIGURE 20.1
Typical RTOS components and relationship to applications.

system, they need to meet DO-178C (or DO-178B).* Figure 20.1 shows that the RTOS kernel is supported by several components, including libraries, board support package (BSP), device drivers, and an API. The RTOS kernel and each of the main supporting components are briefly described in the following subsections.

20.4.1 RTOS Kernel

The kernel is the heart of the RTOS. It provides the basic services (such as the ARINC 653 services described in Section 20.8.1) and is designed to be independent of the underlying hardware, as much as possible.[†] The goal of most RTOS developers is to isolate the kernel from the hardware in order to allow reuse and portability. The hardware-specific code is normally implemented in the BSP and device drivers.

20.4.2 Application Program Interface

The applications access the RTOS services through an API, which functions as the interface between the RTOS kernel and an application.

---

* At time of this writing RTOSs have met DO-178B, but with the publication of DO-178C, they will be transitioning to DO-178C. This should be a relatively easy transition for RTOSs, since most of them do not use object-oriented technology, model-based development, or formal methods.
† There are some parts of the RTOS that are hardware specific, for example, the code that does context switch, but this is kept to a minimum.

The API provides a list of available calls that a programmer can make using the RTOS. The ARINC 653 APEX (APplication EXecutive) is the most common API used in avionics for IMA systems. The POSIX (portable operating system interface) is also employed by some RTOSs. POSIX standards (such as Institute of Electrical and Electronic Engineers [IEEE] 1003.1b) include the following interface definitions for an RTOS: task management, asynchronous I/O, semaphores, message queues, memory management, queued signals, scheduling, clocks, and timers [7]. POSIX is based on UNIX (uniplexed information and computing system). Because there were multiple flavors of UNIX, there are several IEEE standards related to POSIX [4]. As UNIX was not developed for a hard real-time environment, POSIX must be used carefully in avionics. This was one of the motivations for the development of the ARINC 653 APEX. There is some similarity between ARINC 653 and POSIX, since many of the POSIX concepts influenced ARINC 653. The difference between POSIX and ARINC 653 APEX is small at the API level; however, the differences are more significant at the program organization level. APEX was conceived to provide separation between different applications sharing the same processor; i.e., to provide robust partitioning between the APEX partitions. POSIX does not provide the same kind of robust partitioning, but it does provide support for multiple processes to collaborate. Table 20.1 provides a high-level general comparison of ARINC 653 and POSIX.

20.4.3 Board Support Package

The BSP isolates the RTOS from the target computer processing hardware. It allows the RTOS kernel to be portable to various hardware architectures within the same central processing unit (CPU) family. "The BSP initializes the processor, devices, and memory; performs various memory checks; and so on. Once initialization is complete, the BSP can still function to perform low-level cache manipulations" [9], as well as some other hardware access (e.g., flash manipulation or timer access). Much of the BSP code operates in privileged mode and works closely with the RTOS. The BSP is sometimes called a *hardware abstraction layer*, a *hardware interface system*, or *platform enabling software*. It is customized for the specific hardware and is often implemented using C and assembly. The BSP is sometimes developed by the RTOS users (e.g., the avionics developer) using a template from the RTOS supplier—especially when the avionics utilize custom hardware. At other times, the BSP is provided and/or tailored by the RTOS supplier. Depending on the nature of the hardware, the BSP can be a rather large component.

20.4.4 Device Driver

Similar to the BSP, the device driver provides the interface between the RTOS kernel and a hardware device. The hardware devices may be on the processor board or separate. Hardware on the processor board may be handled by

TABLE 20.1

Basic Comparison of POSIX and ARINC 653 Features

POSIX	ARINC 653 APEX
Event-driven execution model.	Cyclic-based execution model to schedule partitions.
Uses a multiprocessing and multithreading execution model.	Same concept as POSIX; however, terminology differs: Uses *partitions* instead of *processes*, and *processes* instead of *threads*.
Does not support temporal partitioning.	Does support temporal partitioning using round-robin scheduling between partitions.
Uses priority preemptive scheduling for processes and threads. Cyclic scheduling can be programmed.	Supports priority preemptive scheduling for threads within a partition, and cyclic scheduling between partitions.
Input/output (I/O) scheduling depends on the device drivers, and may be interrupt-driven or polled.	Many models are possible, with interrupt-driven while the partition doing the I/O is running, or polled I/O to get better control (i.e., I/O on request).
Supports memory segregation.	Same concept as POSIX; however, terminology is different: Uses *spatial partitioning* instead of *memory segregation*. Memory layout, access policies, cache policies, and memory protection are provided using static configuration tables.
Uses sockets.	Uses sampling and queuing ports tied to pseudo ports to communicate off-board. These may be linked to Ethernet sockets, or other protocols.
Uses interprocess communication.	Communicates using sampling and queuing ports, or through shared memory (random access memory) controlled by the memory management system.
Uses timers.	API services can respond to timeouts. Processes can be periodic; delay mechanisms and other mechanisms can be used to synchronize application code with time.
Uses signal management, and requires user to establish error response policies.	Uses health monitoring to manage user policies established in configuration tables or ties in to user-defined error handlers.

Source: Goiffon, S. and Gaufillet, P., Linux: A multi-purpose executive support for civil avionics applications? *IFIP—International Federation for Information Processing*, 156, 719–724, 2004. Also includes input from George Romanski of VEROCEL.

the BSP or a separate driver. The driver is a low-level, hardware-dependent software module that provides the interface between an application (sometimes mediated by higher level RTOS library or kernel functions) and a specific hardware device. A device driver is responsible for accessing the hardware registers of the device and may include an interrupt handler to service interrupts generated by the device. Most avionics systems have multiple devices, such as Ethernet devices, ARINC 664 avionics full-duplexed switched ethernet (AFDX) end systems, RS-232 serial devices, I/O ports, analog-to-digital converters, and controller area network (CAN) databuses.

20.4.5 Support Libraries

Oftentimes, language-specific run-time support is provided as libraries with the RTOS kernel. For example, "in the C language, a number of standard library specifications permit the user to call functions to move memory, compare strings, and use mathematical functions such as *mod, floor,* and *cos*" [9]. These run-time libraries are often packaged with the RTOS (either within the kernel or as a separate library package). The libraries are normally provided by the RTOS supplier as precompiled object code that can be linked together with the applications. That way, only the library functions needed are linked in. Some smart linkers can actually identify which library functions are called by the application and only link those functions into the executable.

20.5 Characteristics of an RTOS Used in Safety-Critical Systems

This section identifies the key characteristics of safety-critical RTOSs used in the aviation domain.

20.5.1 Deterministic

A safety-critical system delivers the right answer in the right order and at the right time. "Determinism is the characteristic of a system which allows the correct prediction of its future behavior given its current state and knowledge of future changes to its environment (inputs)" [10]. Nondeterminism on the other hand "means that a systems' future behavior cannot be correctly predicted. (An unpredictable system cannot be called 'safe.')" [10]. The behavior of an RTOS in a safety-critical domain must be predictable. That is, given a particular input, the RTOS generates the same output. In particular, the outputs are within the bounds defined in the requirements. A deterministic RTOS provides both functional correctness and temporal correctness as defined by the requirements; it also only consumes known and expected (bounded) amounts of memory and time.

20.5.2 Reliable Performance

The RTOS must meet the *horsepower* requirements needed to enable the application to perform its intended functionality. Many factors determine the RTOS performance, including computation times, scheduling techniques, interrupt handling, context switch times, cache management, task dispatching, etc. [4]. Performance benchmarks are typically provided in a data sheet with the RTOS. Performance varies widely depending on the specific hardware, interfaces, clock speed, compiler, design, and operating environment. Thus, when selecting an RTOS, companies almost always run their own benchmarks before deciding which RTOS to purchase.

20.5.3 Compatible with the Hardware

The RTOS kernel must be compatible with the selected processor. Likewise, the BSP must be compatible with the selected processor board and core devices, and the device drivers must be compatible with the system devices.

20.5.4 Compatible with the Environment

The RTOS should be capable of supporting the selected programming language and compiler. Most RTOSs are supported with an integrated development environment, which includes the compiler, linker, debugger, and other tools for successful integration of the applications, RTOS, supporting software, and the hardware.

20.5.5 Fault Tolerant

Fault tolerance is "the built-in capability of a system to provide continued execution in the presence of a limited number of hardware or software faults" [11]. A fault tolerant RTOS plans for application failures and provides mechanisms for recovery or shutdown. The RTOS enables fault management which consists of (1) detecting faults, failures, and errors; (2) correctly identifying faults, failures, and errors when they are detected; and (3) performing the response preplanned for the system [11].

20.5.6 Health Monitoring

Health monitoring is closely related to fault management and is another feature of a fault tolerant RTOS. Health monitoring is a service frequently provided by an RTOS to provide fault management for most (if not all) of the system that uses the RTOS. DO-297 explains that health monitoring is responsible for detecting, isolating, containing, and reporting failures that could adversely affect resources or the applications using those resources [11]. DO-297 focuses on the IMA platform; however, the RTOS plays a key role in the overall health management scheme of the system. The Federal Aviation Administration (FAA) research report on RTOSs in IMA systems states the following:

> Health monitoring is specified in ARINC 653. The health monitor is the RTOS function responsible for identifying, responding to, and reporting hardware, application, and RTOS faults and failures. Health monitoring helps to isolate faults and prevents failures from propagating.

By classifying and categorizing faults and the health management responses, a range of possibilities exists that helps the application supplier and IMA system integrator select appropriate behavior. Configuration tables are used to describe intended recovery of identified faults, such as ignoring the fault, reinitializing a process, restarting (warm or cold) a partition, performing a total system reset, or calling a system-specified routine to take system-specified actions [12].

ARINC 653 identifies a number of error codes that must be detected and handled. "An error can be handled within a process, in a partition, or in the health monitoring module or process" [9].

The objective of the ARINC 653 health monitoring function is to

contain faults before they propagate across an interface boundary. In addition to self-monitoring techniques, application violations, communication failures and faults detected by applications are reported to the RTOS. A recovery table of faults is used to specify the action to be taken in response to the particular fault. This action [is] initiated by the Health Monitor and might include terminating an application and starting an alternative application, together with an appropriate level of reporting [13].*

The recovery action is dependent on the system design and requirements.

20.5.7 Certifiable

Certifiability is a necessary characteristic of an RTOS utilized by safety-critical systems. If the RTOS is to be installed in an aircraft, it will need to satisfy the objectives of DO-178C.† It is worth noting that the RTOS itself is not certified; rather, it is developed to be certified as part of an aircraft, engine, or propeller.‡ Most COTS RTOSs suitable for installation in aviation products have a *certification package* available to support the overall certification. The certification package contains the artifacts to show compliance with DO-178C (or DO-178B) and to support safety and certification. Some of the typical certification challenges for RTOSs are presented later in this chapter (Section 20.7).

20.5.8 Maintainable

Maintainability is a desirable characteristic for all software used in safety-critical systems, including the RTOS. Since the life of the system is likely to be quite long (possibly over 20 years), the RTOS must be maintainable

---

* Brackets added for clarity.
† Most RTOSs that complied with DO-178B will also comply with DO-178C, unless they use object-oriented or related programming techniques, model-based development, or formal methods. However, some attention to DO-178C's guidance on parameter data, data and control coupling, structural coverage, and trace data may be needed.
‡ The FAA certifies aircraft, engines, and propellers—not the parts.

(i.e., able to be modified to accommodate additional applications, equipment, hardware, etc.). The life cycle data and configuration management processes required by DO-178C support maintainability.

20.5.9 Reusable

While not a required characteristic, most RTOSs are designed to be reusable. As noted previously, the kernel is abstracted from the hardware using a BSP and device drivers. The FAA's Advisory Circular (AC) 20-148, entitled *Reusable Software Components*, is typically applied to an RTOS. AC 20-148 provides FAA guidance for packaging a reusable software component (RSC) for reuse in multiple systems. Even if the project does not seek an RSC letter from the FAA, they will probably want to follow the suggestions of the AC in order to develop the RTOS *component* to be as reusable as possible. Reuse is further discussed in Chapter 24.

20.6 Features of an RTOS Used in Safety-Critical Systems

This section identifies the key features of most safety-critical RTOSs used in the aviation domain. These features are closely related to the characteristics noted earlier and provide the means to implement those characteristics. The features most pertinent from a safety perspective are discussed.

20.6.1 Multitasking

Multitasking is a method where multiple tasks, also known as processes, share a common processor. Multitasking creates the appearance that many tasks are running concurrently on a processor when, in fact, the kernel interleaves (for a single core processor) their execution using a scheduling algorithm. When the processor stops executing one task and starts executing another task, this is called a context switch. When context switches occur frequently enough, it appears that tasks are running in parallel. Each seemingly independent program (task) has its own context, which is the processor environment and system resources that the task sees each time it is scheduled to run by the kernel.

20.6.2 Guaranteed and Deterministic Schedulability

Meeting deadlines is one of the most fundamental requirements of an RTOS. For safety-critical systems, the scheduling and timely completion of multiple tasks must be deterministic. For RTOSs used in advanced avionics

(e.g., IMA systems), two kinds of schedulability are normally desired:*
(1) scheduling between partitions and (2) scheduling within partitions. Each
is discussed here.

20.6.2.1 Scheduling between Partitions

The ARINC 653 model requires a round-robin execution sequence between
partitions. This establishes a time frame called the major time frame. Time
slots of fixed duration are defined within the major time frame. However,
the duration of each time slot does not need to be the same. A configuration
table defines which partitions will execute in which time slot within a frame:

> A partition may be allocated to more than one slot within a frame.
> Execution of each partition follows in sequence, starting at the beginning
> of a frame. When the last partition in the frame is executed, the sequence
> is repeated. The time slots are strictly enforced by the MOS [Module
> Operating System]. A clock device is used to ensure the timely switching
> between partitions [9].†

20.6.2.2 Scheduling within Partitions

Many types of scheduling schemes can exist within a partition. Most RTOSs
do their scheduling using priorities and preemption. Each task is assigned a
priority. A higher priority task is executed before all tasks with lower priority.
Preemptive means that as soon as a higher priority task is ready to run, it can stop
the execution of a lower priority task that is currently executing. It's similar to
when you're in a meeting with the boss. If the chief executive officer (CEO) of
the company comes by, your discussion with your boss is preempted until he or
she is finished with the higher priority discussion with the CEO. The following
list summarizes the most common scheduling approaches used in safety-critical
RTOSs. There is a plethora of literature available on each of these processes; only
a brief synopsis is discussed here:‡

- *Cyclic executive.* This technique cycles through a set of processes
  whose execution order has been predetermined. It is a common
  approach even when an RTOS is not used.
- *Round-robin.* The name of the algorithm comes from the round-robin
  principle, where each person has an equal share of something. I once

---

* For nonpartitioned systems, only the second kind of schedulability is needed.
† Brackets added for clarification. In the ARINC 653 specification, two scheduling mechanisms
  are suggested. The Module Operating System (MOS) provides scheduling for partitions, and
  the Partition Operating System (POS) provides scheduling for processes within a partition.
‡ Although somewhat dated, FAA Report DOT/FAA/AR-05/27, entitled *Real-time scheduling
  analysis*, provides some useful information on real-time scheduling in the aviation industry.
  It is available at FAA's website: www.faa.gov.

made it to the 4-H showmanship round-robin competition. Each competitor was given an equal time to show a horse, a steer, a pig, and a lamb. You'll be pleased to know that I earned the Reserved Grand Champion trophy. The round-robin scheduling algorithm is one of the simplest scheduling algorithms for sharing time on a processor. A minor time slice is defined by the system, and all processes are kept in a circular queue. The scheduler goes around the queue, allocating processor resources to each process for a time slice interval. As new processes arrive, they are added to the tail of the queue. The scheduler functions by selecting the first process from the queue, setting a timer to interrupt after one time slice, and then dispatching the process. If the process is not finished at the end of the time slice, it is preempted and added to the tail of the queue. If the process does finish before the end of the time slice, it releases the processor. A context switch occurs every time a process is granted access to the processor. The context switching adds overhead to the process execution time [9].

- *Fixed priority preemptive scheduling.* Each task has a fixed priority that does not change. With fixed priority preemptive scheduling, the scheduler ensures that at any given time the highest priority task is executed out of all the tasks that are currently ready to execute. This approach uses an interrupt to preempt a lower priority task if a higher priority task becomes ready. The advantage of this approach is that it ensures that lower priority tasks don't monopolize the processor time. However, the negative aspect is that it can block a lower priority task from executing indefinitely. Many RTOSs support this scheduling scheme.

- *Rate monotonic scheduling.* The rate monotonic scheduling algorithm is considered the optimal static priority algorithm. It is a priority preemptive algorithm that assigns fixed priorities to tasks in order to maximize schedulability and to ensure that all deadlines are met. Each task is assigned a priority according to its period. The shorter the period (higher frequency), the higher the priority. This scheduling algorithm is implemented in several of the COTS RTOSs for safety-critical systems.

- *Deadline monotonic scheduling.* The deadline monotonic scheduling algorithm is similar to rate monotonic scheduling, except the priority is assigned to the task with the shortest deadline, rather than the shortest period. Thus, the process with the shortest deadline is assigned the highest priority and the process with longest deadline gets the lowest priority.

- *Earliest deadline first scheduling.* This is a dynamic priority preemptive policy. It places tasks in a priority queue, so that whenever a task finishes, the queue is searched for the process closest to its deadline.* This process becomes the next task scheduled for execution. Basically, the scheduler selects the process that has the earliest deadline to run first, which preempts any processes with a later deadline.

- *Least slack scheduling (also known as least laxity first scheduling)*. As with earliest deadline first approach, this is a dynamic priority preemptive policy. The priority is assigned based on the *slack time* (the remaining time to deadline minus the remaining execution time, which can be difficult to precisely predict). Slack time might also be described as the difference between the time until a task's completion deadline and its remaining process time requirement. The process with the smallest slack preempts processes with larger slack.

20.6.3 Deterministic Intertask Communication

Since only one task can be run at a time (for single core processor), there must be mechanisms for tasks to communicate with one another. For many RTOSs, queues and messages provide a means for intertask communication; this approach helps to avoid the situation where a task reads a segment of memory while another is writing to it. There are at least three reasons that intertask communication is needed [4]:

1. *To synchronize or coordinate activities without data transfer.* Generally such tasks are linked by events, including time-related factors, such as time delays or elapsed time. Task synchronization or task cooperation is used to synchronize the shared resources. There is some overlap between task synchronization and task coordination because task synchronized operations may be used to coordinate tasking operations. However, as soon as an interrupt is used to block and release tasks, it is no longer task coordination because the tasks are synchronized with the real-time interrupts.

2. *To exchange data without synchronizing.* There are times when tasks exchange information without the need for synchronization or coordination. This is often accomplished with a data store (e.g., pools or buffers) that incorporates mutual exclusion to ensure data are not corrupted.

3. *To exchange data at carefully synchronized times.* This is the scenario where tasks wait for events and use data associated with those events. For example, task synchronization may be achieved by suspending or halting tasks until the required conditions are met.

---

* This is similar to deadline monotonic scheduling, except earliest deadline first scheduling is dynamic.

20.6.4 Reliable Memory Management

RTOSs support memory allocation and memory mapping and take action when a task uses memory. Most processors include an on-chip memory management unit (MMU) that allows software threads to run in hardware-protected address space [14]. The MMU is responsible for handling accesses to memory. MMU functionality typically includes virtual addresses to physical addresses translation, memory protection, cache control, and bus arbitration. Since the MMU is COTS functionality provided with the processor and the integrity of the MMU is unknown, the MMU typically requires some special attention during the certification effort. Generally, the accuracy of the MMU's functionality, as well as the overall processor functionality, is verified through the detailed testing of the RTOS and its supporting software (e.g., BSP). Some of the common memory issues (such as memory leakage, fragmented memory, and memory coherency) to avoid are discussed later.

20.6.5 Interrupt Processing

Real-time systems usually respond to external events using interrupts through interrupt service routines (ISRs). The RTOS normally saves the context of the interrupted task to return after the interrupt is processed. The interrupt processing can have a significant impact on the system performance. Generally, interrupt processing does the following [15]:

- Suspends the active task.
- Saves the task-related data that will be needed when resuming the task.
- Transfers control to the appropriate ISR.
- Performs some processing in the ISR to determine necessary action.
- Retrieves and saves critical (incoming) data associated with the interrupt.
- Sets required device-specific (output) values.
- Clears the interrupt hardware so the next interrupt can be recognized.
- Transfers control to the next task, as determined by the scheduler.

In order to prevent race conditions (two tasks attempting to change the same data without coordination), RTOSs sometimes disable interrupts while accessing or manipulating internal (critical) operating system data structures. The maximum time that an RTOS disables interrupts is referred to as the interrupt latency. Worst-case interrupt latency times should include "all software overhead that must be endured before the actual ISR is executed" [16]. There is typically a trade-off between interrupt latency, throughput, and processor utilization.* This should be factored in when determining worst-case performance [17]. A key to meeting performance is to have low interrupt latency [18].

Some RTOSs reduce interrupt latency through a technique called *work deferral*. If an ISR causes other RTOS work to be scheduled during an interrupt, the work may be saved on a work queue and invoked later. This reduces the times of critical regions and makes the system more responsive; however, it adds complexity when calculating the worst-case execution time.

20.6.6 Hook Functions

Many RTOSs include what is referred to as *hook functions* (or *callback functions*). A hook function allows a developer to associate application code to a particular function within the RTOS. The hook is executed by the RTOS function (sometimes with elevated priority and access to RTOS resources). These hooks can be used to extend the RTOS for specific user needs, particularly if the proprietary RTOS source code is not available. Hook functions allow customization without modifying the RTOS source code and inadvertently introducing a defect. Hook functions can allow the RTOS user to perform some actions before or after the RTOS responds to some event without the overhead of a separate task creation. Oftentimes, hook functions are used to assist with debugging during development. They may be deactivated or disabled in the final product. Obviously, hook functions must be handled carefully in safety-critical systems because they have the ability to modify the behavior of the RTOS itself [19].

20.6.7 Robustness Checking

An RTOS should be designed to protect itself against certain user errors. Examples of built-in robustness checks include validating parameters passed through the API by an application making a system call, ensuring that a task priority is within the permitted range of the RTOS, or ensuring that a semaphore operation works only on a semaphore object. Robustness checking is complex and verification of the robustness features may require special testing techniques [20].

20.6.8 File System

File system is a way of managing data storage.† Similar to the file system in a desktop environment, the RTOS file system manages and hides the details of the various forms of data storage on the hardware. The file system provides the ability to open, close, read, write, and delete files and directories.

---

* It should be noted that interrupt handling overhead isn't the only overhead that reduces throughput (i.e., the useful computation of results by applications per time period). Context switching, periodic built-in test, health monitoring, etc. also take time and impact throughput.
† ARINC 653's Part 2 includes file system as an extended service.

The implementation is unique to the type of storage media (such as random access memory (RAM), flash memory, erasable programmable read-only memory (EPROM), or network-based media). The file system allows multiple partitions to access the storage media. The RTOS kernel or its supporting library implements the file system to manage the low-level details of the media for the partitions that use it. Avionics applications often use a file system to store and retrieve data. For example, flight management systems and terrain awareness systems may use a file system to access their databases [13].

20.6.9 Robust Partitioning

Many IMA RTOSs support robust partitioning. The definition of *robust partitioning* was included earlier (in Section 20.2). DO-297 section 2.3.3 explains the following:

> The objective of robust partitioning is to provide an equivalent level of functional isolation and independence as a federated system implementation (i.e., applications individually residing on separate Line Replaceable Units (LRU)). This means robust partitioning supports the cooperative coexistence of applications using shared resources, while assuring that any attempted unauthorized or unintended interaction is detected and mitigated. The platform robust partitioning protection mechanisms are independent of any hosted applications, that is, applications can not alter the partitioning protection provided by the platform [11].

The RTOS plays a key role in implementing the robust partitioning—to ensure that the shared resources of time, memory, and I/O are protected. Partitioning is described in more detail in Chapter 21.

20.7 RTOS Issues to Consider

This section summarizes technical and certification issues that often require consideration when developing an RTOS and when implementing it for a safety-critical system. Some issues were noted earlier but are further elaborated here. This is not an exhaustive list of issues that may be encountered on a program but is some of the ones that have traditionally arisen and that are generally most pertinent to safety.

20.7.1 Technical Issues to Consider

20.7.1.1 Resource Contention

As the name implies, resource contention is a conflict over a shared resource, such as processor or memory. Three specific contentions that need to be dealt with in an RTOS are deadlock, starvation, and lockout. Each is described in the following:

- *Deadlock* is a condition where no processes will be completed because they cannot access the resources they require to make progress. The following conditions must be true for a deadlock to occur: (1) mutual exclusion (resources may only be allocated to one process at a time), (2) hold and wait (a process may allocate a resource and wait for others), (3) no preemption (a resource may not be forcibly taken away), and (4) circular wait (processes are holding resources that other processes need). Deadlock is almost impossible to find by testing, but may be found by analysis (formal methods can help with this). Deadlock is typically avoided through design [7] by preventing one of the four conditions in the RTOS architecture [21].

- *Starvation* occurs "when a task does not receive sufficient resources to complete processing in its allocated time" because other tasks are using the needed resources [22].

- *Lockout* is a special condition of starvation where a task is *locked out* because another task is blocked before returning a shared resource [7].

20.7.1.2 Priority Inversion

Priority inversion is a type of deadlock that occurs when a high priority task is forced to wait for the release of a shared resource owned by a lower priority task. "The period of time that a task has a lock on a shared resource is called the task's critical section or critical region" [21]. A famous example of priority inversion is the Mars Pathfinder mission. A few days into the mission, the Pathfinder started to have persistent resets, causing loss of the system for long periods of time. Testing and analysis revealed that the problem was caused by priority inversion. A low priority software task on the Pathfinder shared a resource with a high priority task. The low priority task blocked the shared resource after it was preempted by some medium priority tasks. "When another high priority task discovered the previous high priority task had not completed, it initiated a system reset" [5]. A global default setting in the RTOS allowed the priority inversion.

There are two approaches that are typically used to address the priority inversion issue: (1) priority inheritance protocol or (2) priority ceiling protocol. Some RTOSs provide both protocols and let the user decide the preferred algorithm [14]. Kyle and Bill Renwick cover the advantages and disadvantages of both protocols in their paper entitled *How to use priority inheritance*. They suggest that the "best strategy for solving priority inversion is to design the system so that inversion can't occur" [23]. There is considerable literature available on priority inversion. This section only provided an introduction.

20.7.1.3 Memory Leaks

A memory leak is "an error in a program's dynamic-store allocation logic that causes it to fail to reclaim discarded memory, leading to eventual collapse due to memory exhaustion" [24]. It is typically addressed by avoiding dynamic memory manipulation in safety-critical applications; this is accomplished by locking the memory allocation mechanism after initialization and disabling the freeing of memory.

20.7.1.4 Memory Fragmentation

Memory fragmentation occurs when memory is used inefficiently, leading to poor performance and possibly memory exhaustion. To avoid this, the memory allocation technique needs to be well defined, organized, and controlled. The following practices can help to avoid memory fragmentation: (1) fix the block size allocation, (2) partition and size the allocated memory, (3) use identifiers to track allocated memory, and (4) protect and isolate segments [4].

20.7.1.5 Intertask Interference

Intertask interference occurs when a task can modify the memory of another task or even the operating system itself. This is addressed by separating the RTOS from applications (e.g., using protected modes) and by utilizing the MMU facilities [4]. As noted earlier, memory protection is critical when implementing robust partitioning.

20.7.1.6 Jitter

The cache and pipeline features of a processor improve performance, but they also add uncertainty to task execution times. "This uncertainty is termed jitter, and is practically impossible to quantify analytically" [25]. Jitter depends on the hardware platform, operating system scheduling approach, and the tasks that share the processor. Selective flushing of the cache is a common solution to address cache jitter. The cache is flushed during the partition switch so that the incoming partition has a clear cache memory at the start of its duration. "Flushing means copying all the cache values only present in the cache back to main memory (i.e., they have been updated, and copy-back mode is used). This places the overhead at the start of the partition rather than it being distributed throughout" [9]. The amount of time to perform the flushing varies, depending on the number of values that need to be written to memory [9].

20.7.1.7 Vulnerabilities

Since the RTOS is typically developed separate from the applications that use it, it may have vulnerabilities that the users should note. A software vulnerability analysis is needed to identify the RTOS vulnerabilities and to mitigate them. Any vulnerabilities that are not mitigated by the RTOS design need to be identified for the systems integrator and/or the applications developers to address. Additionally, any assumptions or limitations on the RTOS users should be identified to ensure the RTOS is used properly. Some hazards may require special design or coding practices by the applications. "Other hazards may be mitigated by special analysis or verification technique undertaken during the subsequent states of the verification process" [20].

> A software vulnerability analysis (SVA) can identify areas of potential anomalies, which can be provided as input not only to a robustness or stress-test plan, but also to a system functional hazard analysis or SSA [system safety assessment]. How an SVA is conducted is up to the RTOS developer or applicant [26].*

The SVA is based on a specific RTOS implementation. The FAA research report, *Study of Commercial Off-The-Shelf (COTS) Real-Time Operating Systems (RTOS) in Aviation Applications*, provides a table that can be used as the starting point for an SVA. The table identifies seven functional areas to consider: data consistency, dead or deactivated code, tasking, scheduling, memory and I/O device access, queuing, and interrupts and exceptions [26]. This table is included in Appendix B for convenience.

FAA AC 20-148 section 5.f explains that an RSC developer, such as an RTOS developer, needs to produce the following:

> An analysis of the RSC's behavior that could adversely affect the users' implementation (for example, vulnerabilities, partitioning requirements, hardware failure effects, requirements for redundancy, data latency, and design constraints for correct RSC operation). The analysis may support the integrator's or applicant's safety analysis [27].

An SVA is comparable to a safety assessment of the RTOS functionality.† For a partitioned RTOS, the SVA is normally combined with the partitioning analysis, since many of the vulnerabilities are in the area of partitioning. However, the approach varies for each project; there isn't a standardized packaging scheme for this analysis. Some RTOS developers package it as an SVA, some as a hazard analysis, and others as a partitioning analysis.

20.7.2 Certification Issues to Consider

This section identifies some of the common certification-related scenarios or issues that are encountered when using an RTOS in a safety-critical aviation system.

---

* Brackets added for clarification.
† Bate and Conmy explain typical steps to performing an RTOS failure analysis in their paper *Safe composition of real time software* [5].

20.7.2.1 Creating a Safe Subset

Many of the current RTOSs used in safety-critical applications are based on a commercially available operating system with widespread usage and a proven track record for quality and performance. Most general-purpose RTOSs were developed with *time-to-market* as a priority rather than safety. In order to make the RTOS suitable for safety-critical applications, a new RTOS is created using a subset of the fully featured, general-purpose RTOS. Any functionality that might not be suitable for the safety-critical environment is removed or modified. Identifying the safety issues and removing or modifying them without impacting other RTOS functionality can be challenging and should be carried out with caution. This requires detailed knowledge of the RTOS design and code.

20.7.2.2 User's Manual

Oftentimes, when a safe subset is created, the user's manual is not updated to align with the RTOS subset. This could result in improper understanding or use of the RTOS.

20.7.2.3 Reverse Engineering

In order to comply with DO-178C (or DO-178B) the life cycle data for an RTOS is often reverse engineered from the source code. While doing this, issues with the source code itself may be identified. As will be discussed in Chapter 25, there are a number of issues to consider when reverse engineering.

20.7.2.4 Deactivated Features

Most RTOSs are designed to be used by multiple customers and have features that are not necessarily used or needed on every project. Some RTOSs are designed to be scalable, so the user can select and compile or link only the RTOS functions needed. Other RTOSs may have features that are available but not used. An RTOS built specifically for an application will have the advantage that only features used by the application will be present. Constructing an application-based subset may be difficult as the RTOS features are often interrelated. A more common approach is to define and verify an RTOS irrespective of the specific features used by each application and then to treat unused RTOS functionality as deactivated code. Chapter 17 provides a discussion of deactivated code and some of the things to consider when features are available in the RTOS but are not used by the applications.

20.7.2.5 Complexity

Some RTOSs are extremely complex, containing code with complex interactions and features that may not be needed. The complexity can make it difficult to prove the determinism and safety characteristics. Additionally, data coupling and control coupling analyses may be difficult if complexity is not managed. See Chapters 7 and 8 for design and coding practices to reduce complexity.

20.7.2.6 Disconnect with the System

Because the RTOS is developed separate from the system that will use it, there may be inconsistencies between the system needs and expectations and the RTOS functionality. Additionally, since the RTOS is part of the airborne software, it needs to link into the system or software requirements. The system which implements the RTOS must have requirements (typically software requirements) that trace or link to the RTOS requirements to ensure that the RTOS functionality supports the overall system needs, performs its intended functionality, and does not have any undesired or unused functionality.

20.7.2.7 Code Compliance Issues

If the RTOS was not originally developed for the safety-critical environment and DO-178C (or DO-178B) compliance, the code itself is often not up to the expected standards. In fact, the code may have been developed without any coding standards at all. Many times, the code for commercial RTOSs has limited comments and is so complex that even the original author may not be able to determine what was intended. Additionally, the code is generally written with complex data structures, nesting, and interconnectedness, making it difficult to verify.

20.7.2.8 Error Handling Issues

Error handling is an important feature in embedded systems. The RTOS is often employed to help with the error handling. Interestingly, error handling can account for a substantial portion of the RTOS code. Unfortunately, RTOS error handlers are prone to errors themselves; this is because they may not have been adequately verified since the conditions that activate them can be rare or esoteric. It is not uncommon to have faults reported at the lower levels but not be visible to the API; this may not be a problem with the error handler itself, but rather in code that is supposed to use the error handler.

20.7.2.9 Problem Reporting

Since many of the RTOSs are developed by organizations that are independent of the application developer or system integrator, the problem reports generated by the RTOS developer may not be available to the users (e.g., applications developers, system integrators, avionics manufacturers, or aircraft manufacturers). In the certification world all problem reports need to be visible to the aircraft applicant and confirmed that they don't have a safety impact. This is an ongoing process. Some RTOS developers may not normally share problem reports throughout the life of the RTOS with their customers (this may be a business decision driven by fear that the product will look bad). When choosing to use a COTS RTOS, the user should ensure that problem reports are provided as problems are identified and for as long as the RTOS is on an aircraft.

20.7.2.10 Partitioning Analysis

As noted earlier, many RTOSs include support for robust partitioning. In order to prove the adequacy of the robust partitioning, a partitioning analysis is performed. The RTOS kernel is analyzed, as well as supporting components (such as BSP, device drivers, CPU, MMU, and device hardware), to confirm the adequacy of the partitioning. The partitioning analysis is like a safety analysis of the partitioning implementation. The analysis considers how partitioning violations could occur with shared time, space, and I/O resources; then mitigations are required for each vulnerability. In some projects, the partitioning analysis may be incomplete or inadequate to prove robust partitioning. Chapter 21 discusses partitioning analysis further.

20.7.2.11 Other Supporting Software

Frequently, much of the focus is on the RTOS kernel. However, libraries, BSPs, middleware, device drivers, etc. that connect the RTOS with the hardware or the applications also need to be considered. That is, they also need to comply with DO-178C, since they are part of the safety-critical system.

20.7.2.12 Target Testing

Many times, the RTOS is initially tested using a COTS BSP and COTS hardware. However, the RTOS must be tested with the actual hardware and supporting software that will be installed. This will normally require additional tests and modifications to some of the existing tests that were used with the COTS BSP and hardware. The target-based test not only is important to prove the functionality of the RTOS, BSP, and device drivers, it is also needed to verify that the COTS processor, MMU, etc. work as intended. A key selection criterion when choosing an RTOS is the availability of test suites and the anticipated level of rework necessary to reuse those tests for certification credit.

20.7.2.13 Modifications

While RTOS reuse is a goal, it is rarely fully realized. AC 20-148 provides guidance for RSCs. However, in reality most RTOSs are modified for each specific user. It is important to understand what is modified and what is not (i.e., a thorough change impact analysis is needed). Normally, it is recommended that the entire RTOS be retested when modifying for different users or when integrating with different hardware. Automated test suites tend to make this relatively straightforward.

20.8 Other RTOS-Related Topics

This section considers various topics related to RTOSs, including ARINC 653 overview, tool support, open source RTOSs, multicore processors, virtualization, and hypervisor technology.

20.8.1 ARINC 653 Overview

ARINC 653, *Avionics Application Software Standard Interface*, is a three-part standard that is about three inches thick when printed double-sided. Only a high-level overview is presented here. ARINC 653 specifies the APEX interface between applications and the underlying run-time system. It provides an API standard for a robustly partitioned operating system but allows flexibility for implementing the RTOS.

Paul Prisaznuk explains two key benefits of the ARINC 653 standardized RTOS interface. First, it provides a clear interface boundary for the avionics software development. The avionics software applications and the underlying core software can be developed independently, allowing concurrent development of the RTOS and the applications that use the RTOS. Additionally, the same applications may be portable to other ARINC 653 compliant platforms. Second, the ARINC 653 RTOS interface definition enables the underlying core software and the hardware platform to evolve independent of the software applications that will be hosted on them [13].

The APEX API defines the following functionality for an RTOS: (1) sending and receiving data to/from other software applications or networked systems; (2) managing resources like time, memory, I/O, and displays; (3) handling errors; (4) managing files; and (5) scheduling software tasks or processes.

There are some unique things about the APEX API compared to other APIs. First, it is designed specifically for the aviation community with safety in mind. Second, it is designed for a partitioned computing environment. It includes two-tiered scheduling of applications (between partitions and within partitions) to ensure that one partition will not affect another partition. It also defines interpartition communications to allow safe communication between partitions. Third, it provides a health monitor interface that allows errors and conditions at the computing module level to be communicated to applications, as well as application level errors to be communicated and handled at the module (board) level [3]. Section 20.4.2 and

Table 20.1 provide a high-level comparison of the ARINC 653 APEX features and the POSIX features.

In addition to the API, APEX provides a standardized, configurable operating environment for software applications to facilitate the abstraction of the application software from the underlying hardware. ARINC 653 defines the configuration specification and an assumed operating environment.

The basic concept of ARINC 653 is that an application is given a partition (like a container) to function in. The dimensions of the partition are defined by attributes such as memory size, processor utilization (e.g., in terms of period and duration), interpartition I/O ports and connections (communication channels), and health monitor actions. The attributes of the partition are determined by configuration data that are defined by the system integrator.

ARINC 653 contains three parts. Each part is briefly described in the following.

Part 1 defines the required services in order to meet safety needs, ensure application portability, and allow communication between partitions. Part 1 defines the following core services [3]:

- *Partition management*: allows applications (even with different criticality levels) to execute on the same hardware with no undue influence on one another, spatially or temporally.

- *Process management and control*: includes the resources needed to manage the processes within a single partition.

- *Time management*: allows the management of one of the most critical resources controlled by the operating system (i.e., time). In the APEX philosophy, time is independent of process or partition execution. All time values are related to the core module time and are not relative to the partition or process.

- *Interpartition communication mechanisms*: allow the communication of messages between two or more partitions executing either on the same hardware or on different hardware. Two paradigms are provided: queuing for variable-sized messages and sampling for fixed-size messages.

- *Intrapartition communication mechanisms*: allow the communication of messages between processes within the same partition without the overhead needed for global message passing.

- *Health monitoring functions*: monitors and reports platform and application faults and failures. It also helps isolate faults in order to prevent failures from propagating. In an IMA system the health monitor performs fault monitoring, fault containment, and fault management at both the application level and the system level.

Part 2 extends the Part 1 services to include the following [28]:

1. *File system*: a general-purpose, abstract method for managing data storage, as noted in Section 20.6.8.

2. *Sampling port data structures*: a standardized set of data structures for exchanging parametric data. These help to reduce unnecessary variability in parameter formats, thus reducing the need for custom I/O processing. These data structures also enable portability of applications and improve efficiency of the core software.

3. *Multiple module schedules*: extend the single static module schedule in Part 1 to allow multiple schedules to be defined in the configuration table.

4. *Logbook system*: used to store messages. It retains the stored data after a power failure, so the data can be recovered when power is restored to the module. Each logbook is accessible by only one partition, and the logbook content and status are not altered by partition reset.

5. *Sampling port extensions*: extend Part 1 basic services for sampling ports with the following services:

   a. `READ_UPDATED_SAMPLING_MESSAGE`

   b. `GET_SAMPLING_PORT_CURRENT_STATUS`

   c. `READ_SAMPLING_MESSAGE_CONDITIONAL`

6. *Support access port (SAP)*: a special kind of queuing port that allows access to addressing information when sending and receiving messages.

7. *Name service*: a companion to the SAP services. It allows a partition to retrieve an address based on a name and to retrieve a name based on an address.

8. *Memory blocks*: provide a means for a partition to access blocks of memory within the module's memory space. Access privileges for the partition are defined in the configuration tables. Partitions can be granted read-only or read-write access to a memory block.

Part 3 is a conformity test specification. It describes test assertions and responses necessary to demonstrate conformity to the required services software interface defined in Part 1. This specification is intended to be used to evaluate ARINC 653 compliance.

Most RTOSs comply with parts of ARINC 653 Part 1 and Part 2 but not all of them. At this time, Part 3 is not required for certification, since the APIs are tested as part of DO-178B or DO-178C compliance. It may be used in the future by potential RTOS customers to evaluate RTOS functionality and to confirm ARINC 653 compliance. Additionally, if the ARINC 653 becomes administered by an independent standards body, the conformity testing will become a valuable measure of adherence to the standard.

ARINC 653 is updated on a regular basis to address the needs of the aviation community; therefore, it's important to confirm the version of ARINC 653 being used, as well as the specific services implemented.

20.8.2 Tool Support

Significant tool support is necessary in order to successfully integrate and verify the installed RTOS and to analyze the real-time applications that use the RTOS. An RTOS is usually supported with an integrated development environment. Gerardo Garcia writes: "The development environment has an enormous impact on the quality and speed of development as well as the project's overall success" [18].

Tools are provided to support code development and analysis. Code development tools include an editor, assembler, compiler, debugger, browser, etc. Tools to support run-time analysis include a debugger, coverage analyzer, performance monitor, etc. "The debugging capabilities of your real-time software development tools can mean the difference between creating a successful control system and spiraling into an endless search for elusive bugs" [18]. Debug tools (such as application profilers, memory analysis tools, and execution tracing tools) help to optimize the real-time applications [18].

Tools are also provided to evaluate the RTOS behavior and application performance. RTOS-oriented tools include tools to analyze timing behavior of tasks and ISRs, effects caused by task/task and task/ISR interactions (such as synchronization and preemption), problems related to resource protection features (e.g., priority inversion and deadlocks), and overheads and delays due to intertask communication [4].

Additionally, tools are often provided to help with RTOS configuration. Configuration tools help to configure memory, partitioning constraints, communication mechanism (e.g., buffers and ports), I/O devices, health monitoring parameters, etc. For ARINC 653 compliant platforms, many deployment and implementation details are defined in the configuration tables. Some RTOS or platform developers provide tools to support the configuration; however, this is an area that is still evolving. Horváth et al. claim that despite the complexity of ARINC 653 configurations, current tools are only available for the very low-level design. Tools are lacking at the higher levels to capture configuration process, validate configuration design constraints, record design decisions, trace the configuration data to the requirements, etc. "As a result, verification of configuration tables is a tedious activity" [29].

20.8.3 Open Source RTOSs

An open source RTOS is one that has the source code publicly available, free of charge, for use and/or modification from its original design. Such RTOSs are normally created as a collaborative effort in which programmers improve the code and share the changes within the community. There are a number of open source RTOSs available, such as Linux, which has drawn considerable attention. However, to date I know of no open source RTOS that has the supporting DO-178C data available to support certification. Additionally, the code is regularly updated—normally without attention to safety impact. If one were to use an open source RTOS in a safety-critical system, one would probably either need to implement architectural mitigation (such as wrappers) to limit the impact of the RTOS or reverse engineer the DO-178C life cycle data using a defined code baseline. Some government-funded research is being performed to investigate if it's possible to reap the benefits of open source and the vast toolset available, while still meeting the safety and certification requirements.

Serge Goiffon and Pierre Gaufillet performed a research study considering ARINC 653 and Linux. Their paper points out the Linux is not ready for DO-178B (and now DO-178C) compliance for the following reasons: no development and verification plans exist, the development environment is heterogeneous and complex (distributed over Internet, multiplatform, etc.), there are no universal development standards (requirements, design, or code standards) used, and little-to-no design documentation exists [8]. The paper also identified some steps that are needed to address DO-178B* certification, including reverse engineering the missing data, the addition of ARINC 653 partition scheduling, and APEX API compliance [8]. Getting an open source RTOS ready for use in a safety-critical application that requires DO-178C would be a major undertaking.

20.8.4 Multicore Processors, Virtualization, and Hypervisors

Many avionics organizations are considering the feasibility of multicore processors and the use of virtualization and hypervisor technology. *Virtualization* provides a software environment in which programs (including entire operating systems) can run as if on bare hardware, when in reality they are not running on the hardware but on a layer between called a *virtual machine*. A virtual machine is basically an isolated duplicate of the real machine. A software layer provides the virtual machine environment which is normally called a *virtual machine monitor (VMM)* or a *hypervisor*. The VMM has the following essential characteristics [30]:

1. It provides an environment that appears to be identical to the original machine.

2. Programs running in this environment only have minor decreases in speed.

3. The VMM completely controls the system resources.

---

* The paper was written before DO-178C was published.

Virtualization has gained popularity in the mainstream computing community and is now starting to be offered with COTS RTOSs. It isolates operating systems, applications, devices, and data from each another while running on the same hardware platform. A hypervisor is used to provide virtualization and protection services. In order to tune performance, the size of the hypervisor is relatively small [30].

If virtualization or hypervisor technology is used, DO-332, *Object-Oriented Technology and Related Techniques Supplement to DO-178C and DO-278A*, should be considered. The *related techniques* portion of DO-332 could apply to such technology.

At this time, the use of multicore processors, virtualization, and hypervisors are being considered for installation in aircraft. However, to my knowledge none have been approved for installation in a commercial civil aircraft. Manufacturers and certification authorities are currently performing investigations, identifying the issues, and working to resolve them. I have no doubt that such technology will be used on commercial aircraft in the not-too-distant future.

20.8.5 Security

Many RTOSs are required to meet both safety and security standards, particularly when the RTOS is used in military aviation applications. For the security domain, the Common Criteria* is applied. The Common Criteria has seven evaluation assurance levels (EALs), with EAL 7 being the highest. For RTOSs that are used in safety and security domains, both DO-178C and the Common Criteria are applied.[†]

20.8.6 RTOS Selection Questions

When choosing an RTOS for use in safety-critical systems, there are several aspects to evaluate. Appendix C provides three categories of questions to consider when selecting an RTOS: (1) general RTOS questions, (2) RTOS functionality questions, and (3) RTOS integration questions.

References

1. H. M. Deitel, *Operating Systems*, 2nd edn. (Reading, MA: Addison-Wesley, 1990).
2. W. Stallings, *Operating Systems Internals and Principles*, 3rd edn. (Upper Saddle River, NJ: Prentice Hall, 1998).
3. Aeronautical Radio, Inc., Avionics application software standard interface part 1—Required services, ARINC Specification 653P1–3 (Annapolis, MD: Airlines Electronic Engineering Committee, November 2010).

---

* *Common Criteria* refers to ISO/IEC 15408, *Common Criteria for Information Technology Security Evaluation.*
† In the past, EAL ratings were assigned for the RTOS alone; however, now EAL ratings are based on the entire system.

4. J. Cooling, *Software Engineering for Real-Time Systems* (Harlow, U.K.: Addison-Wesley, 2003).
5. I. Bate and P. Conmy, Safe composition of real time software, *Proceedings of the Ninth IEEE International Symposium on High-Assurance Systems Engineering* (Dallas, TX, 2005).
6. B. L. Di Vito, A formal model of partitioning for integrated modular avionics, NASA/CR-1998-208703 (Hampton, VA: Langley Research Center, August 1998).
7. E. Klein, RTOS design: How is your application affected? *Embedded Systems Conference* (San Jose, CA, Spring 1999).
8. S. Goiffon and P. Gaufillet, Linux: A multi-purpose executive support for civil avionics applications? *IFIP—International Federation for Information Processing*, 156, 719–724, 2004.
9. J. Krodel, Commercial off-the-shelf real-time operating system and architectural considerations, DOT/FAA/AR-03/77 (Washington, DC: Office of Aviation Research, February 2004).
10. K. Driscoll, Integrated modular avionics (IMA) requirements and development, *Integrated Modular Avionics Conference for the European Network of Excellence on Embedded Systems* (Rome, Italy, 2007).
11. RTCA DO-297, *Integrated Modular Avionics (IMA) Development Guidance and Certification Considerations* (Washington, DC: RTCA, Inc., November 2005).
12. J. Krodel and G. Romanski, Real-time operating systems and component integration considerations in integrated modular avionics systems report, DOT/FAA/AR-07/39 (Washington, DC: Office of Aviation Research, August 2007).
13. P.J. Prisaznuk, ARINC 653 role in integrated modular avionics (IMA), *IEEE Digital Avionics Systems Conference* (St. Paul, MN, 2008), pp. 1.E.5-1–1.E.5-10.
14. D. Kleidermacher and M. Griglock, Safety-critical operating systems, *Embedded Systems Programming* 14(10), 22–36, September 2001.
15. W. Lamie and J. Carbone, Measure your RTOS's real-time performance, *Embedded Systems Design* 20(5), 44–53, May 2007.
16. R. G. Landman, Selecting a real-time operating system, *Embedded Systems Programming* 79–96, April 1996.
17. N. Lethaby, Reduce RTOS latency in interrupt-intensive apps, *Embedded Systems Design* 23–27, June 2009.
18. G. Garcia, Choose an RTOS, *Embedded Systems Design* 36–41, November 2007.
19. IAR Systems, How to choose an RTOS, *Embedded Systems Conference* (San Jose, CA, 2011), pp. 1–22.
20. G. Romanski, Certification of an operating system as a reusable component, *IEEE Digital Avionics Systems Conference* (Irvine, CA, 2002), pp. 1–8.
21. S. Ferzetti, Real time operating systems (RTOS), on-line tutorial. http://www.slidefinder.net/r/real_time_operating_systems_rtos/realtimeoperatingsystems/26947789 (accessed December 2011).
22. P. Laplante, *Real-Time Systems Design and Analysis: An Engineer's Handbook* (New York: IEEE Press, 1992).
23. K. Renwick and B. Renwick, How to use priority inheritance, *EE Times*, article #4024970, May 2004. http://www.eetimes.com/General/PrintView/4024970 (accessed December 2011).
24. C. Z. Yang, Embedded RTOS memory management, YZUCSE SYSLAB tutorial. http://syslab.cse.yzu.edu.tw/~czyang (accessed December 2011).

25. F. M. Proctor and W. P. Shackleford, Real-time operating system timing jitter and its impact on motor control, *Proceedings of the SPIE Sensors and Controls for Intelligent Manufacturing II* (Boston, MA, October 2011), Vol. 4563, pp. 10–16.
26. V. Halwan and J. Krodel, Study of commercial off-the-shelf (COTS) real-time operating systems (RTOS) in aviation applications, DOT/FAA/AR-02/118 (Washington, DC: Office of Aviation Research, December 2002).
27. Federal Aviation Administration, *Reusable Software Components*, Advisory Circular 20-148 (December 2004).
28. Aeronautical Radio, Inc., Avionics application software standard interface part 2—Extended services, ARINC Specification 653P2-1 (Annapolis, MD: Airlines Electronic Engineering Committee, December 2008).
29. Á. Horváth, D. Varró, and T. Schoofs, Model-driven development of ARINC 653 configuration tables, *IEEE Digital Avionics Systems Conference* (Salt Lake City, UT, 2010), 6.E.3-1–6.E.3-15.
30. G. Heiser, Virtualizing embedded Linux, *Embedded Systems Design* 18–26, February 2008.

21

Software Partitioning

Acronyms

AFDX	avionics full-duplexed switched ethernet
API	application program interface
BSP	board support package
CAST	Certification Authorities Software Team
COTS	commercial off-the-shelf
CPU	central processing unit
CRC	cyclic redundancy check
DMA	direct memory access
FAA	Federal Aviation Administration
I/O	input/output
IMA	integrated modular avionics
LRU	line replaceable unit
MMU	memory management unit
RAM	random access memory
ROM	read only memory
RTOS	real-time operating system
SFI	software fault isolation
SVA	software vulnerability analysis

21.1 Introduction to Partitioning

Partitioning is the Achilles heel of integrated modular avionics (IMA) systems and advanced avionics. Implementing a robustly partitioned system requires extreme care and caution. This chapter provides an overview of the basic concepts related to partitioning and key actions necessary to implement partitioning in a safety-critical system.

21.1.1 Partitioning: A Subset of Protection

Partitioning is a subset of a broader concept called *protection*. The concepts of protection are described in Certificate Authorities Software Team (CAST)* paper CAST-2, entitled *Guidelines for assessing software partitioning/protection schemes*. The protection concepts in CAST-2 are presented as follows [1]:

- *Two-way protection*—A component X is protected from component Y, and component Y is protected from component X. An example of two-way protection is two components within an avionics unit with no interactions and no shared resources between them.

- *One-way protection*—A component X is protected from component Y, but component Y is not protected from component X. An example of one-way protection is an avionics unit that can receive data from the flight management computer but it cannot send data to the flight management computer. The flight management computer software can impact the avionics unit but not vice versa.

- *Strict protection*—A component X is strictly protected from component Y if any behavior of component Y has no effect on the operation of component X. An example of this type of protection is two components within an avionics unit with no interactions and no shared resources between them. Strict protection can be one-way or two-way.

- *Safety protection*—A component X can be said to be safely protected from component Y if any behavior of component Y has no effect on the safety properties of component X. An example of this is the use of a cyclic redundancy check (CRC) associated with data passed through a nonassured data link, where the only safety property of importance is the corruption of data. Loss of data is not a safety property of interest in this example. Safety protection requires one to identify the safety properties from the safety or hazard analysis. Safety protection can be one-way or two-way.

Protection may be implemented in software, hardware, or a combination of hardware and software. Some examples of how protection has been implemented include encoding/decoding, wrappers, tools, separate hardware resources, and software partitioning. This chapter concentrates on software partitioning, since it is the most common approach implemented in software. It's worth noting that many software documents (such as DO-178C) interchange the terms protection and partitioning.

This chapter is closely related to Chapter 7 on design and Chapter 20 on real-time operating systems (RTOSs).

---

* CAST is a team of international certification authorities who strive to harmonize their positions on airborne software and aircraft electronic hardware in CAST papers.

21.1.2 DO-178C and Partitioning

DO-178C section 2.4.1 describes partitioning as "a technique for providing isolation between software components to contain and/or isolate faults and potentially reduce the effort of the software verification process" [2]. DO-178C goes on to explain that partitioning between software components can be implemented by allocating software components to different hardware resources, or by running more than one software component on the same hardware. With the speed and capacity of processors today, the latter approach is often selected. DO-178C section 2.4.1 provides the following five guidelines regardless of the partitioning approach [2]:

a. A partitioned software component should not be allowed to contaminate another partitioned software component's code, input/output (I/O), or data storage areas.
b. A partitioned software component should be allowed to consume shared processor resources only during its scheduled period of execution.
c. Failures of hardware unique to a partitioned software component should not cause adverse effects on other partitioned software components.
d. Any software providing partitioning should have the same or higher software level as the highest level assigned to any of the partitioned software components.
e. Any hardware providing partitioning should be assessed by the system safety assessment process to ensure that it does not adversely affect safety.

DO-178C has one objective that explicitly mentions partitioning: Table A-4 objective 13, which states: "Software partitioning integrity is confirmed" [2]. This objective references DO-178C section 6.3.3.f, which explains that during review and/or analysis of the software architecture, it must be ensured that partitioning breaches (violations) are prevented. This objective applies to all software levels, when partitioning is used.

There are five DO-178C objectives that have an indirect link to partitioning: Table A-5 objectives 1–5 all reference (either as an objective or as an activity) DO-178C section 6.4.3.a, which explains that "violations of software partitioning" are a typical error type that should be revealed during requirements-based hardware/software integration testing [2]. This means that partitioning is covered by requirements and that it must be tested. This is difficult as partitioning requirements are often *negative*, for example: "It shall not be possible for one partition to modify the data memory of another partition unless that memory has been configured as shared." It would be easy to write a test to show an example or two, but because of various configuration options, showing that the test set is sufficient to verify the requirement can be challenging.

Partitioning also directly supports safety when it is used to separate and/or isolate software functions. It ensures that less critical software functions do not interfere with more critical software functions and that all functions have the time, memory, and I/O resources needed to perform their intended functionality.

21.1.3 Robust Partitioning

Until the late 1990s, partitioning was used rather sparsely to separate two or more software levels running on the same hardware. However, since the late 1990s, it has become a more common technique used in avionics. With the expansion of computing power and the availability of RTOSs that support partitioning, partitioning has become a key characteristic of many modern avionics systems.

Robust partitioning is a cornerstone for IMA. RTCA DO-297 defines IMA as: "A shared set of flexible, reusable, and interoperable hardware and software resources that, when integrated, form a platform that provides services, designed and verified to a defined set of safety and performance requirements, to host applications performing aircraft functions" [3]. Since the IMA platform hosts applications of different software levels, robust partitioning is needed to ensure that each application has the necessary resources and does not interfere with other applications.

Chapter 20 introduced the concept of robust partitioning and mentioned it as related to the RTOS. The concept is further explored in this chapter. DO-297 section 2.3.3 explains the concept of robust partitioning as follows [3]:

> Robust partitioning is a means for assuring the intended isolation of independent aircraft functions and applications residing in IMA shared resources in the presence of design errors and hardware failures that are unique to a partition or associated with application-specific hardware. If a (different) failure can lead to the loss of robust partitioning then it should be detectable and appropriate action taken. The objective of robust partitioning is to provide the same level of functional, if not physical, isolation and protection as a federated implementation. This means robust partitioning should support the cooperative coexistence of core software and applications hosted on a processor and using shared resources, while assuring unauthorized or unintended interference is prevented... Robust partitioning is a means for assuring the intended isolation and independence in all circumstances (including hardware failures, hardware and software design errors, or anomalous behavior) of aircraft functions and hosted applications using shared resources. The objective of robust partitioning is to provide an equivalent level of functional isolation and independence as a federated system implementation (i.e., applications individually residing on separate Line Replaceable Units (LRU)). This means robust partitioning supports the cooperative coexistence of applications using shared resources, while assuring that any attempted unauthorized or unintended interaction is detected and mitigated.

The implementation of robust partitioning draws from the computer security domain, which uses the concepts of data and information flow (i.e., access control, noninterference, and separability), integrity policies, timing channels, storage channels, and denial of service. However, while safety partitioning and software security concepts are related, the two models do not completely coincide since they are driven by different objectives [4,5].

DO-297 section 3.5 lists the following characteristics of robust partitioning [3]:

a. The partitioning services should provide adequate separation and isolation of the aircraft functions and hosted applications sharing platform resources. Partitioning services are the services provided by the platform that define and maintain the independence and separation between partitions. These services ensure that the behavior of functions or applications within a partition cannot unacceptably affect the behavior of functions or applications in any other partition. These services should prevent any adverse effects on the aircraft of the simultaneous undetected corruption of all the functions and applications' partitions sharing the affected resources.

b. The ability to determine in real-time, with an appropriate level of confidence, that the partitioning services are performing as specified consistent with the defined level of safety.

c. Partitioning services should not rely on any required behavior of any aircraft function or hosted application. This implies that all protection mechanisms required to establish and maintain partitioning are provided by the IMA platform.

If robust partitioning is not implemented properly, it can lead to numerous problems, some of which may have safety implications. Examples of partitioning issues are as follows [3,6]:

- Erroneously writing data into the wrong areas; for example, a faulty partition allowing an application to write to a memory location to which another partition assumes it has exclusive access.
- Stealing time from another application.
- Crashing the processor.
- Corrupting the I/O by falsely sending output data which appear to come from a critical function.
- Corrupting input data before the critical function uses it.
- Monopolizing internal communication paths.
- Corrupting a shared flash memory file system.
- Introducing timing jitter on context switch to a new partition (altering the performance of the new partition).

Robust partitioning occurs across the three basic subsystems of any computing platform: memory, central processing unit (CPU), and I/O [7]. Communications between partitions may also be considered as a shared resource [8]; however, this typically overlaps with the other shared resources, so it is not discussed further. Each of the shared resources (memory, time, and I/O) should be addressed by the partitioning design, implementation, and verification. Partitioning of each shared resource is discussed next.

21.2 Shared Memory (Spatial Partitioning)

John Rushby writes: "Spatial partitioning must ensure that software in one partition cannot change the software or private data of another partition (either in memory or in transit) nor command the private devices or actuators of other partitions" [4]. Basically, spatial partitioning prevents a function in one partition from corrupting or overwriting the data space of a function in another partition [9]. Justin Littlefield-Lawwill and Larry Kinnan explain that spatial partitioning assures that shared system resources in one partition "are not consumed in a manner that would result in a denial of service for other partitions requiring access to the same resource" [10]. There are two common approaches to memory protection: (1) using a memory management unit (MMU) or (2) using software fault isolation (SFI).

Hardware-based spatial partitioning is the most prevalent form of spatial partitioning. A hardware MMU is usually provided with the CPU. The details of the MMU operation vary from processor to processor. The MMU ensures that policies expressed in MMU tables provide the desired controls over memory access. Since the MMU is a commercial off-the-shelf (COTS) device without supporting life cycle data, the operating system is used to set up the MMU tables that the processor subsequently uses. The proper functionality (accuracy) of the MMU needs to be confirmed during the certification effort. Chapter 20 explained that this is often confirmed during the RTOS testing. The Federal Aviation Administration (FAA) report entitled *Commercial Off-The-Shelf Real-Time Operating System and Architectural Considerations* [11] provides some additional details for how the MMU and RTOS are used to provide robust partitioning.

If the MMU isn't used, SFI is an alternative [4]. With this approach logical checks are added in the code at each memory access point. The machine code in a partition is examined to determine the destinations of memory references and to check their accuracy.

> Indirect memory references cannot be checked statically, so instructions are added to the program to check the contents of the address register at runtime, immediately prior to its use. The SFI technique imposes

some overhead cost by adding code to the program. It also requires an additional analysis and certification cost on every project. However, it is possible to automate much of the check procedure and to qualify a tool or toolset that can be used on multiple projects [9].

The CAST-2 position paper identifies several areas to consider when implementing memory partitioning, such as loss, corruption, or delay of input or output data, corruption of internal data, program overlays, buffer sequences, external device interaction, control flow defects which affect memory (e.g., incorrect branching into a partition or protected area), etc. [1].

21.3 Shared Central Processing Unit (Temporal Partitioning)

"Temporal partitioning must ensure that the service received from shared resources by the software in one partition cannot be affected by the software in another partition. This includes the performance of the resource concerned, as well as the rate, latency, jitter, and duration of scheduled access to it" [4]. Partitioning in the time domain is closely related to multitasking schedulability, which was discussed in Chapter 20. ARINC 653 enforces strict round-robin scheduling for partitions (durations and periods are specified in a configuration table). Within the partition, other schedulers are used.

The goal of temporal partitioning is to ensure that functions in one partition do not disturb the timing of events in other partitions. Concerns include one partition that monopolizes the CPU, crashes the system, or issues a HALT instruction—leading to denial of services for other partitions. "Other scenarios that can cause a partition to fail to relinquish the CPU on time include simple schedule overruns, where particular parameter values cause a computation to take longer than its allotted time, and runaway executions, where a program gets stuck in a loop" [4]. The approach to temporal partitioning should take these scenarios into account.

Processor interrupts are used in real-time systems to identify an event that needs processor access. Interrupts must be handled carefully to avoid undermining the temporal partitioning. Normally such disruptions are prevented by eliminating interrupts altogether, with the exception of the timer tick used to implement the schedule. However, some parties desire to expand ARINC 653 to provide a deterministic means of handling interrupts [7]. Interrupts are discussed more in Section 21.5.

The CAST-2 position paper identifies areas to consider when implementing temporal partitioning, such as interrupts and interrupt inhibits, loops, frame overrun, counter or timer corruption, pipelining and caching, control flow defects, memory or I/O contention, software traps (such as divide by zero), etc. [1].

21.4 Shared Input/Output

Most systems have multiple I/O ports, devices, and channels, for example, a serial bus, an ARINC 664 AFDX (avionics full-duplexed switched ethernet) end system, a field-programmable gate array device, or a CRC device. Some devices are dedicated to a specific partition and others are shared by multiple partitions. As noted in Chapter 20 on RTOSs, a device driver typically serves as the glue code between the device and the RTOS kernel.

As with other shared resources, I/O resources need to be partitioned. Partitioning concerns for shared I/O are closely related to space and time partitioning. For each I/O device, partitioning for both the time and space domains must be considered.

Addressing partitioning in I/O can be one of the most challenging aspects of design for a partitioned system. ARINC 653 provides sampling and queuing port interface and operation definitions for interpartition I/O; however, I/O to physical devices or intermodule I/O is implemented at the discretion of the RTOS developers or other stakeholders. The ARINC 653 port mechanism (which is based on *pseudo ports* and *pseudo partitions*) can be used as an interface standard for connectivity; however, it may lead to performance issues for the applications [10]. Steve VanderLeest explains that when partitioning I/O devices, not only the end device must be considered, but one must also consider the communication mechanisms connecting the device to the memory and CPU (such as communication busses, direct memory access engines, and intermediate buffers). "The partitioning environment must manage all the salient features of the I/O subsystem" (including latency, bandwidth, control registers, and buffer space) to prevent partitions from affecting one another by an unauthorized means [7].

The approach to address I/O partitioning varies depending on I/O hardware specifics and the low-level device driver. The FAA research report, entitled *Real-Time Operating Systems and Component Integration Considerations in Integrated Modular Avionics Systems Report*, notes:

> Different approaches are possible, such as kernel-centralized I/O control or partition I/O control. Many implementations incorporate some type of RTOS-governed partition or task permission table that permits only certain partitions or tasks to access specific I/O. A health monitor is then employed to identify any undesignated accesses by other tasks or partitions. I/O considerations differ for different RTOSs [8].

Depending on the scenario, the RTOS may implement some constraints or make some assumptions that require special integration steps by the integrator [8]. As with any other assumptions or constraints, these should be documented by the RTOS vendor and/or platform supplier and communicated to the integrator; oftentimes, the platform data sheet includes this type of information.

The use of I/O interrupts should be carefully evaluated in the partitioning scheme. Most CPUs support interrupts from I/O-related events. However, as will be discussed shortly, interrupts can impact partitioning and may not be allowed. Therefore, another approach may need to be established to communicate events that require action.

21.5 Some Partitioning-Related Challenges

Some organizations are tempted to rely on the RTOS alone to implement partitioning. However, partitioning is a system-level characteristic. The processor, board support package (BSP), devices, device drivers, MMU, etc. play a role in the partitioning approach and can disrupt robust time and space partitioning. Some specific areas that pose challenges to partitioning are identified in this section, including direct memory access, cache memory, interrupts, and interpartition communication.

21.5.1 Direct Memory Access

Direct memory access (DMA) is used to transfer a large block of memory in a short period of time, autonomous to the processor. DMA transfers use a DMA engine that may grant exclusive access to the memory bus to perform the block transfer. The memory bus is a shared resource; therefore, it can deny an application its resources if not properly shared. A DMA may violate temporal partitioning when a transfer is initiated by a partition with less execution time remaining than the time needed to complete the DMA transfer. The DMA may also induce memory partitioning violations. One way to address the DMA partitioning issue is to create an application program interface (API) to control the access to DMA rather than allowing the application direct access. Performance may be impacted by the API, but it is still faster than other types of memory transfer, provided the size of the memory blocks transferred is large enough [10].

21.5.2 Cache Memory

Cache memory is an intermediate, high-speed memory that resides between the main memory and the CPU. It contains a copy of frequently accessed memory for rapid access. It significantly improves performance; however, COTS processors do not provide partitioned cache dedicated to specific partitions (although this is changing with some of the new devices planned for future releases). In a partitioned system, the state of the processor is swapped with each partition switch. The state prior to the switch is held in the CPU registers. Most processors provide functionality to perform this swap quickly. In a nonpartitioned system, the state of the cache memory does not need to be saved because there is a single application which does not interfere with itself. However, in a partitioned system, cache is a shared resource [7].

There are a few options to preserve the partitioning when using cache memory. One option to remedy potential cache-induced partitioning violations is to turn off the cache. However, the performance impact is usually too high for this approach, so a deterministic method for using cache must be implemented. To date, cache flushing is the most common solution. The approach is to flush the cache during the partition switch so that the new partition has a clean cache memory when it starts. As noted earlier, "Flushing means copying all the cache values only present in the cache back to main memory (i.e., they have been updated, and copy-back mode is used). This places the overhead at the start of the partition rather than it being distributed throughout" [11]. Flushing adds to the time for context switching and reduces cache performance at the start of a partition while the cache is loading, but it is still more efficient than no cache at all. Since the time taken to perform the flush operations varies, depending on the amount of data written to memory, the worst-case timing for context switching must be considered to ensure that time needs are met. Because each partition time slot starts with an empty cache, the performance of the application can be impacted if the partition duration is too short (so that the cache never gets a chance to fully populate).

Cache write-through mode is another option to address the cache partitioning issues. This is more efficient than the no-cache option, but code execution is slower. The benefit of this option is that cache invalidate is a single fast instruction.

21.5.3 Interrupts

An interrupt is a signal triggered by an asynchronous event. When interrupts are enabled by the processor, the normal operation is suspended in order to service the interrupt event using an interrupt service routine. To prevent such disruptions, interrupts are sometimes eliminated altogether in robustly partitioned systems, except for the timer tick used to implement the schedule [10]. Partitioned RTOSs that comply with ARINC 653 tend to limit interrupts to the system clock. However, some special hardware and/or software techniques may be employed to avoid temporal partitioning violations during interrupt (e.g., disallowing an application access to interrupt-causing hardware, implementing software to poll data related to the interrupt signal, or implementing an API to check if the activity requested can be completed prior to the end of the current minor frame time) [10]. Such techniques must be carefully analyzed for effectiveness, which may not be a trivial exercise.

21.5.4 Interpartition Communication

If each partition was an island that didn't need to communicate with other islands, life would be simpler. However, because partitions may need to communicate with each other, interpartition communication must be designed to support robust partitioning.

> The challenge is to design a partitioning solution that enables the exchange of information between partitioned functions (e.g., interpartition communication) and controls access to other shared resources (such as I/O devices) while keeping the partitioned functions largely autonomous and unaffected by other functions. Inter-partition communication and sharing of I/O devices influences both the space and time aspects of partitioning and protection mechanisms [9].

When addressing interpartition communication, both the memory dimension (primarily focused on transferring data from one partition to another) and the temporal dimension (primarily concerned with synchronization and how one partition invokes services from another) must be considered [4]. Communication between partitions must be restricted to only those that are intended and that are authorized by the system configuration data. The FAA's research report provides suggestions for implementing interpartition communication without violating time or space partitioning [9].

21.6 Recommendations for Partitioning

This section provides some practical recommendations to consider when implementing and verifying partitioning. Each project varies and the technical details must be addressed on a case-by-case basis; however, these recommendations are intended to provide a starting point.

Recommendation 1: Keep in mind that robust partitioning is a system-level concern. As explained in Section 21.5, robust partitioning is not implemented by software alone. It is a system-level concern that requires early and ongoing collaboration between systems, hardware, and software teams. Likewise, robust partitioning isn't implemented by the RTOS alone. The RTOS can play a significant role in enforcing robust partitioning, but it is not the only component involved.

Recommendation 2: Proactively design for robust partitioning. Robust partitioning requires diligent design—it does not just happen. There are multiple dimensions to consider and numerous technical challenges. However, addressing these challenges proactively during the development phases significantly reduces surprises during the integration and verification phases. The following suggestions are offered:

1. Consider the common issues noted earlier (such as interrupts, cache memory, I/O challenges) and address them as part of the design solution.
2. Document partitioning goals, for example:
   a. *The system will implement robust memory protection*: An application will always receive its allocated memory resources, without interference from other applications.
   b. *The system will implement robust time protection*: An application will always receive its specified execution time without interference from other applications.
   c. *The system will implement robust resource protection*: An application will always receive its physical and logical allocated resources without interference from other applications.
3. Define robust partitioning requirements to meet the goals.
4. Identify all shared resources so they can be properly utilized to support robust partitioning.
5. Document partitioning details in the design. In particular, ensure that all components and their interactions are clearly identified.
6. Identify all requirements (at each hierarchical level) that support or impact partitioning. For example, a requirements attribute may be assigned to distinguish partitioning-related requirements. This also assists with the subsequent partitioning analysis.
7. Identify multiple means to prevent failure propagation.
8. Use containment boundaries to limit the effect of failures.
9. Reference the guidance of DO-297 section 3.5.1 (entitled *Design for robust partitioning*) [3], which provides guidelines to consider when designing a robustly partitioned IMA platform. The DO-297 concepts are applicable, even if the system is not classified as IMA.

Recommendation 3: During design and design reviews ensure that vulnerabilities are addressed. As the design is documented and reviewed, consider probing questions that would uncover partitioning violations. Table 21.1 provides example questions to help uncover data-related and control-related vulnerabilities.

Recommendation 4: Consider using an ARINC 653 compliant RTOS. As previously noted, the RTOS alone doesn't address all of the partitioning issues; however, it is a good start. ARINC 653 provides guidelines to help RTOS and platform developers think through the partitioning challenges.

TABLE 21.1

Example Questions to Help Identify Partitioning Vulnerabilities

Data-Related Vulnerabilities	Control-Related Vulnerabilities
• Can partitioning be violated by data flow?	• Can partitioning be violated by control flow?
• Can shared data be inappropriately used?	• Can functions be inappropriately invoked within a feature or between features?
• Can messages be sent or received improperly?	• Can interrupts cause erroneous behavior?
• Can function parameters be inappropriately used?	• Can hardware faults or failures impact data integrity or execution order?
• Can configuration data be invalid?	• Can transitions between modes be improperly implemented?
• Can data be incorrectly passed?	• Can resources be inappropriately allocated?
• Can data be improperly initialized?	• Can deactivated code be inadvertently activated?
• Can global data be read or written to improperly?	• Can initialization sequence be incorrect?
• Can global data be erroneously written to by unintended functions?	• Can inappropriate responses to exceptions occur?
• Can global data be uninitialized or improperly reinitialized?	• Can fault handlers behave inappropriately (e.g., miss faults or failures, or handle faults or failures improperly)?
• Can hardware registers be inappropriately used?	• Can memory overlaps occur?
• Can the linker incorrectly assemble data or code?	• Can improper hardware addresses be read from or written to?
• Can data become stale or invalid?	• Can inappropriate responses occur during resets?
• Can data drop out?	• Can synchronization be impacted by miscompares or erroneous waits?
• Can improper responses to data miscompares occur?	• Can improper context switching cause erroneous data or timing?
• Can unexpected floating-point values occur?	• Can any unexpected exceptions be generated?
	• Can functions be executed at incorrect rates or times?

Recommendation 5: Utilize other development and verification activities to prevent redundant efforts. Partitioning should be considered throughout the system and software development. Other development or verification activities can be utilized to support the robust partitioning design and verification, including the following:

- *Build upon the data and control coupling activities.* The data and control coupling analyses provides a good starting point for robust partitioning. Data and control coupling and software partitioning are related, but are distinct concepts. Partitioning is a way to provide isolation between independent software components and thus provides protection of unintended coupling between independent partitioned components. Data and control coupling is an intentional interaction between components, including coupling between separately partitioned components. Software partitioning does not guarantee avoidance of data or control coupling problems; conversely, data and control coupling problems do not imply that the partitioning mechanism is flawed. For partitioned systems, data and control coupling analyses, as well as partitioning analysis, are required. There may be some synergy between the two objectives.

- *Use robustness testing to confirm partitioning is not violated.* When partitioning is clearly documented in the requirements and design, it drives the testing effort, which provides a good start toward proving the robust partitioning claims. The partitioning analysis may identify the need for additional testing; however, the requirements-based tests provide a foundation to begin building confidence in the partitioning strategy.

- *Utilize the RTOS software vulnerability analysis (SVA).* If an RTOS is used, the RTOS vendor may have an SVA available to use as input to the partitioning analysis data. The SVA was discussed in Chapter 20.

Recommendation 6: Test the partitioning mechanism. The main purpose of verifying the partitioning mechanism is to ensure that no erroneous behavior in one partition can contribute to misbehavior or failure of any other partition and to ensure that partitioning of shared resources are not violated. DO-248C discussion paper #14 (entitled *Partitioning aspects in DO-178C/DO-278A*) provides some suggestions for verifying the partitioning mechanism:

> Elements of partitioning integrity can be verified by exercising the partitioning mechanisms using special test scenarios, simulations, and/or analysis techniques. Test scenarios should be written to stimulate the partitioning mechanism by injecting errors or violation attempts to bypass time and space constraints. An example of additional analysis would be the calculation of worst-case execution times to assess temporal performance... A verification test suite (containing normal range test cases and abnormal or out of range test cases for all requirements of the partitioning mechanism) should be established. Robustness of the partitioning mechanism may be demonstrated by use of a requirements-based test suite (satisfying requirements-based test coverage) [12].

Some projects also implement *rogue partition testing*, where a rogue partition is developed to try to deliberately violate the partitioning.

Recommendation 7: Perform a partitioning analysis. DO-297 promotes an activity called *partitioning analysis*. The purpose of the analysis is to demonstrate that no application in a partition can affect the behavior of applications in any other partition in an adverse manner. The partitioning analysis is similar to the system safety assessment, where all potential sources of failure are considered and mitigated. All vulnerabilities should be identified, classified, and mitigated [11]. The engineer(s) performing the analysis must have detailed understanding of the system, hardware, and software architecture. The chief architect is often the ideal person to do this analysis. Following are some common tasks performed as part of the partitioning analysis:

1. *Gather data to support the analysis,* including the preliminary system safety assessment, system requirements and design, software requirements and design, hardware architecture, BSP and device driver design data, processor data sheet and/or user manual, device user manuals, interface specifications, configuration tool requirements and design (if applicable), etc.

2. *Identify the robust partitioning claims* that will be analyzed, for example,*

   a. The system will implement robust memory protection: An application will always receive its allocated memory resources, without interference from other applications.

   b. The system will implement robust time protection: An application will always receive its specified execution time without interference from other applications.

   c. The system will implement robust resource protection: An application will always receive its physical and logical allocated resources without interference from other applications.

3. *Identify potential vulnerabilities which could violate each claim.* All potential sources of error should be systematically identified and considered, including resource limitations, scheduling tasks, I/O, interrupt error sources, etc. A traceability analysis that traces all shared resources to the modules, components, and applications that use those shared resources helps confirm that all potential vulnerabilities have been considered [13]. The potential partitioning violations of shared memory devices (such as read only memory [ROM], random access memory [RAM], cache, queues, and onboard chip registers) should be analyzed. Likewise, the effects of hardware failures on shared and nonshared hardware components should also be analyzed. DO-297 section 3.5.2.5 identifies some common potential sources of design errors that could impact the partitioning [3]:

   a. Interrupts and interrupt inhibits (software and hardware).
   b. Loops (for example, infinite loops or` indirect non-terminating call loops).

---

* This same example was provided earlier in the chapter but is repeated for completeness.

    c. Real-time correspondence (for example, frame overrun, interference with real-time clock, counter/timer corruption, pipeline and caching, deterministic scheduling).

    d. Control flow (for example, incorrect branching into a partitioned or protected area, corruption of a jump table, corruption of the processor sequence control, corruption of return addresses, unrecoverable hardware state corruption (for example, mask and halt)).

    e. Memory, input, and/or output contention.

    f. Sharing of data flags.

    g. Software traps (for example, divide by zero, unimplemented instruction, specific software interrupt instructions, unrecognized instruction, and recursion termination).

    h. Hold-up commands (i.e., performance hedges).

    i. Loss of input or output data.

    j. Corruption of input or output data.

    k. Corruption of internal data (for example, direct or indirect memory writes, table overrun, incorrect linking, calculations involving time, corrupted cache memory).

    l. Delayed data.

    m. Program overlays.

    n. Buffer sequence.

    o. External device interaction (for example, loss of data, delayed data, incorrect data, protocol halts).

4. *Identify potential vulnerabilities that are mitigated by design.* The majority of the vulnerabilities are mitigated by the design—especially when robust partitioning is proactively considered during the development phases. Each vulnerability mitigated by design should trace to the requirement(s) that demonstrate the mitigation (i.e., there should be a mapping between vulnerabilities and requirements that provide the mitigations). Additionally, each mitigation should be verified during testing.

5. *Identify potential vulnerabilities that are mitigated by process.* Some potential vulnerabilities may be mitigated by process (e.g., the use of a qualified tool to verify accuracy of configuration data or a design standard that confirms certain memory assignments).

6. *Identify potential vulnerabilities that cannot be mitigated by design or process.* These will need to be communicated to the integrator and possibly the application developer. Such communication is normally documented in the data sheet and appropriate user information, so the integrator or application developer can take the appropriate action.

7. *Coordinate with safety and systems personnel throughout the partitioning analysis.* The partitioning analysis is essentially an extension of the safety assessment and should be closely coordinated with the system safety personnel.

8. *Ensure that all potential vulnerabilities have been mitigated*, particularly those that were not addressed by design or process. The goal is to minimize mitigation activities to be taken by the integrator or applications; however, in some cases there may be special actions required. For example, the integrator may have to perform some special verification steps or impose special restrictions through the configuration files.

References

1. Certification Authorities Software Team (CAST), Guidelines for assessing software partitioning/protection schemes, Position Paper CAST-2 (February 2001).
2. RTCA DO-178C, *Software Considerations in Airborne Systems and Equipment Certification* (Washington, DC: RTCA, Inc., December 2011).
3. RTCA DO-297, *Integrated Modular Avionics (IMA) Development Guidance and Certification Considerations* (Washington, DC: RTCA, Inc., November 2005).
4. J. Rushby, Partitioning in avionics architectures: Requirements, mechanisms, and assurance, DOT/FAA/AR-99/58 (Washington, DC: Office of Aviation Research, March 2000). Also published as NASA/CR-1999-209347 (Hampton, VA: Langley Research Center, March 2000).
5. B. L. Di Vito, A formal model of partitioning for integrated modular avionics, NASA/CR-1998-208703 (Hampton, VA: Langley Research Center, August 1998).
6. K. Driscoll, Integrated modular avionics (IMA) requirements and development, *Integrated Modular Avionics Conference for the European Network of Excellence on Embedded Systems* (Rome, Italy, 2007), p. 440.
7. S. H. VanderLeest, ARINC 653 hypervisor, *IEEE Digital Avionics Systems Conference* (Salt Lake City, UT, 2010), pp. 5.E.2-1–5.E.2-20.
8. J. Krodel and G. Romanski, Real-time operating systems and component integration considerations in integrated modular avionics systems report, DOT/FAA/AR-07/39 (Hampton, VA: Langley Research Center, August 2007).
9. V. Halwan and J. Krodel, Study of commercial off-the-shelf (COTS) real-time operating systems (RTOS) in aviation applications, DOT/FAA/AR-02/118 (Washington, DC: Office of Aviation Research, December 2002).
10. J. Littlefield-Lawwill and L. Kinnan, System considerations for robust time and space partitioning in integrated modular avionics, *IEEE Digital Avionics Systems Conference* (Orlando, FL, October 2008).
11. J. Krodel, Commercial off-the-shelf real-time operating system and architectural considerations, DOT/FAA/AR-03/77 (Washington, DC: Office of Aviation Research, February 2004).
12. RTCA DO-248C, *Supporting Information for DO-178C and DO-278A* (Washington, DC: RTCA, Inc., December 2011).
13. J. Krodel and G. Romanski, *Handbook for Real-Time Operating Systems Integration and Component Integration Considerations in Integrated Modular Avionics Systems*, DOT/FAA/AR-07/48 (Washington, DC: Office of Aviation Research, January 2008).

22

Configuration Data

Acronyms

ATP   acceptance test plan
CC1   control category #1
CC2   control category #2
DAL   development assurance level
EASA   European Aviation Safety Agency
FAA   Federal Aviation Administration
IMA   integrated modular avionics
XML   eXtensible Markup Language

22.1 Introduction

Many safety-critical systems are designed to be configurable. Configuration data provide a flexible, yet controlled, way to configure and reconfigure a system. The data used to configure the system are referred to as *configuration data* in this chapter. DO-178C and the certification authorities' policy and guidance use a variety of other terms to describe these *configuration data*, including configuration files, databases, airborne system databases, parameter data items, adaptation data items, etc.

This chapter explains some of the terminology, guidance, and recommendations related to configuration data. This chapter does not discuss aeronautical databases (e.g., navigational or terrain databases). They are discussed in Chapter 23.

22.2 Terminology and Examples

The European Aviation Safety Agency (EASA) Certification Memorandum CM-SWCEH-002 refers to configuration data as *configuration files* and defines them as:

> Files embedding parameters used by an operational software program as computational data, or to activate/deactivate software components (such as, to adapt the software to one of several aircraft/engine configurations). The terms "registry" or "definition file" are sometimes used for a configuration file. Configuration files such as symbology data, bus specifications or aircraft/engine configuration files are segregated from the rest of the embedded software for modularity and portability purposes [1].

Section 15.2 of Federal Aviation Administration (FAA) Order 8110.49 (Change 1) refers to configuration data as *airborne system databases*. The Order explains that airborne system databases are:

> Used by an airborne system and approved as part of the type design of the aircraft or engine. These databases may influence paths executed through the executable object code, be used to activate or deactivate software components and functions, adapt the software computations to the aircraft configuration, or be used as computational data. Airborne system databases may consist of script files, interpretive languages, data structures, or configuration files (including registries, software options, operating program configuration, aircraft configuration modules, and option-selectable software) [2].

DO-178C builds upon the certification authority guidance by introducing the concept of *parameter data item* and *parameter data item file*, defined as follows [3]:

- *Parameter data item*—A set of data that, when in the form of a Parameter Data Item File, influence the behavior of the software without modifying the Executable Object Code and that is managed as a separate configuration item. Examples include databases and configuration tables.
- *Parameter Data Item File*—The representation of the parameter data item that is directly usable by the processing unit of the target computer. A Parameter Data Item File is an instantiation of the parameter data item containing defined values for each data element.

Configuration data are separate from the executable object code. Sometimes the configuration data are included as part of the airborne software part number and therefore are included in the DO-178C software life cycle data (i.e., they are called out in the Software Accomplishment Summary and Software Configuration Index of the airborne software). Other configuration data may have a separate part number than the airborne software in order to configure the software for the specific aircraft needs. In this situation, the configuration data may have their own life cycle data (including a separate Software Accomplishment Summary and Software Configuration Index) and a separate part number. Configuration data may also be field loadable or user modifiable.

Configuration data come in many formats, including text files, binary data, XML (eXtensible Markup Language) data, lookup tables, databases, etc. Some examples of configuration data include data used to do the following:

- Define symbology for a display system.
- Specify databus parameters.
- Enable or disable preprogrammed functions (e.g., option-selectable software).
- Configure a data network.
- Program integrated modular avionics (IMA) platform resource allocation (such as memory, time, and shared resources).
- Identify aircraft-specific options (such as a personality module).
- Calibrate sensors or actuators.
- Configure a real-time operating system.

DO-178C section 2.5.1 explains the following [3]:

Parameter data items may contain data that can:
a. Influence paths executed through the Executable Object Code.
b. Activate or deactivate software components and functions.
c. Adapt the software computations to the system configuration.
d. Be used as computational data.
e. Establish time and memory partitioning allotments.
f. Provide initial values to the software component.

Configuration data are normally developed and controlled by one of the following entities:

- *Equipment/avionics manufacturer's systems or systems integration team.* The configuration of the system may be established by the systems or systems integration team. For example, when determining the systems communications requirements, the systems team may identify the network connections, and the integration team may develop the configuration data to implement the connections.

- *Equipment/avionics manufacturer's software team.* The software team may package configuration data as part of the equipment's software or as a separate configuration item. An example of configuration data that is part of the software package is a lookup table to set gains. An example of configuration data that is packaged separately is a resource allocation file for an application.

- *Equipment/avionics manufacturer's production process.* In some situations, the configuration data are established as part of the production process rather than the engineering and software development process. For example, the hardware configuration data such as processor serial number and equipment serial number may be entered as part of the production process. Another example is calibration data that are entered prior to acceptance testing and shipping to the customer.

- *Aircraft manufacturer's installation process.* Sometimes, the aircraft manufacturer is responsible for some configuration data. For example, the aircraft manufacturer may use a configurable aircraft personality module to configure the aircraft to meet customer's specific needs.

- *User's modification process.* Some configuration data may be user modifiable. For example, an airline may use configuration data to identify which parameters they will collect for maintenance purposes.

22.3 Summary of DO-178C Guidance on Parameter Data

DO-178B did not provide unique guidance for configuration data. However, as noted earlier, DO-178C uses the terms *parameter data item* and *parameter data item file* to describe configuration data. DO-248C explains that a parameter data item is a software component that contains only data (and no executable object code). A parameter data item is also a configuration item. Airborne software may consist of one or more executable object code configuration items and one or more parameter data configuration items. A parameter data item *file* is an instantiation of the parameter data item containing defined values for each data element [4].

Parameter data items are assigned the same software level as the software component that uses the parameter data. Parameter data item files are treated very similar to executable object code in DO-178C; they are driven by the requirements and must be verified, either separately or as part of the airborne software. DO-178C sections 7 and 8 explain that parameter data item files need to be under configuration management and assessed during software quality assurance activities, just as the executable object code is. Likewise, DO-178C section 9 identifies parameter data item files as type design data along with the executable object code.

The following DO-178C objectives, explicitly reference parameter data item files [3]:

- *DO-178C Table A-3 objective 7:* "Executable Object Code and Parameter Data Item Files, if any, are produced and loaded in the target computer." This objective was expanded from DO-178B to DO-178C to address the integration of the parameter data into the target computer.

- *DO-178C Table A-5 objective 8*: "Parameter Data Item File is correct and complete." This objective is new to DO-178C. It ensures that each element in a parameter data item file satisfies its requirements, has correct values, and is consistent with other data elements.

- *DO-178C Table A-5 objective 9*: "Verification of Parameter Data Item File is achieved." This ensures that all elements of each parameter data item file were covered during verification.

22.4 Recommendations

Configuration data cover a wide variety of scenarios. Each scenario has its own unique needs to be addressed on the specific project. However, this section provides some general recommendations to keep in mind. Many of these recommendations are based on DO-178C and/or certification authorities' policy and guidance.

Recommendation 1: Determine how the configuration data will be used and what policy and/or guidance applies. Depending on how the configuration data are used, guidance on field-loadable software (see Chapter 18), user-modifiable software (see Chapter 19), or deactivated code or option-selectable software (see Chapter 17) may apply. The DO-178C guidance on parameter data items and any special guidance on configuration data from the certification authority should be considered. At the time of this writing, the certification guidance related to configuration data is still evolving. The publication of DO-178C clarifies what is expected; however, because of the newness of the parameter data item guidance, there may be some additional clarification from the certification authorities. In general, the configuration data are treated as a special class of software; therefore, most of the guidance applicable to airborne software also applies to the configuration data—with a few adjustments here and there, as will be discussed later.

Recommendation 2: Determine the software level or development assurance level (DAL) that applies. The safety assessment considers the potential conditions that will transpire if the configuration data are erroneous or corrupted. The configuration data are normally treated like software in the safety assessment process; therefore, they are assigned a software level. However, if the configuration data are established at the system level, a DAL may apply. The configuration data are usually assigned the same software level (or DAL) as the software (or system) component that uses the configuration data, or the highest level if several levels are present, such as in an IMA system.

DO-178C explains that software level for a parameter data item is the same as the software level of software component that uses the parameter data. However, the safety assessment will take precedence. If architectural mitigation is applied, there could be a situation where the configuration data level is lower, but this would probably be quite rare.

Recommendation 3: Develop the appropriate life cycle process for the configuration data. While configuration data are similar to software and much of the guidance related to airborne software applies, there are some differences. Most projects apply the objectives of DO-178C to configuration data. However, for configuration data there will often only be one level of requirements needed (such as high-level requirements) and the structural coverage objectives are satisfied differently, since there are no conditions or decisions in configuration data. DO-178C's guidance, objectives, and activities for parameter data items and parameter data item files explain how to apply DO-178C to configuration data.

Recommendation 4: Develop plans to document the process. The planning for the configuration data may be included as part of the software plans or it may be separate. Typically, if there is a significant amount of configuration data or if the configuration data will be regularly modified separate from the software that uses it, it makes sense to separate the configuration data and its supporting life cycle data from the executable software's data. When the configuration data are packaged separate from the airborne software, the five plans identified in DO-178C may be compressed into one or two plans for configuration data, since the processes may not be as complex as the airborne software and the team may be significantly smaller. In more than one project, I've seen the plans condensed into two plans: (1) a Plan for Software Aspects of Certification which is submitted to the certification authority and (2) another plan that details the development, verification, configuration management, and quality assurance of the configuration data. Many times, the Software Configuration Management Plan and the Software Quality Assurance Plan for the airborne software may also apply to the configuration data, so those plans are merely referenced from the second plan. Regardless of how the plans are packaged, DO-178C section 4.2.j explains that the following additional information should be considered in the plans when using *parameter data items* [3]:

1. The way that parameter data items are used.
2. The software level of the parameter data items.
3. The processes to develop, verify, and modify parameter data items, and any associated tool qualification.
4. Software load control and compatibility.

Recommendation 5: Define requirements for configuration data. Configuration data should be requirements driven, just like any other data that impact aircraft functionality. There is often just one level of requirements needed to define the functionality of configuration data. DO-178C suggests that configuration data requirements be captured as high-level software requirements;

however, there have been a few situations where system-level requirements or a configuration file design specification have been used to capture the configuration data requirements. The packaging is somewhat flexible; however, regardless of the packing scheme, it should be remembered (1) requirements to specify the configuration data are necessary, and (2) the configuration data must trace to those requirements.

Recommendation 6: Trace the configuration data to its requirements. As with source code, the configuration data must trace to its requirements. As will be discussed later, the trace data are particularly important for configuration data, since an analysis of the trace data is sometimes used to help ensure there are no unintended configuration data.

Recommendation 7: Develop standards to ensure that the configuration data are in the proper format. In order for configuration data to be readable by the executable object code, they need to be in a well-defined format. Standards defining the format of data, allowed values, data types, etc. are typically needed to guide the configuration data developer and to provide consistency and accuracy. There may be some scenarios where certain constant values are always needed, for example, some network settings for a specific device driver. Such situations may be handled by requirements or development standards. Note that standards should also be considered when verifying the configuration data (i.e., verify that the configuration data meet the development standards). When tools are used, the tools tend to enforce the data format; however, there may still be a need for data formatting standards.

Recommendation 8: Design the executable software to be configurable. The software that will interact with the configuration data should be designed to use the data. The software requirements should specify the structure, attributes, allowable ranges, etc. of configuration data to be used. Oftentimes, the executable software will perform some kind of validity check to ensure that the configuration data are within the preapproved range and have not been corrupted.

Recommendation 9: Verify the configuration data. Verification of configuration data involves several activities including the following:

1. *Ensure that the requirements for the configuration data are accurate, complete, consistent, verifiable, and correct.* EASA's CM-SWCEH-002 treats this as a *validation* step [1]. However, whether it is called validation or verification, it needs to take place. This activity is typically accomplished by review and analysis. For complex configuration data, such as the network configuration for an aircraft, simulations are often used to verify the correctness and completeness of the data. For simpler configuration data, a review may be adequate. Some configuration data may be verified by a combination of reviews and analyses.

2. *Verify that configuration data trace to the appropriate requirements.* It is important to ensure that all configuration data requirements

are implemented and that all configuration data are requirements driven. Accurate and complete traceability helps to ensure this. It's important not just to make sure the trace is there, but also to make sure that it is the *right* trace.

3. *Ensure that configuration data complies with the requirements.* Requirements-based testing, as well as review and analysis, should be applied to configuration data to ensure that the requirements are satisfied. Sometimes the configuration data will be verified along with the executable object code. At other times, it may be verified separately. DO-178C section 6.6 explains that the following must apply if configuration data (DO-178C calls it *parameter data item files*) are verified separate from the executable object code that uses the data [3]:

   • The Executable Object Code has been developed and verified by normal range testing to correctly handle all Parameter Data Item Files that comply with their defined structure and attributes.
   • The Executable Object Code is robust with respect to Parameter Data Item Files structures and attributes.
   • All behavior of the Executable Object Code resulting from the contents of the Parameter Data Item File can be verified.
   • The structure of the life cycle data allows the parameter data item to be managed separately.

4. *Ensure that all configuration data have been covered during verification.* To ensure that there are no unintended configuration data, there should be an activity to ensure that all configuration data have been verified. This is similar to structural coverage analysis that is performed for software. However, since structural coverage criteria doesn't normally apply to databases, another approach must be developed. The approach may be similar to the DO-254 concept of elemental analysis, used for electronic hardware to ensure that all *elements* are tied to requirements that have been verified. For our purposes, the configuration data could be treated as an *element*; the trace data may be used to support the analysis and to confirm that every element of the configuration data has been tested. Regardless of the approach, the goal is to ensure that all of the configuration data have been properly verified and that there is no extraneous or dead data.

5. *Use a qualified verification tool to check for consistency where possible.* Since configuration data can be very tedious and somewhat ineffective to verify manually (particularly if it is in bitmap format), tools can be very effective. For example, a qualified tool may be used to confirm that the system does not send messages to a sending communications port, memory segments that are not shared do not

overlap, the sum of all resource requests is not greater than those provided by the system, and so on.

6. *Integrate the configuration data with the executable object code into the target computer.* The integration may take place during the testing, particularly if the executable object code and configuration data are verified together. However, there may be situations when the configuration data and executable object code are verified separately. If this is the case, there will need to be an additional integration step to ensure that the executable object code and configuration data work together as intended on the specific target computer.

Recommendation 10: Ensure compatibility of the executable object code and configuration data. Perform verification to ensure that the specific configuration of the executable object code and configuration data are compatible. Activities should be in place to ensure that the configuration data are not corrupted or unacceptable by the executable object code (e.g., noncompliant with the agreed upon format and ranges). Additionally, the compatibility needs to be documented in the appropriate configuration index.

Recommendation 11: Apply configuration management to the configuration data. Configuration management and configuration control apply to the configuration data, as well as their supporting life cycle data. For example, configuration files should have configuration identification, changes to configuration data requirements and files should be handled via change request or problem report, configuration data should only be changed by authorized personnel, etc. Additionally, all aspects of control category #1 or #2 (CC1 or CC2) apply to the configuration data and their supporting life cycle data—just as it does for the airborne software. See Chapter 10 for information on DO-178C's configuration management expectations.

Recommendation 12: If the production process develops configuration data, ensure it is a controlled process. In my experience, configuration data entered during production are not as common as the configuration data developed by engineering during the development process; however, they may be needed for some systems. The production process must be well defined and repeatable. Additionally, the process should include an independent verification of the data entry (particularly for more critical data, such as level A and B). In some scenarios, the acceptance test plan (ATP) may be traced to the configuration data requirements to ensure that the ATP will identify any erroneous data. In other scenarios, an independent person may manually verify data (e.g., review the serial numbers of hardware and firmware parts) and sign the production record.

Recommendation 13: Produce the necessary documentation (life cycle data). As with airborne software, the configuration data must have the required life cycle data to support their approval. Plans, development data, verification data, configuration management data, and quality assurance data are all needed.

In general, the same life cycle data identified in DO-178C section 11 and Annex A for airborne software compliance are needed for configuration data approval. However, as noted earlier, the data may be packaged differently (e.g., plans may be merged, there may only be one level of requirements, and parameter data item guidance in DO-178C section 11 applies).

Recommendation 14: Update documentation and re-verify when modifying the configuration data. When configuration data are updated, a change impact analysis should be performed (see Chapter 10) and all impacted data updated. There may also be a need for additional verification to ensure that the updated configuration data are still compatible with the software that uses it.

Recommendation 15: If tools are used, ensure that necessary qualification is considered. Tools are often used to generate and verify configuration data, since tools can eliminate some of the trivial errors that humans may create. If the output of a tool is not verified, it will likely require qualification. See Chapter 13 for information on tool qualification.

Recommendation 16: Ensure that loading instructions are documented and approved. Since configuration data may be loaded separate from the software, it is important to ensure that the instructions for loading and updating the configuration data are well documented, repeatable, unambiguous, and approved. The loading instructions should be included in or referenced from the Software Configuration Index.

References

1. European Aviation Safety Agency, Software aspects of certification, Certification Memorandum CM-SWCEH-002 (Issue 1, August 2011).
2. Federal Aviation Administration, *Software Approval Guidelines*, Order 8110.49 (Change 1, September 2011).
3. RTCA DO-178C, *Software Considerations in Airborne Systems and Equipment Certification* (Washington, DC: RTCA, Inc., December 2011).
4. RTCA DO-248C, *Supporting Information for DO-178C and DO-278A* (Washington, DC: RTCA, Inc., December 2011).

23

Aeronautical Data

Acronyms

AC	advisory circular
AMDB	airport mapping database
AMM	airport moving map
AR	authorization required
ASCII	American standard code for information interchange
DAL	development assurance level
DQR	data quality requirement
EUROCAE	European Organization for Civil Aviation Equipment
FAA	Federal Aviation Administration
FAQ	frequently asked question
ICAO	International Civil Aviation Organization
ISO	International Standards Organization
LOA	letter of acceptance
MASPS	minimum aviation system performance standards
RNAV	area navigation
RNP	required navigation performance
TAWS	terrain awareness and warning system
TSO	Technical Standard Order
UML	Unified Modeling Language
XML	eXtensible Markup Language

23.1 Introduction

Aeronautical data are "data used for aeronautical applications such as navigation, flight planning, flight simulators, terrain awareness and other purposes, which comprises navigation data and terrain and obstacle data" [1]. An *aeronautical database* is "a collection of data that is organized and arranged for ease of electronic storage and retrieval in a system that supports airborne or ground-based aeronautical applications" [1].

Aeronautical data are treated differently than the configuration data that were discussed in Chapter 22. The configuration data discussed in Chapter 22 are approved as part of the aircraft's type design data. However, aeronautical data are often not part of the type design data. Instead, loading such data is frequently treated as a maintenance action that is identified in the aircraft's Instructions for Continued Airworthiness. Title 14 of the U.S. Code of Federal Regulations Part 43.3 allows this approach and treats such updates as *preventative maintenance*. There are at least a couple of reasons that aeronautical data are treated differently than configuration data.

First, aeronautical data typically require frequent updates that are not practical to implement under the type certification process. For example, navigation databases are updated every 28 days. To go through the DO-178C software approval process and supplemental type certification or amended type certification every 28 days is virtually impossible. Terrain databases are updated less frequently (perhaps three or four times per year), which is still too frequent for the DO-178C process to be viable.

Second, the source of the data for such databases is often a government organization rather than an avionics or aircraft manufacturer. The International Civil Aviation Organization (ICAO) Annex 15 places requirements on the ICAO contracting states around the world that are responsible for compiling and transmitting the aeronautical data through Aeronautical Information Publications [2]. "Each Contracting State must take all necessary measures to ensure that the aeronautical information/data it provides is adequate, of required quality (accuracy, resolution and integrity) and provided in a timely manner for the entire territory that the State is responsible for" [1]. In the past, ICAO requirements were primarily applicable to navigational data. Recently, terrain data requirements have also been included in the ICAO requirements.

As noted in Chapter 22, DO-178C does not generally apply to aeronautical data. However, RTCA DO-200A, entitled *Standards for Processing Aeronautical Data*, does apply to this kind of data. The Federal Aviation Administration (FAA) recognizes DO-200A in Advisory Circular (AC) 20-153A, *Acceptance of Aeronautical Data Processes and Associated Databases*, and in various other documents (including Technical Standard Order (TSO)-146c, TSO-151b, AC 20-138B, AC 90-101A, Order 8110.55A, and Order 8110.49).

This chapter surveys DO-200A, AC 20-153A, and several other FAA and industry documents related to aeronautical data in order to provide a high-level overview of expectations for aeronautical data.

23.2 DO-200A: Standards for Processing Aeronautical Data

RTCA's DO-200A (as well as its European Organization for Civil Aviation Equipment (EUROCAE) equivalent, ED-76) is recognized as *the standard* for ensuring the quality of aeronautical data. DO-200A is referenced in the following FAA documents, just to name a few:

- AC 20–153A—*Acceptance of Aeronautical Data Processes and Associated Databases*
- Order 8110.55A—*How to Evaluate and Accept Processes for Aeronautical Database Suppliers*
- Order 8110.49—*Software Approval Guidelines*
- AC 20–138B—*Airworthiness Approval of Positioning and Navigation Systems*
- AC 90–101A—*Approval Guidance for Required Navigation Performance (RNP) Procedures with Authorization Required (AR)*
- TSO-C151b—*Terrain Awareness and Warning System (TAWS)*
- TSO-C146c—*Stand-Alone Airborne Navigation Equipment Using the Global Positioning System Augmented By The Satellite Based Augmentation System*

This section provides an overview of DO-200A. The next section examines the FAA's AC 20-153A and Order 8110.55A, which define the process for obtaining a letter of acceptance for a DO-200A compliant process.

DO-200A is considered the minimum standard and guidance to address the processing quality assurance and data quality management of aeronautical data used for navigation, flight planning, terrain awareness, flight simulators, etc. The output of the DO-200A compliant process is a database that is distributed to the user for implementation in their equipment.

DO-200A identifies requirements and recommendations to provide the appropriate assurance level for the aeronautical data. DO-200A defines *assurance level* as "the degree of confidence that a data element is not corrupted while stored or in transit. This can be categorized into three levels: 1, 2, and 3; with 1 being the highest degree of confidence" [1]. As with the DO-178C software levels, DO-200A assurance levels are determined by the potential impact of corrupted data on safety. DO-201A (discussed in Section 23.5.1) defines the assurance levels for aeronautical information related to area navigation (RNAV). Assurance levels for applications of data not covered by DO-201A need to be determined by the end user or application provider, based on the safety impact [2].

Table 23.1 shows the three DO-200A assurance levels and their relationship to the ICAO criticality levels and the failure condition categories. The related development assurance levels (DALs) or software levels are also shown. The ICAO classifications of critical, essential, and routine are defined as follows [3]:

- *Critical data (Assurance Level 1)*—The data, if erroneous, would prevent continued safe flight and landing or would reduce the ability to cope with adverse operating conditions to the extent that there is a large reduction in safety margins or functional capabilities. There is a high probability when trying to use corrupted critical data that an aircraft would be placed in a life threatening situation;

TABLE 23.1

DO-200A Assurance Levels

DO-200A Assurance Level	Related Requirement on State-Provided Data (ICAO)	Failure Condition Category	DAL or SW Level
1	Critical	Catastrophic or hazardous/ severe-major	A, B
2	Essential	Major or minor	C, D
3	Routine	No safety effect	E

- *Essential data (Assurance Level 2)*—The data, if erroneous, would reduce the ability to cope with adverse operating conditions to the extent that there is a significant reduction in safety margins. There is a low probability when trying to use corrupted essential data that an aircraft would be placed in a life threatening situation;
- *Routine data (Assurance Level 3)*—The data, if erroneous, would not significantly reduce airplane safety. There is a very low probability when trying to use corrupted routine data that an aircraft would be placed in a life threatening situation.

DO-200A uses the concept of an aeronautical data chain to explain the path that the aeronautical data takes: "An aeronautical data chain is a series of interrelated links wherein each link provides a function that facilitates the origination, transmission, and use of aeronautical data for a specific purpose" [1]. There are normally five major links, including origination, transmission, preparation, application integration, and end use [1]. There may be multiple organizations involved in a single link (e.g., there may be *sublinks* in the preparation and transmission phases), or a single organization may perform multiple types of links (e.g., one company may originate, prepare, and transmit data). The originator is "the first organization in an aeronautical data chain that accepts responsibility for the data. For example, a State or RTCA DO-200A/EUROCAE ED-76-compliant organization" [1]. An end user is "the last user in an aeronautical data chain. Aeronautical data end-users are typically aircraft operators, airline planning departments, air traffic service providers, flight simulation providers, airframe manufacturers, systems integrators, and regulatory authorities" [1]. Figure 23.1 illustrates a typical aeronautical data chain.

At each link in the aeronautical data chain, the data should satisfy the following seven quality characteristics based upon the intended function that will use the data: (1) accuracy, (2) resolution, (3) assurance level (based on the safety assessment), (4) traceability (to the origin of the data), (5) timeliness (to support need for valid and current data), (6) completeness

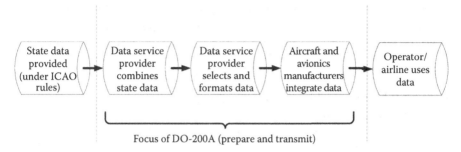

FIGURE 23.1
Typical aeronautical data chain.

(all necessary data provided), and (7) format (consistent with the data's intent, including transmission resolution). These data quality requirements are defined based upon the intended function supported by the data.

DO-200A focuses on the *preparation* and *transmission* links in the aeronautical data chain. While origination, application integration, and end use links are mentioned in order to provide context, they are considered beyond the scope of DO-200A.

The aeronautical data preparation functional link includes the following four phases:

1. *Assemble*—collecting data from suppliers
2. *Translate*—changing the information format (often combined with the assemble phase)
3. *Select*—choosing desired data from the collection of assembled aeronautical data
4. *Format*—transforming the data into the format acceptable for the next functional link

The aeronautical data transmission functional link includes two phases:

1. *Receive*—accepting, verifying, and validating data
2. *Distribute*—last phase of the processing model which becomes part of the transmission link

At each phase of the processing, the data are verified and any issues are documented in an error report. Corrective action is taken as needed. If the data from a *trusted source*, such as an ICAO member state, are found to have an error, the error must be reported to the trusted source. However, it is often difficult to get the trusted source to immediately correct the data. Oftentimes, once the data are confirmed to be erroneous, they are corrected by the organization that applies the DO-200A compliant process.

DO-200A section 2 defines the guidance for each organization in order to ensure that the data satisfy the seven quality characteristics mentioned earlier. Each organization in the data chain is responsible for the following:

- Developing a compliance plan
- Defining data quality requirements
- Defining the data processing procedures
- Ensuring that data alteration is not performed unless properly coordinated with the originator
- Maintaining configuration management
- Ensuring the skills and competencies of personnel
- Performing tool assessment and qualification as needed
- Developing and implementing quality management procedures

Depending on the assurance level, the amount of required validation and verification varies. DO-200A defines validation and verification as follows [1]:

- *Validation*—The activity whereby a data element is checked as having a value that is fully applicable to the identity given to the data element, or a set of data elements that is checked as being acceptable for their purpose.
- *Verification*—The activity whereby the current value of a data element is checked against the value originally supplied.

In general, for assurance level 1, validation by application (i.e., applying the data under test) and verification are required; for assurance level 2, validation by application is not required, but verification is; and for assurance level 3, validation and verification are recommended but not required [1].

23.3 FAA Advisory Circular 20-153A

Just as the previous section provided a high-level overview of DO-200A, this section includes a summary of FAA AC 20-153A.

AC 20-153 (the predecessor to AC 20-153A) only applied to navigation databases. In 2010, the FAA expanded the AC to apply to other types of aeronautical data, including terrain, obstacle, and airport map databases. Each of the databases covered by AC 20-153 is briefly explained as follows. The AC may be applied to other databases; however, those should be closely coordinated with the certification authority.

- *Navigation database*—"Any navigation data stored electronically in a system supporting navigation applications. Navigation data is information intended to be used to assist the pilot to identify the aircraft's position with respect to flight plans, ground reference points and navaid fixes ... as well as items on the airport surface" [2].

- *Terrain database*—"Any data stored electronically in a system supporting terrain applications. Terrain data includes the natural surface of the earth excluding man-made obstacles" [2].

- *Obstacle database*—"Any data stored electronically in a system supporting obstacle applications. Obstacle data includes any natural or manmade fixed object which has vertical significance in relation to adjacent and surrounding features and which is considered as a potential hazard to the safe passage of aircraft" [2].

- *Airport map database*—"Any navigation data stored electronically in a system supporting airport map applications. Airport map data is information intended to be used to assist the pilot to identify the aircraft's position with respect to items on the airport surface" [2].

AC 20-153A provides guidance to aeronautical service providers, equipment or avionics manufacturers, and/or operators necessary to obtain a letter of acceptance (LOA). An LOA is a letter granted by the FAA, acknowledging compliance with AC 20-153A and DO-200A for aeronautical data processing. "The LOA formally documents that a supplier's databases are being produced pursuant to RTCA/DO-200A, or for some established systems, RTCA/DO-200" [2]. AC 20-153A identifies two types of LOAs:

- *Type 1 LOA*: Type 1 acceptance letters are primarily for data suppliers that are data service providers. "A Type 1 LOA provides recognition of a data supplier's compliance with RTCA/DO-200A with no identified compatibility with an aircraft system... This acceptance letter may be issued to data suppliers, operators, avionics manufacturers, or others" [2].

- *Type 2 LOA*: "Type 2 acceptance letters are based on requirements that ensure compatibility with particular systems or equipment and are for data suppliers that are avionics manufacturers/application integrators... Type 2 data suppliers have additional requirements to ensure the delivered database is compatible with the DQRs [data quality requirements] necessary to support the intended function approved for the target application" [2].*

---

* Brackets added for clarification.

AC 20-153A explains the LOA application process and how to apply DO-200A and additional requirements for Types 1 and 2 LOA applications. The AC also provides additional notes about DO-200A processes. Guidance is provided for operators, equipment and avionics manufactures (i.e., application providers), and data service providers. The AC also explains that contracting states are encouraged to follow DO-200A, as well as other applicable data quality guidelines (e.g., DO-201A, DO-272B, and DO-276A).

AC 20-153A section 14 explains that data obtained from a DO-200A compliant source or from a contracting state may be trusted. However, data from other suppliers must be verified and validated for the appropriate assurance level. Per the AC, DO-272B (section 3.9) provides acceptable techniques for verification and validation of airport map data [4].* Likewise, per the AC, DO-276A (sections 6.1.4 and 6.1.5) provides acceptable verification and validation techniques for terrain and obstacle data [5].

AC 20-153A section 17 provides a correction to DO-200A's error detection probabilities. The AC points out that for assurance level 1, the error detection must have a probability of undetected corruption of less than or equal to 10^{-9} (rather than 10^{-8} as identified in DO-200A). Likewise, for assurance level 2, the error detection must have a probability of undetected corruption of less than or equal to 10^{-5} (rather than 10^{-4} as identified in DO-200A). These probabilities better align with XX.1309 requirements at the aircraft level [2].†

AC 20-153A Appendix 3 provides a compliance matrix for aeronautical database suppliers to use to help ensure that they satisfy the DO-200A and AC 20-153A requirements prior to applying for an LOA.

In addition to AC 20-153A, the FAA has published Order 8110.55A, *How to Evaluate and Accept Processes for Aeronautical Database Suppliers* [6]. The Order provides guidelines for FAA Aircraft Certification Offices to evaluate and accept aeronautical processes and to grant LOAs. While this Order is primarily for FAA staff, it provides some insight for aeronautical database providers on what to expect during FAA audits and how to prepare for such audits.

23.4 Tools Used for Processing Aeronautical Data

Because of the volume of data and the need for repeatability, aeronautical data processing is tool intensive. Typically, a combination of integrity checks (such as cyclic redundancy checks) and qualified tools are used to ensure the integrity of the aeronautical data as they go through the data chain. Sometimes, tools are run in parallel and results compared to avoid the need

---

* It should be noted that DO-272B has been superseded by DO-272C since the publication of AC 20-153A.
† XX may be Part 23, 25, 27, or 29 of Title 14 of the U.S. Code of Federal Regulations.

for qualification. All tools must be identified and assessed. Tools whose output is not verified may require qualification. DO-200A and AC 20-153A reference DO-178B section 12.2 criteria for tool qualification. DO-178C's section 12.2 is also acceptable. As explained in Chapter 13, DO-178C section 12.2 invokes DO-330, *Software Tool Qualification Considerations*. DO-330 was written to be applicable to multiple domains, including the aeronautical data domain. DO-330 frequently asked question (FAQ) D.7 may provide valuable information for aeronautical data processing tools in some scenarios. This FAQ describes how the output of an unqualified tool may be verified by a qualified tool to ensure the accuracy of the unqualified tool.

23.5 Other Industry Documents Related to Aeronautical Data

DO-200A and AC 20-153A reference several other industry documents. A brief summary of these documents is provided to complete the survey of aeronautical database guidance. The RTCA documents are presented first (in sequential order), followed by two ARINC documents.

23.5.1 DO-201A: Standards for Aeronautical Information

DO-201A compiles general and specific requirements for aeronautical data with emphasis on RNAV operations in RNP airspace. General requirements and standards for aeronautical data are provided, including accuracy, resolution, calculation conventions, naming conventions, and the timely dissemination of the finished data [3]. Specific operational requirements and standards are also given; these are to be considered when developing procedures in the enroute, arrival, departure, approach, and aerodrome environments. As noted earlier, DO-201A defines the assurance levels for aeronautical information specific to RNP. Assurance levels for applications of data not covered by DO-201A need to be determined by the end user or application provider based on the safety assessment process. The EUROCAE equivalent to DO-201A is ED-77.

23.5.2 DO-236B: Minimum Aviation System Performance Standards: Required Navigation Performance for Area Navigation

DO-236B contains minimum aviation system performance standards (MASPS) for RNAV systems operating in an RNP environment. The standards are for designers, manufacturers, and installers of avionics equipment, as well as service providers and users of such systems for worldwide operations [7]. DO-236B references DO-201A and DO-200A compliance for navigation databases used in RNAV systems. Additionally, DO-201A provides guidance on requirements supporting RNAV and the RNP operations described in DO-236B. The EUROCAE equivalent to DO-236B is ED-75B.

23.5.3 DO-272C: User Requirements for Aerodrome Mapping Information

DO-272C provides minimum requirements applicable to the content, origination, publication, and updating of aerodrome (airport) mapping information. It also provides guidance to assess compliance to the requirements and to determine the necessary level of confidence. It defines a minimum set of data quality requirements which may be used for airport map displays [2,8]. The EUROCAE equivalent to DO-272C is ED-99C.

23.5.4 DO-276A: User Requirements for Terrain and Obstacle Data

DO-276A defines the minimum user requirements applicable to terrain and obstacle data. It includes the minimum list of attributes associated with terrain and obstacle data and describes associated errors that may need to be addressed [5]. The EUROCAE equivalent to DO-276A is ED-98A.

23.5.5 DO-291B: Interchange Standards for Terrain, Obstacle, and Aerodrome Mapping Data

DO-291B* is used in conjunction with DO-272C and DO-276A, which specify user requirements for terrain, obstacle, and aerodrome database content and quality. DO-291B sets data interchange format requirements for the data generated in compliance with DO-272C and DO-276A. DO-291B was based on the International Standards Organization (ISO) 19100 (geographic information) series of standards as applied to terrain, obstacle, and aerodrome mapping databases used in aviation. The standard specifies requirements for scope, identification, metadata, content, reference system, data quality, data capture, and maintenance information [9]. The EUROCAE equivalent of DO-291B is ED-119B. DO-291B is also related to ARINC 816, which is described in Section 23.5.7.

23.5.6 ARINC 424: Standard, Navigation System Database

"ARINC 424 has been the industry standard for exchanging navigation data between data suppliers and avionics vendors for more than 30 years" [10].

---

* AC 20-153A only references DO-291A, since the AC was published before DO-291B was released.

It serves as the air transport industry's recommended standard for preparing airborne navigation system reference data files.

> The data on these files are intended for merging with airborne navigation computer operational software to produce media for use by such computers on board aircraft... It enables data base suppliers, avionics systems, and other users of the data bases to fly and flight plan procedures as prescribed by procedure designers [11].

The original version of ARINC 424 was published in 1975. At this time, the latest version of ARINC 424 is 424-19. However, a committee is actively working on ARINC 424A. The current ARINC 424 format defines fixed-length text files with 132 characters per line. These files are converted into binary images and loaded into a specific flight management system. Because of target dependencies, each type of flight management computer has a different binary image. This leads to the scenario where a single airline with multiple aircraft types and flight management systems may have dozens of database packages, even though the raw data used to create the database packages is identical. The committee's goal for ARINC 424A is to develop an open standard database which is directly readable by the flight management computer—without the need for the flight management computer-specific binary. This allows an airline to obtain data from one source. ARINC 424A proposes using an Unified Modeling Language (UML) model which will output the same ASCII (American standard code for information interchange) as the traditional ARINC 424. However, the UML model will also be used to automatically generate an eXtensible Markup Language (XML) format which can be converted to a binary XML (for smaller file sizes and easier parsing). The committee is striving to keep the current ARINC 424 format, while at the same time enabling more standardized resultant files that can be used on multiple flight management systems [10].

23.5.7 ARINC 816-1: Embedded Interchange Format for Airport Mapping Database

ARINC 816-1 defines a format for databases used for airport moving maps (AMMs). AMMs have the potential to improve safety by reducing runway and taxiway incursions caused by a loss of crew situational awareness. ARINC 816-1 simplifies data handling and provides features which benefit airport moving map displays [12].

ARINC 816-1 is related to DO-272C and DO-291B, which were described earlier. Figure 23.2 illustrates the relationship between the three documents. DO-272C defines the content, quality, and processing requirements for airport mapping databases (AMDBs). DO-291B defines the data interchange format requirements for the databases, enabling the exchange of AMDBs between data originators and integrators. ARINC 816-1 defines a database

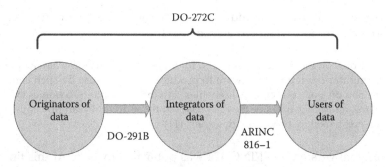

FIGURE 23.2
Relationship between DO-272C, DO-291B, and ARINC 816-1.

standard for embedded avionics systems. The standardized format allows end users to be able to choose between different database integrators independent of an avionics supplier. This allows database integrators to convert airport data directly into the end system specification [13].

References

1. RTCA/DO-200A, *Standards for Processing Aeronautical Data* (Washington, DC: RTCA, Inc., September 1998).
2. Federal Aviation Administration, *Acceptance of Aeronautical Data Processes and Associated Databases*, Advisory Circular 20–153A (September 2010).
3. RTCA/DO-201A, *Standards for Aeronautical Information* (Washington, DC: RTCA, Inc., April 2000).
4. RTCA DO-272B, *User Requirements for Aerodrome Mapping Information* (Washington, DC: RTCA, Inc., April 2009).
5. RTCA DO-276A, *User Requirements for Terrain and Obstacle Data* (Washington, DC: RTCA, Inc., August 2005).
6. Federal Aviation Administration, *How to Evaluate and Accept Processes for Aeronautical Database Suppliers*, Order 8110.55A (November 2011).
7. RTCA DO-236B, *Minimum Aviation System Performance Standards: Required Navigation Performance for Area Navigation* (Washington, DC: RTCA, Inc., October 2003).
8. RTCA DO-272C, *User Requirements for Aerodrome Mapping Information* (Washington, DC: RTCA, Inc., September 2011).
9. RTCA DO-291B, *Interchange Standards for Terrain, Obstacle, and Aerodrome Mapping Data* (Washington, DC: RTCA, Inc., September 2011).
10. C. Pschierer, J. Kasten, J. Schiefele, H. Lepori, P. Bruneaux, A. Bourdais, and R. Andreae, ARINC 424A—A next generation navigation database specification, *IEEE Digital Avionics Systems Conference* (Orlando, FL, 2009), 4.D.2-1–4.D.2.10.

11. Aeronautical Radio, Inc., Navigation system database, ARINC Specification 424-19 (Annapolis, MD: Airlines Electronic Engineering Committee, December 2008).
12. C. Pschierer and J. Schiefele, Open standards for airport databases—ARINC 816, *IEEE Digital Avionics Systems Conference* (Dallas, TX, 2007), 2.B.6-1–2.B.6-8.
13. Aeronautical Radio, Inc., Embedded interchange format for airport mapping database, ARINC Specification 816-1 (Annapolis, MD: Airlines Electronic Engineering Committee, November 2007).

24

Software Reuse

Acronyms

AC   Advisory Circular
CAST   Certification Authorities Software Team
COTS   commercial off-the-shelf
FAA   Federal Aviation Administration
PDS   previously developed software
PSAC   Plan for Software Aspects of Certification
RSC   reusable software component
RTOS   real-time operating system
SOUP   software of unknown pedigree
U.S.   United States

24.1 Introduction

Software reuse is an important subject because the majority of software projects, at least in the aviation world, are derivatives of existing systems. New or clean-sheet avionics, electrical systems, airframes, and engines are actually rather rare. DO-178C encourages a well organized and disciplined process that not only supports the initial certification and compliance with the regulations but also supports ongoing modifications and maintenance.

It is also important to discuss software reuse because there have been some rather well-publicized consequences of unsuccessful software reuse. The explosion of the Ariane Five rocket is one example. The software used on the Ariane Five was originally intended for the Ariane Four and worked properly on that platform. However, the launch characteristics of the Ariane Four and Five rockets were different. Improper reuse of the Ariane Four software caused the Ariane Five to explode [1]. In her book *Safeware*, Nancy Leveson explains that there is a myth that software reuse increases safety. She provides a number of examples of safety-related problems that arose from software reuse (three of which follow). First, the Therac-25 medical device reused parts from its predecessor, the Therac-20. An error existed in the Therac-20 software, but it had no serious consequences on the Therac-20 operation, except an occasional blown fuse. Unfortunately, when utilized on the Therac-25, the error led to massive radiation overdoses and the death of at least two people. Software reuse was not the sole reason for the Therac-25 problem, but it was a significant contributing factor. Second, air traffic control software used in the United States (U.S.) was reused in Great Britain. In this case, the British users failed to account for the longitudinal differences between the U.S. and Great Britain. Third, problems arose when software written for an aircraft in the northern hemisphere and above sea level was reused in the southern hemisphere or below sea level [2].

Software must be reused with extreme caution. This chapter examines reuse by discussing previously developed software (PDS) and commercial off-the-shelf (COTS) software. The terms reuse, PDS, COTS software, and software component are interrelated. COTS is a subset of PDS, and reuse implements PDS components. Perhaps some definitions will clarify the relationship:

- *Software reuse*: There are greatly ranging opinions on and definitions of software reuse. I prefer to look at it as a process of implementing or updating systems using existing software assets. Assets may be software components, software requirements, software design, source code, and other software life cycle data (including plans, standards, verification cases and procedures, and tool qualification data). Software reuse may occur within an existing system, across similar systems, or in widely differing systems.

- *Software component*: DO-178C defines *component* as: "A self-contained part, combination of parts, sub-assemblies, or units that perform a distinct function of a system" [3]. However, I prefer the following, more descriptive definition: "an atomic software element that can be reused or used in conjunction with other components; ideally, it should work without modification and without the engineer needing to know the content and internal function of the component. However, the interface, functionality, pre-conditions and post-conditions, performance characteristics, and required supporting elements must be well known" [4].

- *COTS software*: Commercially available software components that are not intended to be customized or enhanced by the user, although they may be configured for user-specific needs [3].

- *PDS*: "Software already developed for use. This encompasses a wide range of software, including COTS software through software developed to previous or current software guidance" [3].

Together, these definitions indicate that software reuse often occurs using PDS that is packaged as a component. The PDS may be commercially available

TABLE 24.1

Examples of Previously Developed Software

	Non-COTS	COTS
DO-178[ ]	• An avionics application developed to DO-178A. • DO-178B-compliant flight control software to be modified for use in a similar system. • DO-178C-compliant battery management software to be installed in a new aircraft.	• A real-time operating system (RTOS) with DO-178B or DO-178C data available. • A DO-178C-compliant board support package for a specific RTOS and microprocessor.
Non-DO-178[ ]	• An electrical power system developed using U.S. Department of Defense Military Standard 498. • Flight management system software for military aircraft developed using the United Kingdom's Defence Standard 00-55. • Brake system software for an automobile. • Device driver for a controller area network databus.	• An operating system without DO-178[ ] data (e.g., Windows). • A compiler-supplied library. • Databus software used in automotive market. • Communication stack compliant with Open Systems International communication protocol.

(COTS software), developed in-house from a past project, developed using DO-178[ ],* developed using some other guidance or standard (e.g., a military or automotive standard), or developed using no guidance at all. Table 24.1 provides some examples of PDS considering four categories. Some PDS is utilized without change; while some PDS requires modification for use in the new system or environment.

In order for software to be truly reused (rather than salvaged), it must be designed to be reusable. While this is not *required* per any aviation standard, it is a practical reality and a best practice. Therefore, this chapter discusses how to plan and design for reuse. The focus then shifts to examining how to reuse PDS. A brief survey of software service history, which is closely related to PDS, concludes the chapter.

24.2 Designing Reusable Components

Software reuse doesn't just happen. A successful reuse effort requires planning and careful design decisions. In order for components to be successfully reused, they must be designed to be reusable. Designing for reuse normally

---

* DO-178[ ], indicates DO-178, DO-178A, DO-178B, or DO-178C.

increases the initial development time and cost, but the return on investment is recovered when the component is reused without significant rework.

The Ariane Five, Therac-25, and other previously discussed examples provide a brief examination of cases where software reuse was not carefully evaluated and implemented. As software becomes more complex and more widely used, the concerns of software reuse in safety-critical systems also increase. Reuse can be a viable option; however, it must be evaluated and implemented with caution. If software is designed to be reusable in the first place, it can help avoid such incidents.

In general, the software industry, particularly the aviation software industry, is rather immature when it comes to developing for reuse. In the 10 years since I completed my master's thesis on the subject of reuse and arriving at the same conclusion, there have been some advances in the reuse arena for components like real-time operating systems and library functions. Unfortunately, designing for reuse still seems to be a rather elusive goal in the aviation industry. Frequently, a project starts out with the intent to design for reuse. However, once schedule and budget constraints are imposed, the reuse goals are abandoned.

Following are 16 recommendations to consider when designing for reuse. These interrelated recommendations summarize vital programmatic and technical concepts for effective reuse.

Recommendation 1: Secure management's long-term commitment to reuse. Designing a component to be reusable will take longer and cost more than a component that is not designed for reuse. The magnitude of the schedule and cost increase varies depending on the processes used, the organization's experience and domain expertise, the type of component being developed, management's commitment, and a variety of other factors. Some estimate that a reusable component costs two to three times as much as the same component for one-time use [5]. It is important for management to understand this reality and to make the long-term commitment. For reuse to be successful, upper-level management must champion it. Most reuse failures come from management's lack of commitment to support the effort when the going gets tough.

Recommendation 2: Establish and train a reuse team. A well-trained team with the primary objective of developing reusable software is more successful than a development-as-usual team. It may be a small dedicated team, or it may be composed of people who only contribute part of their time to the effort. However, to ensure accountability, it is important to have an identified team and to ensure that all members are properly trained.

Recommendation 3: Evaluate existing projects to identify issues that prevent reuse. It is beneficial to evaluate existing software projects within your company to identify what issues prevent reuse. Design practices, coding practices, hardware interfaces, development environment, and compiler issues may prevent reuse. By compiling a list of issues that prevent reuse, the team can begin to develop strategies to overcome these issues.

Recommendation 4: Try a pilot project. Rather than trying to change the organization overnight, it is often best to start with a small pilot project. That project can be the basis for identifying reuse practices and training the reuse team. A small, successful project builds experience and confidence.

Recommendation 5: Document reuse practices and lessons-learned. Before launching into a reusable component development effort, draft practices for the team to follow. After the first project or two, those practices should be updated. Ideally, practices and procedures are continually refined as lessons are learned from real-life experience. Eventually, the practices can be established as company-wide recommendations or procedures.

Recommendation 6: Identify intended users during the component development and support them throughout. Since the users are the component customers, it's important to identify who the component users are in order to ensure that their needs are met, any conflicting needs are identified, and that they are informed of any problems that may arise when developing and maintaining the component. It is also important to design the component to be user friendly, which includes such characteristics as (1) easy identification (users should be able to easily see if the component meets their needs), (2) ease of use (users should be able to quickly learn how to use the component), and (3) usability by integrators with a wide range of experience (novices to experts) [6].

Recommendation 7: Implement a domain engineering process. A sound domain engineering process is essential to successful software reuse. A domain is "a group or family of related systems. All systems in that domain share a set of capabilities and/or data" [7]. Domain engineering is the process of creating assets through domain analysis, design, and implementation that can be managed and reused. The domain engineer suggests an architecture that meets the majority of the application needs and is suitable for future reuse. It is important to distinguish the concepts of *domain engineering* and *reuse engineering*. *Domain engineering* is developing for reuse; whereas *reuse engineering* is developing with reuse. First-rate domain engineering is critical to being able to reuse entire components or design portions of those components, and not just salvage code.

Recommendation 8: Identify and document the usage domain. A *usage domain* is a declared set of characteristics for which the following can be shown:

- The component is compliant to its functional, performance, and safety requirements.
- The component meets all the assertions and guarantees regarding its defined allocateable resources and capabilities.
- The component performance is fully characterized, including fault and error handling, failure modes, and behavior during adverse environmental effects [8].

It is important to document the usage domain because when reusing the component, a usage domain analysis will need to be performed to evaluate the component reuse in the new installation. The usage domain analysis evaluates the subsequent reuse of the component to ensure that (1) any assumptions made by the component developer are addressed, (2) the impact of the component on the installation is considered, (3) the impact of the new environment on the component is considered, and (4) any interfaces with the component in the new installation are consistent with the component's usage domain and interface specification [8].

Recommendation 9: Create small, well-defined components. Steve McConnell writes: "You might be better off focusing your reuse work on the creation of small, sharp, specialized components rather than on large, bulky, general components. Developers who try to create reusable software by creating general components rarely anticipate future users' needs adequately... 'Large and bulky' means 'too hard to understand,' and that means 'too error-prone to use'" [5]. In order for a component to be reusable, its functionality must be clear and well documented. The functionality is defined by a high-level *what* the component does—not *how* it is implemented. The functionality should have a single purpose, only provide functions related to its purpose, and be properly sized (i.e., it is not too small for use and not too large to become unmanageable). Additionally, in order to allow smooth and successful integration, a software component must have a well-defined interface. An interface defines how the user interacts with the component. A successful interface has consistent syntax, logical design, predictable behavior, and a consistent method of error handling. A well-defined interface is complete, consistent, and cohesive—it provides what the user needs to make the component work [9].

Recommendation 10: Design with portability in mind. Portability is a desirable attribute for most software products because it enhances the value of a software package both by extending its useful life and by expanding the range of installations in which it can be used [10]. There are two types of portability: *binary portability* (porting the executable form) and *source portability* (porting the source language representation). Binary portability is clearly desirable, but is usually possible only across strongly similar processors (e.g., same binary instruction set) or if a binary translator is available. Source portability assumes availability of source code, but provides opportunities to adapt a software unit to a wide range of environments [11]. In order to obtain binary or source portability, the software must be designed for portability. In general, incorporating portability calls for design strategies such as the following [10]:

1. Identify the minimum needed environmental requirements and assumptions.
2. Eliminate unnecessary assumptions throughout the design.

3. Identify required interfaces specific to the environment (such as procedure calls, parameters, and data structures).

4. For each interface, do one of the following:

a. Encapsulate the interface in a suitable module, package, object, etc. (i.e., anticipate the need to adapt the interface for each target system).

b. Identify a standard for the interface, which will be available in most target environments. And, follow this standard throughout the design.

Recommendation 11: Design the component robustly. Bertrand Meyer writes: "The component must not fail, when it is used properly" [6]. Since the component will have multiple users, it must be designed robustly. The component should anticipate unexpected inputs and address them (e.g., using error handling capabilities). Robustness must be considered when developing component requirements and design.

Recommendation 12: Package the data to be reusable. Federal Aviation Administration (FAA) Order 8110.49 chapter 12 discusses the reuse of software life cycle data in the aircraft certification environment [18]. In order for data (and the software itself) to be reusable, the software needs to be packaged for reuse. Often, this means having a full set of DO-178C life cycle data for each component. Order 8110.49 provides guidelines for reuse within a company; however, FAA Advisory Circular (AC) 20-148 discusses the concept of software component reuse across company boundaries (e.g., reuse of an RTOS on multiple avionics systems). AC 20-148 provides guidance for how to document a component to be reusable. Whether seeking an FAA reuse acceptance letter or not, the component can still be packed to be reusable from program to program. The Order and AC provide packaging suggestions.

Recommendation 13: Document the reusable component well. Since it is often unknown who will use the component in the future, it is important to thoroughly document the component. This involves creating documentation to ensure proper use and integration of the component (e.g., interface specification and user's manual), as well as data to support certification and maintenance. AC 20-148 requires the creation of a data sheet that explains information needed by the user of the reusable software component (RSC) in order to ensure proper usage of the component. AC 20-148 (section 6.i) requires the following data in the data sheet: component functions, limitations, analysis of potential interface safety concerns, assumptions, configuration, supporting data, open problem reports, characteristics of the component (such as worst-case timing, memory, throughput), and other relevant information that supports the use of the component [12].

Recommendation 14: Document the design rationale. In order to effectively reuse a software component, its design rationale must be well documented.

Both the design decisions of the component itself and the design decisions of the system that first integrates the component are important. The documented design decisions help determine where a component can "appropriately and advantageously be reused" [13]. Additionally, the documented design decisions for the system using the component can also be helpful to determine if a component is suitable for reuse in another system. Dean Allemang divides design rationale into two categories: (1) internal rationale (the relation of parts of the design to other parts of the same design—the way in which a component interacts with other components), and (2) external rationale (the relation of parts of the design to parts of different designs—use of component in multiple systems) [13].

Recommendation 15: Document safety information for the user. During the development and subsequent reuse of a component, safety must be considered. It is essential that the component developer defines the failure conditions, safety features, protection mechanisms, architecture, limitations, software levels, interface specifications, and intended use of the component. All interfaces and configurable parameters must be analyzed to describe the functional and performance effects of these parameters and interfaces on the user. The analysis documents required actions by the user to ensure proper operation. Additionally, per AC 20-148, section 5.f, a RSC developer must "produce an analysis of the RSC's behavior that could adversely affect the users' implementation (as in vulnerabilities, partitioning requirements, hardware failure effects, requirements for redundancy, data latency, and design constraints for correct RSC operation). The analysis may support the integrator's or applicant's safety analysis" [12].

Recommendation 16: Focus on quality, not quantity. Since reusable components may have multiple users, it is important to ensure that each component works as required. McConnell writes: "Successful reuse requires the creation of components that are virtually error-free. If a developer who tries to employ a reusable component finds that it contains defects, the reuse program will quickly lose its shine. A reuse program based on low-quality programs can actually increase the cost of developing software... If you want to implement reuse, focus on quality, not quantity" [5]. When developing AC 20-148, the FAA emphasized the fact that the first approval of a reusable component requires a high level of certification authority oversight and involvement to ensure it functions correctly.

24.3 Reusing Previously Developed Software

Three aspects of PDS are considered in this section: (1) a process to evaluate PDS for inclusion in a safety-critical system, particularly when required to meet the civil aviation regulations, (2) special considerations when reusing

PDS that was not developed using DO-178[ ],* and (3) additional factors to evaluate when the PDS is also COTS software.

24.3.1 Evaluating PDS for Use in Civil Aviation Products

To better understand the evaluation process for the use of PDS in a civil aviation product (an aircraft, engine, or propeller) a flowchart is provided (see Figure 24.1). Although it may not cover all situations that arise; it should cover the majority of those that pertain to civil aviation projects. Such an approach could also be applied to other domains but may require some modification. Each block in the flowchart is numbered and is described in the following:

1. *Determine if the PDS was approved in a civil aircraft installation.* If the PDS was previously approved in a certified civil aircraft, engine, or propeller, it may be suitable for reuse without rework or re-verification. If it was not previously approved, it will need to show that it meets DO-178C or an equivalent level of assurance.

2. *Determine if the installation and the software level will be the same.* If the software is being installed in the same system and used the same way (e.g., a traffic collision and avoidance system that is being installed in another aircraft without change to the hardware or software) and the safety impact is the same, the PDS is probably suitable for reuse without rework or re-verification. A similarity analysis at the system level is normally needed to evaluate the use of the software, confirm the safety considerations, and support the reuse claim (this information would likely go in the system-level plans). If the installation changes or the software level is inadequate, further evaluation is needed (see Block 4).

3. *Document the intent to use the PDS in the plans.* The intent to use the PDS and the results of the similarity analysis need to be documented in the plans. Depending on the situation, this may be the system-level plans (e.g., if the entire system is reused) or in the Plan for Software Aspects of Certification (PSAC) (if only a software component is reused). The plans must explain the equivalent installation and adequacy of the software level, as explained in Items 1 and 2.

4. *Was the software originally approved using DO-178[ ]?* This process block's purpose is to identify if DO-178C or its predecessors were followed during the original development. If not, a gap analysis is performed to identify the gaps, and an approach for filling those gaps is identified (Blocks 5 and 6).

5. *Perform a gap analysis.* A gap analysis involves an assessment of the PDS and its supporting data against the DO-178C objectives.

---

* DO-178[ ] could be DO-178, DO-178A, DO-178B, or DO-178C.

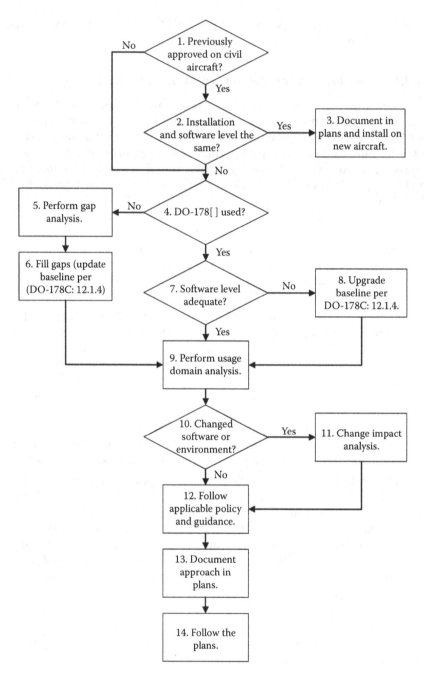

FIGURE 24.1
Process for evaluating PDS for civil aviation project.

For software of unknown pedigree (SOUP) this analysis may not be possible because the data is normally unavailable. Such software can usually only be approved to level D or E.

6. *Fill the gaps.* Once the gaps have been identified, they need to be filled. DO-178C section 12.1.4, "Upgrading a Development Baseline," provides guidance on this task. Depending on the original development, this may be a significant effort and may lead one to rethink the reuse plan. Some alternative approaches that may be used to fill the gaps are discussed later in this chapter (see Section 24.3.2). If the PDS is a COTS component, see Section 24.3.3 for additional recommendations to identify and fill the gaps.

7. *Determine if the software level is adequate.* Determine the necessary software level for the new installation and determine if the software level of the PDS is adequate. For software not developed per DO-178[ ],* the answer is *NO* and a gap analysis is needed (see Blocks 5 and 6). If the software was developed per DO-178 or DO-178A, the equivalent levels for DO-178B and DO-178C are shown in Table 24.2.

8. *Upgrade the baseline.* If the software level of the PDS is not adequate, the data (and possibly the software itself) will need to be updated to address the additional DO-178[ ] objectives. DO-178C section 12.1.4 describes this process. Normally, this involves additional verification activities. If the software was not developed using DO-178B or DO-178C, this upgrade may also require a gap analysis (Blocks 5 and 6) to bring it up to DO-178C compliance. The need for such an analysis will depend on the software level increase desired and the ability of the existing data to support safety needs and should be closely coordinated with the certification authority.

TABLE 24.2

Software Level Equivalence

DO-178B/C SW Level Required by the Installation	Legacy System Software Level per DO-178/DO-178A		
	Critical/Level 1	Essential/Level 2	Nonessential/Level 3
A	Yes/analyze	No	No
B	Yes	No/analyze	No
C	Yes	Yes	No
D	Yes	Yes	No
E	Yes	Yes	Yes

Source: Federal Aviation Administration, *Software Approval Guidelines*, Order 8110.49, Federal Aviation Administration, Washington, DC, Change 1, September 2011.

---

* DO-178[ ] may be DO-178, DO-178A, DO-178B, or DO-178C.

9. *Perform a usage domain analysis.* For a new installation of the PDS, this analysis ensures the following, as a minimum:

  a. The software that will be used is the same as the originally approved software.

  b. There are no adverse effects on safety or operational capability of the aircraft.

  c. Equipment performance characteristics are the same.

  d. Any open problem reports are evaluated for impact in the new installation. If problem reports are not available, this could prohibit the reuse of the PDS.

  e. Ranges, data types, and parameters are equivalent to original installation.

  f. Interfaces are the same.

  Some of this analysis is already addressed by other blocks; however, Block 9 is intended to gather the information in one place and to more thoroughly evaluate the intended reuse. This analysis could identify required changes to the software. If the PDS will be modified, this analysis may be performed in conjunction with the change impact analysis and documented as part of the change impact analysis. However, if the software and its environment do not change, this usage domain analysis is included in the plans (normally the PSAC or system certification plan):

10. *Determine if the software or its environment changed.* The usage domain analysis (Block 9) is used to determine if there are any changes to the PDS installation. Additionally, if the development environment (e.g., compiler, compiler settings, version of code generator, linker version) or software changes, a change impact analysis is needed and appropriate software life cycle data will need to be modified. DO-178C sections 12.1.2 and 12.1.3 discuss changes to installation, application, and development environment.

11. *Perform a change impact analysis.* As noted in Chapter 10, modified software requires a change impact analysis to analyze the impact of the change and plan for re-verification.* As discussed in Chapter 10, the change impact analysis is also used to determine if the software change is *major* or *minor.* If the PDS is classified as *legacy software* (software developed to an older version of DO-178 than is required by the certification basis), this may impact how the software change is implemented. Per Order 8110.49, for *minor changes,* the process originally used for the legacy software may be used to implement the change (e.g., a minor change to a DO-178A compliant PDS may use the DO-178A process). However, if the change is classified as a *major change,* the change to the PDS and all subsequent changes to

---

* DO-178C sections 12.1.1.c and 12.1.1.d also require a change impact analysis

that PDS must be to the version of DO-178 required by the certification basis (probably DO-178C for new projects).*

12. *Follow applicable certification policy and guidance.* In addition to the DO-178C section 12.1 guidance, most certification authorities have policy or guidance related to PDS and/or reuse (e.g., FAA Order 8110.49 and AC 20-148). Also, project-specific issue papers (or equivalent) may be raised when the existing guidance doesn't fit the project-specific scenario.

13. *Document approach in plans and obtain agreement with certification authority.* The plan to use PDS must be documented in the PSAC (or possibly a system-level certification plan). The PSAC describes the PDS functionality; its original pedigree; any changes to installation, software, or development environment; and results of change impact analysis and/or usage domain analysis. If the PDS was not developed to DO-178[ ], details of the gap analysis and alternative approach(es) should be explained. The PSAC should also explain any necessary changes and how they will be verified. Additionally, the PSAC explains how software configuration management and software quality assurance considerations for the software will be addressed per DO-178C section 12.1.5† and 12.1.6.‡ Once the PSAC is completed, it is submitted to certification authority. (Most certification authorities will also want to see the change impact analysis. As noted in Chapter 10, the change impact analysis is often included in the PSAC.) It is important to obtain certification authority agreement before proceeding too far into the reuse effort.

14. *Follow the plans and implement per agreed process.* Once the certification authority approves the plans (this could take a couple of iterations), the approved plans must be followed. All necessary rework, verification, and documentation should be performed. The change impact analysis and plans may require updates as the project evolves.

24.3.2 Reusing PDS That Was Not Developed Using DO-178[ ]

As noted in Figure 24.1 Blocks 5 and 6, if the software was not developed to DO-178[ ] then a gap analysis is performed to identify DO-178C objectives that are not satisfied. Depending on the original development, filling these gaps might be a trivial task or it could be extremely time consuming. For SOUP it may prove more efficient to start over than to try to reconstruct

---

* At this time, Order 8110.49 (change 1) is in effect [18]. The approach for legacy software could change when Order 8110.49 is updated to recognize DO-178C.

† DO-178C section 12.1.5 ensures that a change management process is implemented, including traceability to previous baseline and a change control system with problem reporting, problem resolution, and change tracking.

‡ DO-178C section 12.1.6 ensures that software quality assurance has evaluated the PDS and any changes to the PDS.

nonexistent data. For PDS that has some data available, filling the gaps may be a viable path. PDS that has an excellent track record, but not in civil aviation (e.g., a COTS RTOS, a military application, or an automotive component) may be successful using this approach.

The gap analysis examines the available data against what is required for DO-178C to determine how much additional work is required. In my experience, for levels A and B software, the gap tends to be rather wide. If source code is not available, level D is generally the highest one can go.* If there are very little data available but source code is available, service history and reverse engineering tend to be the most common options.

DO-278A section 12.4 and DO-248C section 4.5 identify typical alternative approaches used to ensure that PDS has the same level of confidence as would be the case if DO-178C objectives had been satisfied initially. Table 24.3 summarizes some of the more common approaches. In most scenarios, the PDS gaps are filled by a combination of these approaches. Service history is discussed later in this chapter and reverse engineering is examined in Chapter 25.

24.3.3 Additional Thoughts on COTS Software

As noted earlier, COTS software is a special kind of PDS. Examples of COTS software are included in Table 24.1. DO-278A section 12.4 provides some specific guidance for acquiring and integrating COTS into a safety-critical system. DO-178C does not provide equivalent guidance, but the DO-278A approach can be applied to a DO-178C project. DO-278A promotes the concept of a *COTS software integrity assurance case* that is developed to ensure that the COTS software provides the same level of confidence as software developed to DO-278A or DO-178C. The integrity assurance case for DO-178C compliance includes the following information [14]:

- Claims about the integrity of the COTS software and which DO-178C objectives are not met by the case.
- Environment where the COTS software will be used.
- Requirements that the COTS software satisfies.
- Identification, assessment, and mitigation of unneeded capabilities in the COTS software.
- Explanation of DO-178C objectives that are satisfied with the existing COTS software data.
- Identification of DO-178C objectives not satisfied and explanation of how an equivalent level of confidence will be achieved using alternative approaches.
- Explanation of strategies to be taken.

---

* FAA Order 8110.49 chapter 8, provides some insight into how to comply with level D objectives [18].

TABLE 24.3

Summary of Alternative Approaches for PDS

Approach	Description	Limitations
Service history/ experience	DO-178C defines product service history as: "A contiguous period of time during which the software is operated within a known environment, and during which successive failures are recorded" [3]. This is discussed in more detail later in this chapter.	• The environment where history exists and the intended environment may not be the same. • It may be difficult to prove that problem reporting and collection is adequate. • The PDS may have changed several times, making it difficult to prove history. • The hours of service experience may not be adequate. • This is a very subjective approach and is difficult to obtain certification authority approval. Most likely, other approaches will need to supplement the service history/experience claim.
Process recognition	If the PDS was approved or independently accredited using a defined process, it may be possible to claim credit for some or all of the DO-178C objectives. Example processes include a Military Standard 498 (MIL-STD-498) or Department of Defense Standard 2167A (DOD-STD-2167A) compliant process, or a Software Engineering Institute Capability Maturing Model Integration® assessed process.	• The process typically won't address all DO-178C objectives. • The data may not be available for the user to assess.
Reverse engineering	"Reverse engineering is the process of developing higher level software data from existing software data. Examples include source code from object code (or executable object code), or developing high-level requirements from low-level requirements" [17]. This topic is discussed in Chapter 25.	• Reverse engineering is a subjective approach that is discouraged by certification authorities. • It is difficult to go from lower levels of abstraction to higher levels of abstraction (e.g., when low-level requirements are reverse engineered from source code, the requirements often mimic the code). • The thought process of the original developers may not be understood. • Reverse engineering requires a very experienced team to do it right.

(continued)

TABLE 24.3 (continued)

Summary of Alternative Approaches for PDS

Approach	Description	Limitations
Restriction of functionality	Restriction of functionality is the process of restricting the PDS use to a subset of its overall functionality. Some ways to do the restriction are run-time checks, deactivated code, built-in restrictions, or wrappers [17]. Most of the RTOSs used for aviation are a subset of the fully featured, general purpose RTOS. Any nondeterministic or potentially unsafe functionality is removed.	• It can be difficult to prove that undesired code won't be exercised. • Protective mechanisms may require additional verification.
Architectural mitigation	The system may be architected in order to protect from any undesired behavior of the software. Some common approaches are software partitioning and safety monitoring. Partitioning is used to isolate the software interaction with other software. Monitors may be used to detect potential failures.	• Without knowledge of all the PDS functionality, it may be difficult to prove any undesired effects are mitigated.
Additional testing	Additional and extensive testing may be developed to ensure that the software performs as intended and does not have adverse effects. Some additional tests might include the following [14]: • *Exhaustive input tests* for a simple software function with isolated inputs and outputs. • *Extensive robustness testing* to ensure any out-of-range, off-nominal, or erroneous inputs do not cause undesired behavior. • *System-level testing* to ensure the software performs as intended at the system level.	• It is difficult to prove no unintended functionality exists in the software—particularly for larger or more complex software.

TABLE 24.3 (continued)

Summary of Alternative Approaches for PDS

Approach	Description	Limitations
	• *Long-term soak testing* which occurs by running the system for a long period of time while exposing it to a wide range of normal and abnormal inputs—followed by functional testing without resetting the system. • *Using system for training* which exposes it to multiple independent users.	

Source: RTCA DO-278A, *Guidelines for Communications, Navigation, Surveillance, and Air Traffic Management (CNS/ATM) Systems Software Integrity Assurance*, RTCA, Inc., Washington, DC, December 2011; RTCA DO-248C, *Supporting Information for DO-178C and DO-278A*, Washington, DC, RTCA, Inc., December 2011.

- Identification of all data to support the case, including additional software life cycle data that will be generated using alternative methods.
- List of assumptions and justifications used in the integrity assurance argument.
- Description of processes for verifying that all uncovered objectives identified in the gap analysis are satisfied.
- Evidence that the COTS software has the same integrity as would be the case had all objectives been met during the original development.

Additionally, the following COTS-specific concerns need to be addressed when determining the feasibility of COTS software in a safety-critical system:

1. The availability of the supporting life cycle data.
2. The suitability of the COTS software for safety-critical use. Many COTS software components were not designed with safety in mind.
3. The stability of the COTS software. Evaluate the following factors:
   a. How often has it been updated?
   b. Have patches been added?
   c. Is the complete set of life cycle data for each update provided or available?
   d. Will the supplier provide notification when updates are available, explain why they occurred, provide problem reports that were fixed (to support change impact analysis), and provide updated data to support the change?

4. The technical support available from the supplier. Consider the following:

    a. If issues are discovered while using the COTS software, is the supplier willing (and obligated) to fix them?

    b. Is the supplier willing to support certification?

    c. Is the supplier willing (and obligated) to disclose known issues throughout the entire life of the COTS software usage (perhaps issues identified by other users)?

    d. What will happen if the supplier goes out of business or decides to no longer support the software?

5. The configuration control of the COTS software. Consider the following:

    a. Is there a problem reporting system in place?

    b. Are changes traced to the previous baseline(s)?

    c. Is the COTS software and its supporting life cycle data uniquely identified?

    d. Is there a controlled release process?

    e. Are open problem reports provided?

6. Protection of the COTS software from viruses, security vulnerabilities, etc.

7. Tools and hardware support for the COTS software.

8. Compatibility of the COTS software with the target computer and interfacing systems.

9. Modifiability or configurability of the COTS software.

10. Ability to deactivate, disable, or remove unwanted or unneeded functionality in the COTS software.

11. Adequacy of the supplier's quality assurance.

24.4 Product Service History

This section briefly explains product service history and factors to consider when proposing it as an alternative method for PDS.

24.4.1 Definition of Product Service History

As noted in Table 24.3, DO-178C defines *product service history* as follows: "A continuous period of time during which the software is operated within a known environment, and during which successive failures are recorded" [3].

This definition includes the concepts of problem reporting, environment (including operational and target computer environments), and time.

Significant effort has been expended by both the industry and the certification authorities to identify a feasible way to use product service history for certification credit. The Certification Authorities Software Team (CAST)* published a paper on the subject (CAST-1); the FAA sponsored research in this area (resulting in a research report and a handbook); the one page of text in DO-178B grew to four pages in DO-178C; and DO-248C includes a seven-page discussion paper on the topic. However, despite the effort to clarify the subject, it is still difficult to implement.

DO-178C section 12.3.4 explains that a product service history case depends on the following factors [3]: (1) configuration management of the software, (2) effectiveness of problem reporting activity, (3) stability and maturity of the software, (4) relevance of product service history environment, (5) length of the product service history, (6) actual error rates in the product service history, and (7) impact of modifications. The section goes on to provide additional guidance on the relevance of the service history, the sufficiency of the accumulated service history, the collection and analysis of problems found during service history, and the information to include in the PSAC when posing service history as an alternative method.

Product service history may be applied to software that was not developed to DO-178[ ] or to software that was developed to a lower level of DO-178[ ] than is required for the desired higher level system.

24.4.2 Difficulties in Seeking Credit Using Product Service History

To date, it has been virtually impossible to make a successful claim using product service history alone. However, it has been successfully used to supplement other alternatives (e.g., reverse engineering or process recognition).

As noted in the FAA's research report, entitled *Software Service History Report*, authored by Uma and Tom Ferrell, the definition of product service history in DO-178C is "very similar to the IEEE [Institute for Electronic and Electronic Engineers] definition of reliability, which is 'the ability of a product to perform a required function under stated conditions for a stated period of time' " [15].† Neither DO-178C nor the certification authorities encourage software reliability models because historically they have not been proven accurate. Because of the similarity between reliability and service history, it is also difficult to make a satisfactory claim using product service history.

---

* CAST is a team of international certification authorities who strive to harmonize their positions on airborne software and aircraft electronic hardware in CAST papers.
† Brackets added for clarity.

Another factor that tends to make product service history difficult to prove is that the data collected during the product's history is often inadequate. Companies normally do not plan up front to make a product service history claim, so the problem reporting mechanism may not be in place to collect the data.

24.4.3 Factors to Consider When Claiming Credit Using Product Service History

When claiming credit for service history, the following need to be addressed:

1. *The service history relevance must be demonstrated.* Per DO-178C section 12.3.4.1, to demonstrate the relevance of the product service history the following should occur [3]:

    a. The amount of time that the PDS has been active in service must be documented and adequate.

    b. The configuration of the PDS and the environment must be known, relevant, and under control.

    c. If the PDS was changed during the service history, the relevance of the service history for the updated PDS needs to be evaluated and shown to be relevant.

    d. The intended usage of the PDS must be analyzed to show the relevance of the product service history (to demonstrate software will be used in the same way).

    e. Any differences between the service history environment and the environment in which the PDS will be installed must be evaluated to ensure the history applies.

    f. An analysis must be performed to ensure any software that was not used in production (e.g., deactivated code) is not seeking service history credit.

2. *The service history must be sufficient.* In addition to showing the relevance of the service history, the amount of service history must be shown to be sufficient to satisfy the system safety objectives, including the software level. The service history must also satisfactorily address the DO-178C gaps that it is intended to fill.

3. *In-service problems must be collected and analyzed.* In order to make a claim of service history, it must be demonstrated that problems that occurred during the PDS service history are known, documented, and acceptable from a safety perspective. This requires evidence of an adequate problem reporting process. As noted earlier, this can be a difficult task to carry out.

The FAA's *Software Service History Handbook* identifies four categories of questions to ask when proposing a product service history claim [16].

These are excellent questions. If one can satisfactorily answer these, it might be feasible to claim service history. If not, it will be difficult to make a successful case to the certification authorities. The categories of questions are noted here:

- 45 questions related to problem reporting
- 11 questions about the operation (comparing the similarity of operation of the PDS within the previous domain as compared with the target domain)
- 12 questions about the environment (assessing the computing environment to assure that the environment in which the PDS was hosted during the service history is similar to the proposed environment)
- 19 questions about time (evaluating the service history time duration and error rates using the data available from product service history)

For convenience, the specific questions are included in Appendix D. For additional information on product service history, consult DO-178C (section 12.3.4) [3], DO-248C (discussion paper #5) [17], and the FAA's *Software Service History Handbook* [16].

References

1. R. Reihl, Can software be safe?—An ADA viewpoint. *Embedded Systems Programming* December 1997.
2. N. Leveson, *Safeware: System Safety and Computers* (Reading, MA: Addison-Wesley, 1995).
3. RTCA DO-178C, *Software Considerations in Airborne Systems and Equipment Certification* (Washington, DC: RTCA, Inc., December 2011).
4. A. Lattanze, A component-based construction framework for DoD software systems development, *CrossTalk* November 1997.
5. S. McConnell, *Rapid Development* (Redmond, WA: Microsoft Press, 1996).
6. B. Meyer, Rules for component builders, *Software Development* 7(5), 26–30, May 1999.
7. J. Sodhi and P. Sodhi, *Software Reuse: Domain Analysis and Design Process* (New York: McGraw-Hill, 1999).
8. RTCA DO-297, *Integrated Modular Avionics (IMA) Development Guidance and Certification Considerations* (Washington, DC: RTCA, Inc., November 2005).
9. A. Rhodes, Component based development for embedded systems, *Embedded Systems Conference* (San Jose, CA, Spring 1999), Paper #313.
10. J. D. Mooney, Portability and reuse: Common issues and differences, Report TR 94-2 (Morgantown, WV: West Virginia University, June 1994).
11. J. D. Mooney, Issues in the specification and measurement of software portability, Report TR 93-6 (Morgantown, WV: West Virginia University, May 1993).

12. Federal Aviation Administration, *Reusable Software Components*, Advisory Circular 20-148 (Washington, DC: Federal Aviation Administration, December 2004).

13. D. Allemang, Design rationale and reuse, *IEEE Software Reuse Conference* (Orlando, FL, 1996).

14. RTCA DO-278A, *Guidelines for Communications, Navigation, Surveillance, and Air Traffic Management* (CNS/ATM) systems software integrity assurance (Washington, DC: RTCA, Inc., December 2011).

15. U. D. Ferrell and T. K. Ferrell, Software service history report, DOT/FAA/AR-01/125 (Washington, DC: Office of Aviation Research, January 2002).

16. U. D. Ferrell and T. K. Ferrell, *Software Service History Handbook*, DOT/FAA/AR-01/116 (Washington, DC: Office of Aviation Research, January 2002).

17. RTCA DO-248C, *Supporting Information for DO-178C and DO-278A* (Washington, DC: RTCA, Inc., December 2011).

18. Federal Aviation Administration, *Software Approval Guidelines*, Order 8110.49 (Washington, DC: Federal Aviation Administration, Change 1, September 2011).

25

Reverse Engineering

Acronyms

CAST   Certification Authorities Software Team
COTS   commercial off-the-shelf
EASA   European Aviation Safety Agency
FAA   Federal Aviation Administration
LAL   less abstract level
LLR   low-level requirement
MAL   more abstract level
PDS   previously developed software
RE   reverse engineering
SQA   software quality assurance

This chapter defines reverse engineering, identifies some issues related to it, and provides high level recommendations for how to reverse engineer the life cycle data required to satisfy DO-178C objectives. This chapter is closely related to Chapter 24, since reverse engineering is recognized in DO-178C as an alternative method for generating life cycle data, particularly for previously developed software (PDS). As with other subjects throughout this book, the topic is covered with a focus on safety-critical software in the civil aviation domain. The concepts may also apply to other safety-critical domains.

Note that much of this chapter concentrates on reverse engineering software life cycle data (requirements and design) from source code, since this tends to be the most common application of reverse engineering in the aviation software industry. However, the concepts may be applied to other scenarios, such as starting with object code or documenting missing system requirements.

25.1 What Is Reverse Engineering?

The definitions of reverse engineering vary. DO-178C defines it as: "The process of developing higher level software data from existing software data. Examples include developing Source Code from object code or Executable Object Code, or developing high-level requirements from low-level requirements" [1].

DO-178C section 12.1.4.d explains that reverse engineering may be used as an approach to upgrade a baseline: "Reverse engineering may be used to regenerate software life cycle data that is inadequate or missing in satisfying the objectives of this document [DO-178C]. In addition to producing the software product, additional activities may need to be performed to satisfy the software verification process objectives" [1].*

Roger Pressman defines reverse engineering for software as: "The process of analyzing a program in an effort to create a representation of the program at a higher level of abstraction than source code. Reverse engineering is a process of design recovery" [2].

The Certification Authorities Software Team (CAST)† paper CAST-18 explains reverse engineering as follows:

> Reverse engineering is an approach to generating software life cycle data that did not originally exist, cannot be found, is inadequate, or is not available in order to satisfy the applicable DO-178B/ED-12B objectives. However, it is not just the generation of the relevant software life cycle data, but a process of assuring that the data is correct, the software functionality is understood and documented, and the software functions (performs) as intended and required by the system. It involves recovery of requirements and design, as well as conducting the relevant verification activities to the appropriate level to ensure the integrity of the software, to ensure all software life cycle data is available and correct, and that an appropriate level of design assurance is achieved [3].‡

The Federal Aviation Administration (FAA) research report, entitled *Reverse Engineering Software and Digital Systems*, by George Romanski et al. states:

> Reverse Engineering (RE) is a class of development processes that start with detailed representations of an implementation, and apply various techniques to produce more generalized, less detailed representations. The goal is to have more abstract representations that can be used to understand and reason about the structure and the intent of the more detailed representations [4].

---

* Brackets added for clarification.
† CAST is a team of international certification authorities who strive to harmonize their positions on airborne software and aircraft electronic hardware in CAST papers.
‡ CAST-18 was written before DO-178C was published and hence references DO-178B.

25.2 Examples of Reverse Engineering

There are numerous examples where reverse engineering may be or has been applied, including the following:

- Commercial off-the-shelf (COTS) software, such as real-time operating systems or vendor-supplied libraries.
- Software originally developed to another standard (e.g., military or automotive standard), such as an engine controller, a controller area network bus driver, or a flight control component.
- Open source software, such as the runtime libraries for the GNAT open source compiler for the Ada programming language, Linux operating system, or Xen hypervisor.
- Existing software, which after years of maintenance has become *too fragile* to upgrade or fix.

25.3 Issues to Be Addressed When Reverse Engineering

The certification authorities have identified common issues surrounding reverse engineering in CAST-18. The CAST-18 focuses on reverse engineering from source code. These issues have also been noted in project-specific FAA issue papers, when reverse engineering is proposed. The common issues are explained here [3,5].*

Issue 1: Lack of a well-defined process. The process of reverse engineering software artifacts must be organized and well defined. Too often, reverse engineering is used to compensate for poor development practices and does not follow a documented process. If reverse engineering is used, the processes, activities, transition criteria, and strategies for satisfying the DO-178C objectives must be documented in the plans and standards.

Issue 2: Failure to justify how DO-178C objectives are satisfied. Certification authorities have frequently observed that when companies propose reverse engineering as a life cycle model, they do not adequately explain how the DO-178C objectives will be satisfied [3]. Per CAST-18: "Reverse engineering should be used cautiously and only in well-justified cases (i.e., for a project that has been used in a number of applications and has shown itself to be of high integrity). The use of reverse engineering in new software development is strongly discouraged by the certification authorities" [3].

---

* Appendix C of the FAA's reverse engineering research report identifies potential mitigations for the most common issues [4].

Issue 3: Lack of access to experts and original developers. When life cycle data are missing, it is often necessary to access the original developers in order to understand their thought process. Source code can be difficult to decipher, especially if it has few or no comments and is written for optimized performance. Per CAST-18, the most successful reverse engineering projects are those with access to the original development team, particularly when clarification is needed for ambiguous or difficult areas [3].

Issue 4: Complex or poorly documented source code. Unless strict coding standards are enforced, the source code for many COTS products is difficult to read. Having examined source code for several COTS real-time operating systems, I know firsthand how challenging the source code can be. The source code is often filled with complex data structures and pointers; and to top it off, it contains minimal comments. The source code issues may happen because the code was not developed for a safety-critical environment or because it was a quick prototype that was never intended to be used as the final product. Poorly documented source code makes it difficult to assess the code's intended function and to ensure that the reverse engineered requirements and design are adequate. CAST-18 summarizes it well: "A thorough understanding of the code is essential to successful reverse engineering. Poorly documented or complex code is not a good candidate for reverse engineering" [3].

Issue 5: Abstraction difficulties. When reverse engineering design and requirements from source code, it is difficult to achieve the appropriate level of abstraction. Pressman explains: "Ideally, the abstraction level should be as high as possible" [2]. It is quite difficult to go from low levels of abstraction (such as code) to higher levels of abstraction (such as requirements). There are several consequences of not achieving the right level of abstraction—two are noted. First, when not performed properly, the design and requirements closely resemble the source code. The testing performed against such requirements provides very little value, since it does not evaluate the intent of the software (the *what*) but rather its implementation (the *how*). Essentially such testing just proves that the code is the code—it doesn't prove that the code does what it is supposed to. Second, without proper abstraction, unwanted functionality may exist in the source code which is not visible at the system level. When there is a large gap in the granularity between the system requirements and the software requirements, it is difficult to confirm the completeness of the system-level requirements.

This is a good time to raise the *pseudocode alert*. Reverse engineered projects often represent the low-level requirements (LLRs) as pseudocode that looks almost exactly like the code. That is bad enough, but to make it even worse, such projects sometimes attempt to use the testing against the pseudocode to satisfy the DO-178C structural coverage analysis objectives. This makes the structural coverage analysis virtually useless. See Chapters 7 and 8 for more information on pseudocode.

Issue 6: Traceability difficulties. Traceability is closely related to the abstraction issue. If the levels of abstraction are not properly established, there are two potential tracing issues that seem to evolve. Both are symptoms of the developers not really understanding what the software does. First, the tracing can be brute-forced; that is, traces are added because the code or design has to trace to something. Links are added based on similar words rather than a solid understanding of the software. Second, requirements may be identified as *derived* to avoid the need to trace. Both of these tracing issues can mask unwanted functionality that exists in the code.

Issue 7: Certification liaison problems. The certification liaison process is often not well executed in reverse engineering projects. Sometimes, reverse engineering is not identified in the plans and is not coordinated with the certification authorities [3]. Far too many times, I've reviewed a set of plans that indicate the project is using a waterfall life cycle model only to find when I assess the actual data, that it is really reverse engineered, without a plan.

25.4 Recommendations for Reverse Engineering

In order to proactively address the noted issues, the following recommendations are offered. Some of these recommendations are not necessarily unique to reverse engineering.

Recommendation 1: Evaluate and justify appropriateness of reverse engineering. Before launching into an reverse engineering effort, it is important to evaluate the appropriateness of the approach. Mature, well-proven code that has extensive service experience (such as the examples included in Section 25.2 earlier) may be a suitable candidate for reverse engineering. However, it may be difficult to justify reverse engineering for prototyped code used to validate the requirements. Some projects attempt to use their rapid prototype code *as is*, and write the design and requirements to match. This approach is not recommended, since the maturity and stability of the code and its design are uncertain. As explained in Chapter 6, prototype code may be used as input to the requirements, design, and final code. However, the use of prototype code as an input to other processes and life cycle data should be tempered since this code may not have a clean, robust, or even safe architecture. In my experience, reverse engineering from prototyped code typically costs more and takes longer than if the code were simply rewritten.

Recommendation 2: Be honest when reverse engineering is being utilized. Many projects that I've reviewed over the last 10 years have used some form of reverse engineering; however, most never admit it in their plans. If reverse engineering is going to be used effectively, it must be planned and properly implemented. It should be clear why it is being used, how it will be implemented, and how it will satisfy the DO-178C objectives.

Recommendation 3: Perform a complete gap analysis. Before committing to a reverse engineering effort, it is important to do a complete gap analysis on the existing data to determine which DO-178C objectives are satisfied and which are not. Often, a quick survey is performed and the schedule is built on the results of that survey. But, a few months into the project, it is discovered that there are many more holes than were originally identified. To avoid this risk, assemble a team of qualified engineers (a *Tiger Team* or an *A-Team*) to thoroughly analyze and identify both the data and process gaps. The team should include talented and experienced engineers who possess technical experience through the entire life cycle, domain knowledge, and DO-178C experience.

Recommendation 4: Document the life cycle, processes, and transition criteria. Reverse engineering is a life cycle and it should be documented in the plans as any other life cycle is. The phases of the development and verification effort should be defined, including the inputs, entry criteria, activities to be performed, outputs, and exit criteria for each phase. For example, Figure 25.1 shows a generic process for a reverse engineering phase. The figure shows a less abstract level (LAL) that is used as input to develop a more abstract level (MAL). It illustrates that the LAL is reviewed prior to developing the MAL (to ensure the quality of the LAL and compliance to a set of standards, such as the coding standards). Once the MAL is created, it is also reviewed. And, then both the MAL and

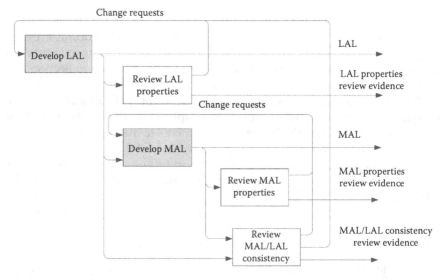

FIGURE 25.1
Generic life cycle for different abstraction layers. (From G. Romanski et al., Reverse engineering software and digital systems, Draft report to be published by FAA/DOT Office of Aviation Research, Washington, DC, October 2011. Used with permission of the author.)

LAL are reviewed together. It should also be noted that the generic process shows the existence of change requests. The process does not assume that the LAL is perfect, hence avoiding the *code is king* phenomenon (i.e., assuming the code is perfect).

Recommendation 5: Document detailed procedures and standards. Implementing a successful and repeatable reverse engineering process requires detailed procedures, checklists, and standards. For example, if the reverse engineering project is starting with source code, the original coding standards may be inadequate or missing altogether. Therefore, appropriate coding standards will need to be generated and the code reviewed against those standards. There may also be specific procedures developed for managing changes to the code. In particular, changes will need to be coordinated with all appropriate entities and approved prior to implementation.

Recommendation 6: Coordinate plans and get agreement. As with any project seeking approval by the certification authorities, the plans and standards should be reviewed, internally approved, coordinated with the certification authority, and approved by the certification authority. A reverse engineering effort may require additional coordination given the earlier mentioned challenges. Be sure to take that into account. As noted in Chapter 5, the earlier the plans are coordinated and agreed with the certification authority, the better.

Recommendation 7: Coordinate with multiple stakeholders. Depending on the nature of the project, there may be multiple stakeholders, including the systems team, the original developers of the software, the customer (or maybe even multiple customers), the reverse engineering team, etc. It is important to ensure that all stakeholders are identified, informed, and performing their tasks as expected. Projects with multiple stakeholders should do the following [4]:

- Clearly identify roles and responsibilities.
- Identify and coordinate the processes used.
- Describe, coordinate, and verify the configuration management between stakeholders.
- Coordinate problem and fault tracking between all stakeholders.
- Control and track information flow between the stakeholders.
- Ensure that the stakeholders have necessary expertise to carry out their responsibilities.
- Ensure that communication between stakeholders is unencumbered.

Recommendation 8: Gather all existing data and apply software configuration management, including change control and problem reporting. The existing artifacts should be baselined and under change control prior to implementing the reverse engineering process. For example, the source code and user's manuals, requirements documents, or design data used as input to the development effort should be captured, baselined, and controlled. Any changes should be handled through the change control process (using problem reports and/or change requests).

Recommendation 9: Involve software quality assurance (SQA) and certification liaison personnel. As with all projects, it's important to involve SQA and certification liaison personnel. For a reverse engineering effort, this is particularly important. Because reverse engineering is considered a higher risk solution, the certification authorities often provide project-specific guidance (e.g., FAA issue papers or European Aviation Safety Agency (EASA) certification review items). Early and continual involvement of SQA and certification liaison personnel helps to proactively address these concerns.

Recommendation 10: Use a technically strong team with domain expertise. Reverse engineering is not a job for junior engineers. Because it involves the creation of more abstract data from less abstract data, reverse engineering requires engineers with technical expertise and domain knowledge. It would be difficult (or maybe impossible) to have a flight management system engineer successfully reverse engineer a real-time operating system. Likewise, an otherwise competent engineer who hasn't experienced the full development life cycle on multiple projects is not a good candidate for such an effort. In my experience, the success of reverse engineering projects is directly proportional to the experience of the engineers implementing them. Good reverse engineering requires a multidisciplined team with strong cognitive ability and good communication skills in order to generate the life cycle data. Tools can help, but the success of the project will depend heavily on the quality of the engineers.

Recommendation 11: Consult with the original developers. If it is possible, it is extremely valuable to coordinate with the original software developers. Some may still be available and their insight can make the difference between success and failure. They may not be able to take an active role in the project, but even a small amount of quality time with them throughout the project is beneficial. This communication significantly improves the understanding of the software and the quality of the artifacts. Since there may be few opportunities to consult with the original developers, use the time wisely. Generate a list of specific questions and ensure the answers are fully understood. It should be noted that the FAA's research on reverse engineering considers access to the subject matter experts (preferably the original developers) as a necessity [4].

Recommendation 12: Strive to thoroughly understand the software functionality. When reverse engineering, it is important to comprehend the software's functionality and behavior. In fact, reverse engineering can be seen as a behavioral discovery process [4]. It is imperative to consider the big picture (*what* the software does)—not just the low level details.

Recommendation 13: Think top-down. Having evaluated dozens of projects, I can typically identify a reverse engineered effort, even if the plans indicate otherwise, because the top-down view is incomplete. That is, when doing the top-down requirements threads (from system requirements to software requirements to design to code), there are disconnects and missing implementation details. The bottom-up view might look good (e.g., all of the code traces to LLRs and all the LLRs trace to high-level requirements), but the top-down view is *rough*. The main problem I find is that the traces do not show that the requirements are fully implemented. Even when developing the data bottom-up, it is important to keep the top-down view in mind. The FAA's research report on reverse engineering explains the need to "emphasize that going bottom up can't guarantee that all of the desired features at the system level will exist" [4]. There needs to be some "means of introducing real system intent" and establishing the "completeness of the reverse engineered perceived system intent" [4].

Recommendation 14: Evaluate robustness. When reverse engineering, particularly from code, it is important to evaluate the robustness of the software. The PDS or prototyped code may not have been developed to be robust in the presence of unexpected or abnormal inputs. Any missing robustness functionality should be identified in a problem report and properly addressed.

Recommendation 15: Develop appropriate levels of abstraction. As previously discussed, reverse engineering involves developing MALs of data from LALs of data (e.g., design from code, and high-level requirements from design). Developing MALs of abstraction is difficult. It is hard in a forward engineering project and even tougher when reverse engineering. It is challenging to go from more detail to less. "Careful consideration should be given to the difference between abstraction levels, to ensure that there is sufficient intellectual value added to demonstrate a thorough understanding of the two representations being traced and verified" [4]. This is where experienced engineers help. When an inexperienced or unqualified engineer does the reverse engineering, the LLRs look like the code.

The FAA research report states it well:

> The difference in the abstraction levels of the MAL and the LAL should strike a balance between the difference between each level of abstraction, and the number of levels of abstraction. The MAL should provide a representation that specifies the intended behavior of the LAL.
>
> 1. A re-statement of the intended behavior at the same or similar level using a different notation is only useful to check that the transformation was correct, not the intended behavior.
> 2. If the abstraction level between the MAL and LAL is too large, there will be less confidence that the process is repeatable with the same or equivalent results.
>
> This issue also applies to forward engineering, but there may be a temptation for engineers to develop low-level requirements from code that specify how the code works rather than the intended behavior. This may then result in the difference between the abstraction level of the low-level requirements and the high-level requirements being too large [4].

Recommendation 16: Proactively trace the data. Bidirectional traceability is important for both forward and reverse engineered projects. It is especially important in reverse engineered projects where some higher level artifacts (e.g., system or software requirements) exist, and the intent is to develop consistent lower level artifacts (such as design and code). That is, lower level data is being reverse engineered to be compliant with higher level data. The bidirectional tracing will help to ensure consistency in both the top-down and bottom-up views.

Recommendation 17: Look for errors and plan for changes to code. One of the most frequent blunders in reverse engineering is to treat the code as *king* or as *golden* (i.e., to consider the code *perfect*). A good reverse engineering process realizes that the code may have issues. It may have unneeded functionality, lack robustness, be poorly commented, or be overly complex. It could even have some logic or functional errors. A survey performed as part of the FAA-sponsored reverse engineering research indicated that most of the problems raised during reverse engineering effort were raised against the source code (71%) [4]. The FAA's report also noted that the majority of the source code errors are found using manual processes: "The striking observation is that the most prolific error detection method when performing RE [reverse engineering] is manual analysis, which is performed as a development process and producing LLRs [low-level requirements] and establishing the traceability between LLRs [low-level requirements] and source code" [4].* The FAA research report also warned that "if a working code base is reverse engineered, the fact that the code base is working could lead to a false sense of confidence that the software is correct which might result in a less thorough investigation or care in developing the RE artifacts" [4]. The reverse engineering process should look for errors, weaknesses, complexities, ambiguities, etc. in the code that might impact safety, and then proactively address the issues (e.g., create a safe subset of the code).

Recommendation 18: Look for holes. When putting together all of the various abstraction levels, it's important to look for and identify missing pieces. That is, search for errors of omission—functionality that should be in the code (and requirements) but is not. Also, look for inconsistencies between the various abstraction levels, including missing or erroneous traces, incomplete system or software requirements, erroneous design, and unused code. It is more challenging to identify what is not there, than it is to review what is there. This is another case where experienced engineers can provide great insight—having seen multiple projects, they know what to expect and what may be missing.

Recommendation 19: Document problems as they are identified. When reverse engineering, it is important to document potential and actual problems when they are observed. Otherwise, they may be forgotten or overlooked. There are several ways to do this. One is to keep an investigation list, which is periodically analyzed. Legitimate issues then get rolled into problem reports.

---

* Brackets added for clarity.

Another approach is to open a problem report for each issue. If it ends up not being a problem, the issue can easily be cancelled or closed with no action. As noted in Chapters 8 and 10, it is important to identify the person who discovered the problem, as well as the engineer who is most knowledgeable about the topic.

Recommendation 20: Address the issues. Issues that are classified as valid problems must be addressed. Some common issues found are uninitialized variables, pointer issues, inconsistent data definitions, inconsistent use of data (types, units), incorrect algorithms, inadequate integration of modules, immature startup design (warm and cold), failure to address all functional scenarios, incorrect order of events, missing built-in tests, and data passed but not used [6].

Recommendation 21: Validate requirements. DO-178C assumes that system requirements allocated to software are validated. Therefore, DO-178C doesn't require requirements validation (ensuring the requirements are complete and correct). However, when the system requirements are reverse engineered, they require validation—to ensure they are the right requirements. The system requirements should be validated before reviewing the software requirements, design, and code. Additionally, any software requirements that get classified as *derived* (do not trace to system requirements) need to be validated.

Recommendation 22: Perform forward verification. In order to satisfy the DO-178C objectives, any reverse engineered development data need to be forward verified. That is, bottom-up development is acceptable; however, top-down verification is still needed. In other words, review high-level software requirements against system requirements, review low-level software requirements against high-level software requirements, review source code against LLRs, etc. Forward verification also confirms the top-down consistency.

Recommendation 23: Forward engineer once a solid requirements baseline is established. Once a solid baseline with supporting life cycle data exists, the project should be forward engineered. It is highly discouraged to continue with the reverse engineering once requirements are in place.

Recommendation 24: Know when to stop. Not all reverse engineering projects will be successful. If the code is not mature or stable at the beginning of the reverse engineering effort, considerable rework to the code may result. This can become expensive and unmanageable. The FAA's research report states:

> If the product is unstable, and many changes to the code base are required then the costs become very high. Developing requirements to software that is being updated uses up a lot of resources as work is repeated and through the spread of impact, the problem becomes larger and unmanageable. If the product is unstable, then the RE processes should be stopped and informal debugging steps should be taken before continuing the reverse engineering processes [4].

I would go a step further to suggest that there may be some situations where the code should be discarded and restarted. Sometimes, it is faster to rewrite clean code than trying to salvage broken or fragile code.

Recommendation 25: Be advised that reverse engineering is hard. End-to-end reverse engineering is not easy. It requires considerable effort and due diligence. In some ways, it may even be more difficult than forward engineering, because it can be tougher to abstract up than to decompose down.

Recommendation 26: Document lessons learned. As with all experiences in life, it is important to document the lessons learned in order to not learn them again. Throughout the project, it's advisable to keep a lessons learned list and to perform ongoing assessments of what does and does not work.

References

1. RTCA DO-178C, *Software Considerations in Airborne Systems and Equipment Certification* (Washington, DC: RTCA, Inc., December 2011).
2. R. S. Pressman, *Software Engineering: A Practitioner's Approach*, 4 edn. (New York: McGraw-Hill, 1997).
3. Certification Authorities Software Team (CAST), Reverse engineering in certification projects, Position Paper CAST-18 (June 2003, Rev. 1).
4. G. Romanski, M. DeWalt, D. Daniels, and M. Bryan, Reverse engineering software and digital systems, Draft report to be published by FAA/DOT Office of Aviation Research (Washington, DC, October 2011).
5. L. Rierson and B. Lingberg, Reverse engineering of software life cycle data in certification projects, *IEEE Digital Avionics Systems Conference* (Indianapolis, IN, 2003).
6. C. Dorsey, Reverse engineering within a DO-178B framework, *Federal Aviation Administration National Software Conference* (Danvers, MA, 2001).

26

Outsourcing and Offshoring Software Life Cycle Activities

Acronyms

FAA	Federal Aviation Administration
PSAC	Plan for Software Aspects of Certification
SOW	statement of work
TSO	Technical Standard Order
U.S.	United States

26.1 Introduction

"Sourcing refers to the act of transferring work, responsibilities, and decision rights to someone else" [1]. Outsourcing is the act of sourcing work to an outside organization; that is, to an external entity. Outsourcing software development and verification is not new, but it is definitely more prevalent now than in the past. There are many aspects to consider when outsourcing; however, it is not feasible to examine them all here. This chapter concentrates on outsourcing some or all of the safety-critical software development and verification to one or more teams. This chapter investigates why outsourcing is promoted, potential risks and challenges of outsourcing and offshoring, and recommendations for addressing the risks.

This chapter concentrates on the topic of outsourcing from a safety and certification perspective. It is primarily written for software technical leaders who make decisions about outsourcing and manage outsourced resources for safety-critical software; however, it is also applicable to developers, verifiers, quality assurance, and certification personnel. Many programmatic aspects are discussed because they impact the software life cycle and the end product. Political, legal, and contractual aspects of outsourcing are not examined. Please note that this chapter considers outsourcing from a

United States (U.S.) perspective. The concepts should apply if you live elsewhere; however, there are some specific concerns for U.S.-based companies to consider when offshoring civil aviation products.

From my avionics experience, outsourcing tends to fall into three categories: (1) products (such as line replaceable units or entire systems), (2) software or electronic hardware applications, or (3) labor services (e.g., portions of software development and/or verification). From a U.S. perspective there are three locations where outsourcing occurs: (1) within the U.S., (2) outside of the U.S. in countries with bilateral agreements with the Federal Aviation Administration (FAA), and (3) outside of the U.S. in countries without bilateral agreements with the FAA. The location is important for several reasons, including experience, domain expertise, legal framework, and certification requirements. The FAA, in particular, considers the location of the outsourced organization because it directly affects who is responsible for the certification oversight. Projects performed in the U.S. are overseen by the FAA. For activities performed in countries with bilateral agreements with the FAA, the FAA may request the international certification authority to perform the oversight.* In countries without bilateral agreements with the FAA, there is no local certification authority for the FAA to legally approve or accept; therefore, it could lead to additional resources for the FAA to oversee such activities. This can be considered an *undue burden* to the U.S. government, since the FAA is not a fee-for-service organization. If the outsourcing effort is classified as an *undue burden*, the resulting work may not be accepted by the FAA.

Experience shows that software activities performed offshore require careful consideration. Therefore, the certification authorities consider offshore projects to be higher risk. The FAA doesn't distinguish between outsourced, offshored, and subcontracted work—entities that perform such work are simply referred to as *suppliers*. Even companies that own off-site facilities inside or outside the U.S. need to address the sourcing issues.

For the rest of this chapter, the term *outsourcing* will include subcontracted or subsidized work performed onshore (within your country), near-shore (in a country that shares borders with your country), or offshore (in a country that does not share borders with your country). These terms are independent of who owns the organization performing the work. Some argue vehemently that work done by an owned subsidiary is not considered *outsourcing*. I agree that the business arrangement is different and that the risk should be lower; however, many of the issues are similar. Therefore, I include them in the *outsourcing* category. From a certification perspective, the business arrangement isn't as important as the quality of the work. History shows us that many offshored projects have

---

* The bilateral agreement must cover the kind of work being delegated. Partial or limited bilateral agreements may not cover software or advanced avionics.

quality issues, some of which are described later. As companies learn to better address the quality challenges, this perception may change.

26.2 Reasons for Outsourcing

There are numerous reasons for outsourcing; some are economic and some are technical. A list of the common reasons is noted in the following:

- *It helps address engineering shortage.* There is a shortage of software engineers in the western world (e.g., U.S., Canada, and Europe). To meet project demands, outsourced engineers are used to augment a company's existing team and to help with project sequencing. This allows a company to take on additional work and to allocate internal resources/staff to the most critical projects.

- *The labor costs are lower in some countries.* The decreased labor costs are also appealing to business managers. However, in my experience, the cost tends to be proportional to the quality of work produced. As my dad is so fond of saying: "You get what you pay for." Due to inexperience, inefficiencies, and poor oversight, I've seen a *cheap* engineer end up costing over three times more than an experienced *high-rate* engineer. When looking at labor rates, business managers frequently forget that outsourced teams require oversight; the oversight ratio of in-country engineers to offshore engineers ranges from 1:1 to 1:30, depending on the nature of the work being outsourced and the team performing the work. Once an offshore team becomes more experienced, their labor rates tend to rise, so they are no longer *cheap.* Managers who think they will save a bundle with lower outsourced labor rates should be forewarned: there are no free lunches in safety-critical software development.

- *Some companies have specialized skills.* Rather than hiring or developing expertise in a new or noncore area (e.g., operating systems), it may be more practical to outsource the work to an organization that specializes in that area. This allows a company access to expertise, knowledge, and capabilities outside its own boundaries [1].

- *It allows around-the-clock work.* By outsourcing to teams in significantly different time zones, it can allow for around-the-clock productivity. There are advantages and disadvantages to this. An around-the-clock schedule means many late-night and early-morning meetings or teleconferences, which can burn out the team. Running an around-the-clock operation also takes a global infrastructure that needs to be developed in advance of the project launch.

- *It is mandated.* Some outsourcing (particularly to offshore countries) is mandated by upper-level management for other business considerations or by aircraft contracts. The mandates are usually accompanied with a certain percentage or amount of work that must be outsourced to certain countries or organizations. Unfortunately, these are often the projects that struggle the most to effectively outsource.

When done properly, outsourcing can be a win–win situation for everyone involved. However, I've seen enough outsourcing fiascos (both within and outside the U.S.) to know that it requires considerable tender loving care and constant oversight to make it successful.

There is no magic formula or a one-size-fits-all solution for outsourcing. The remainder of this chapter documents some of the cautions and suggestions for how to tackle the issues.

26.3 Challenges and Risks in Outsourcing

Outsourcing endeavors vary significantly—all the way from a few tasks outsourced to a domestic supplier, to the entire system outsourced to an offshore entity in a country without bilateral agreements, with a lot of variants between these two extremes. This section examines some of the common challenges. Recommendations to mitigate these challenges are discussed in the next section. The first nine challenges are relevant to domestic and offshore teams. The last eight items are more applicable to offshore work.

Please note that throughout this section, the outsourced team is called the *supplier*, regardless of the type of work performed or the nature of the business relationship. The supplier is the organization selected to do some or all of the work for a safety-critical software project. Also, the company that utilizes the outsourcing team (the supplier) is called the *customer* throughout this section. The *customer* is assumed to be ultimately responsible for the DO-178C compliance.

Challenge 1: Lack of experience. If the supplier doesn't have the appropriate software engineering, management, certification, and safety experience, success is unlikely. Years ago in my late 30s, I visited an offshore supplier for a level A project. I felt like a grandmother. Nearly every team member was right out of college. I respect youth, but it needs to be balanced with and guided by experience. There's a common saying that goes: "If you want to win, you don't enter mules in the Kentucky Derby." To win the Kentucky Derby, you need a thoroughbred. Likewise, talented and experienced engineers are needed to develop and verify high quality software.

Challenge 2: Lack of domain knowledge. If an engineer lacks experience in a domain, the quality of the work suffers. While the engineer may be incredibly bright, he or she may not have the knowledge to make informed decisions. Outsourcing safety-critical embedded software development and verification to a company that has telecommunication experience is probably not a good fit for most aviation software, since such engineers generally lack an understanding of real-time, deterministic, safety-driven software, as well as the discipline and rigor required by DO-178C.

I recently consulted on a level A project in which the software lead was doing his first avionics and certification project. He had worked on farm equipment in the past. He was a bright engineer; however, he didn't understand flight controls, safety, DO-178B objectives, or certification expectations. Despite some people's thinking, avionics require domain experience, a commitment to excellence, and a perfectionist mentality. Engineers cannot be moved around like pieces on a game board, if quality is the desired outcome.

Challenge 3: Unstable workforce. High staff turnover can be an issue for some suppliers. Turnover may be due to low morale, inadequate pay, lack of reward opportunities, the country's growth dynamics, etc. It takes years to develop a skilled engineering team with the right people in the right positions. If the staff continually churns, retraining and starting at the bottom (or close to it) of the learning curve are constant. This results in lower quality work.

Challenge 4: Unrealistic expectations. A number of efforts fail because of unrealistic expectations. Many projects use what I call *success-based planning*. This occurs when project managers don't plan for the learning curves, transition times, probable performance rates, failed tests, audit findings, or schedule dependencies that invariably happen. As a result, projects are set up for failure. Expecting high quality software with a short schedule from an untested team is not realistic. To put it another way, expecting too much too fast is foolish. Schedule pressures often lead to bad decisions (e.g., throwing more engineers at the problem rather than letting the right and qualified engineers do their job successfully). Likewise, it's unrealistic to expect an inexperienced team to produce quality products (steep learning curves aren't always possible to climb). Good quality requires adequate resources, a reasonable schedule, and experienced staff.

Challenge 5: Rapid growth. Success can lead to disaster. Some suppliers grow too fast. In an effort to meet the demands of multiple customers or projects, they may spread themselves too thin, pull in less experienced people to help, or have experienced engineers jump between projects. It is a challenge to get the right people in the right place at the right time. One company I consult with is proposing doubling or tripling their workforce every year for the next 5 years, in a country with little safety-critical software experience. It does not seem like a reality-based plan to me.

Challenge 6: Overselling the company and its abilities. Some software suppliers have smooth-talking marketers but are unable to deliver the goods. I have attended several marketing pitches where the marketing representative threw around DO-178B or DO-178C and certification jargon like he knew what he was talking about. But once I started asking questions, it became obvious that the pitch was like a television advertisement. Buyers beware! Deceptive marketing isn't just for used cars. Even conscientious suppliers can sometimes oversell their abilities.

Challenge 7: Changing processes. Some companies perform well when they follow their own process. However, when the processes change (perhaps due to the needs of the customer or domain), they may not adapt well. In my experience, the larger the team, the more difficult it is to change a process. It's difficult to train everyone and enforce the change, especially if the team members don't understand why the change is needed.

Challenge 8: Poor selection process. When determining what work will be done and where, it is important to have an objective assessment. Oftentimes, teams are forced to select certain suppliers (perhaps based on location or marketing relationships) rather than the best supplier. Failure to properly assess the skills, depth, and record of the supplier leads to problems down the road. I once witnessed a project that had three potential suppliers bid for the work. They chose the one who quoted the lowest estimate, even though there were clear signs that the low bidder would be unable to deliver the work with quality and on time. The endeavor ended up costing at least 10 times the original bid of the low bidder, which equated to about three times the bid of the more qualified supplier. And, incidentally, it was 2 years late.

Challenge 9: Poor quality. It is common to have challenges getting quality work from outsourced suppliers. This can be a challenge within a single company too, but it's often more difficult to identify the quality issue in another organization until later in the process or life cycle (and possibly too late to change the direction). Quality is more than just tracking metrics. Without proper technical oversight, it could be late in the project before one realizes the requirements are bad, the design is flawed, and the code doesn't work. Sometimes, even experienced companies can have a quality lapse; perhaps they get stretched thin, hire a new program manager that isn't ready for the job, or succumb to the schedule pressure.

Challenge 10: Language issues. Communication with some offshoring suppliers can be a challenge. If a team doesn't have a solid grasp of English,* it's difficult to write unambiguous requirements, accurately test the requirements,

---

* English is the international language used by the aviation community. Since most aircraft are certified for international use, English is typically required, particularly for projects seeking certification by the FAA.

write useful problem reports, communicate issues, etc. I've consulted on a couple of projects where a translator was needed to communicate. This can be very time consuming and error-prone. Combine the language issues with inexperience and/or a lack of safety culture and the probability of success drops significantly. While on the topic of language, it should be noted that the FAA requires certification-related data to be in English and accessible in the U.S. [2].

Challenge 11: Cultural issues. It is important to understand cultural differences when outsourcing. There can be cultural differences even within the U.S., but the greater challenges are with international projects. Here are some examples of cultural aspects to be aware of:

- Different cultures are often motivated differently. What works in the U.S. to motivate a team may not work in Europe, Asia, or South America.

- In some countries it's considered disrespectful to point out negative things about someone or something. However, when doing software verification, the goal is to discover errors. Without a proper verification mentality, one could end up with the most successful test program ever, but the system still doesn't work.*

- Not all countries have the same safety standards or value for human life. The evaluation of software defects requires a zero-defect mentality for safety-critical software. In one instance, an offshore company that I consulted with didn't understand the need to test or fix errors in level A software.

- Some cultures may not respect or have a concept of excellence. In some cultures, it may even be considered wrong to have pride in your work. However, developing safety-critical software requires a perfectionist mentality and attention to detail. If the engineers are satisfied with mediocre or *good enough* work, it'll be difficult to prove that it's safe.

Every culture has its strengths and weaknesses. It's important to understand the cultural background of all entities, provide cultural training for all team members involved, and allow time for diversity integration.

Challenge 12: Ethical issues. Not all cultures adhere to the same ethical standards. In some cultures it's acceptable to lie to get ahead, to steal as long as you don't get caught, and to offer bribes. Some countries are suspected of introducing security vulnerabilities and backdoors in the products they develop.

---

* Recall from Chapter 9 that a truly successful test program is one that finds errors—not one that passes all tests.

Challenge 13: Education issues. Not all engineering programs are created equally. In some schools across the world, 50% is considered passing. Just because someone has a degree in engineering doesn't mean they can engineer. Be sure to check out the quality of education, as well as the experience, of the supplier's engineers.

Challenge 14: Hidden expenses. There can be, and probably will be, unexpected expenses when offshoring. Business class airline tickets and safe hotels add up fast. Despite the advances in video conferencing and network-based meetings, there is still a need for in-person meetings and oversight. Additionally, setting up an infrastructure (equipment, high-speed network access, etc.) can be costly.

Challenge 15: Losing experienced engineers. Despite best efforts to put a positive spin on outsourcing and to encourage teamwork, sometimes experienced engineers are lost when outsourcing occurs. The customer's engineers may see the outsourcing arrangement as giving away their job or they may want to design and code rather than oversee a supplier. Burnout due to early-morning and late-night meetings can also be a factor.

Challenge 16: Certification challenges. As mentioned earlier, certification authorities have seen issues with outsourcing (particularly offshoring), so they normally require special justification for and oversight of offshore suppliers. If work is performed in a country without bilateral agreements, there needs to be justification to ensure there is no undue burden. Some suggestions for how to do this are examined in the next section.

Challenge 17: Post-project amnesia. I've consulted on projects that I considered to be certification fiascos. Yet, the mere fact that they made it to the end, without bankrupting the company, is somehow twisted to be viewed as a success. I've seen poor-performing outsourced companies rewarded with additional work because management seems to forget the prior chaos that took place. The supplier's marketing department may even proclaim they were key to the project's success (without mentioning the hassles they caused). Unless specific action is taken to correct the issues from the past, the issues will likely happen again. As I've heard it said: "The definition of insanity is doing the same thing again and expecting different results." Unfortunately, when it comes to outsourcing, particularly offshoring, I've seen enough insanity to make me a skeptic.

26.4 Recommendations to Overcome the Challenges and Risks

Despite the challenges identified in the previous section, outsourcing can be successful. It can provide great benefits when done properly. This section presents some strategies and recommendations to handle the challenges of outsourcing. As noted earlier, the focus is on safety and certification. An organized process is essential to developing software that performs its intended function. Therefore, some of the practical programmatic practices that impact the overall software quality when outsourcing are covered.

Even as a youngster in rural America, I had a desire to see the world. Engineering has opened the doors of the world for me. I enjoy meeting, working with, and eating with people from all over the world. It is enlightening to interact with such diverse, interesting, and talented people across the globe. It takes some time to grow a project or a team to reap the benefits of outsourcing, but once you do it, it can be extremely fulfilling. These recommendations are intended to help you and your team experience such fulfillment.

As with the last section, the outsourced team is called the *supplier* and the company that utilizes the outsourcing team is called the *customer* throughout this section.

Recommendation 1: Determine readiness for outsourcing. Some companies outsource because they see everyone else doing it. Without the proper preparation and groundwork, outsourcing can be a dismal failure. Before diving into an outsourcing arrangement, it's important to evaluate the customer's outsourcing readiness. In his book, entitled *Software Without Borders*, Steve Mezak provides a 20-question *outsourcing readiness test* to evaluate an organization's readiness for outsourcing. The test examines outsourcing experience, technology, business situation, and management approach [3].* Such an evaluation can help identify if outsourcing is a wise decision and where additional help may be needed.

Recommendation 2: Ensure top-level management's commitment. It is essential to have both the short-term and long-term support of executive-level management. Without this, failure may be declared at the first sign of trouble. Management must be provided a complete and accurate understanding of the needs, risks, options, and expectations. Both the customer and supplier teams should be careful not to promise what can't be delivered. It's okay to be optimistic, but the optimism ought to be balanced with an accurate reflection of reality. A well-informed and fully committed management can open doors, provide support, and give vision to the project and to the overall enterprise.

Recommendation 3: Use a systematic and organized approach. Some organizations jump into an outsourcing arrangement with both feet before they even have a plan—it's the *ready-fire-aim approach* to software development. One doesn't build a house without a blueprint and one cannot effectively outsource part, or even all, of the company's work without an organized approach. The *Outsourcing Handbook* by Mark Powers et al. explains the need for a process-driven approach to outsourcing and proposes an outsourcing life cycle with

---

* The test can be found at www.softwarewithoutbordersbook.com.

seven phases. The seven phases are cyclical (unless the exit option is exercised), and each phase is performed independent from the phase that follows it. It is important to address each phase well and not to rush through any one of the phases. Each phase is briefly discussed here [1]:

1. *Strategic assessment.* This entails making a business case to identify the benefits of an outsourcing arrangement (including an analysis of core competencies and areas suitable to outsource).

2. *Needs analysis.* This is performed on a particular project to identify the project-specific needs.

3. *Supplier assessment.* This involves assessing suppliers in order to determine which one best fits the specific needs. This includes not just program management assessments, but a detailed technical review of the engineering competence, depth, infrastructure, and quality. "Choosing the right vendor [supplier] is much like choosing a good partner; the chances are that if you make the right decision from the onset, you will have a potentially lasting relationship, while choosing the wrong vendor [supplier] could damage and thwart a well-intentioned outsourcing project" [1].*

4. *Contract and negotiation management.* This is the process of negotiating the outsourcing agreement and getting the contract in place. Quality and transition criteria need to be established for each deliverable and milestone. I have seen projects pay for deliverables only to discover they were inadequate and needed to be completely reworked.

5. *Project initiation and transition phase.* This is the phase where work is transitioned to the supplier. This phase sets the tone for the rest of the project; therefore, it's important to address issues quickly. This is a particularly critical phase for the technical work. Projects should plan for the appropriate level of technical oversight to ensure communication, issues reconciliation, and acceptability of deliverables.

6. *Relationship management.* As the project moves into a routine, it is important to keep the customer–supplier relationship healthy. This phase includes evaluating the relationship, resolving problems, managing communications, sharing knowledge, and managing the process.

7. *Continuance, modification, or exit strategy.* At the end of each project, it's necessary to evaluate the outsourcing relationship to determine if the arrangement should be continued, modified, or exited. If the decision is to continue, lessons learned should be evaluated and improvements implemented. If the decision is to modify the process, a clear strategy for modification is needed. If the decision is to exit, that too requires a well-planned strategy.

---

* Brackets added for clarification.

Recommendation 4: Be well informed. Make sure knowledgeable people are available to implement the outsourcing arrangement. If new to outsourcing, it is advisable to get some help. Perhaps an outsourcing strategist or another division of the enterprise can help. There are a multitude of details to consider (e.g., legally binding contracts, import and export laws, intellectual property protection, cultural sensitivity training [for offshore projects], visa applications, equipment transfer, network infrastructure). It is extremely beneficial to get some expert help the first time through and to have a dedicated team to learn all they can from those experts in order to establish a sound foundation.

Recommendation 5: Know your own organization. Prior to launching into an outsourcing arrangement, it is important to know your own organization's characteristics. Identify strengths, weaknesses, specific needs, organizational maturity, technical expertise, process maturity, etc. From a technical perspective, it is important to examine core and noncore competencies. "Core competencies are the combinations of special skills, proprietary technologies, knowledge, information and unique operating processes and procedures that are integrated into the organization's products and services and are unique differentiators for the organization's customers" [1]. Noncore competencies are competencies that do not differentiate the organization from competitors and that do not directly impact the organization's products and services. Noncore competencies are required for daily operations and indirectly impact the organization's services and products [1]. Most companies choose to outsource noncore competencies first.

Recommendation 6: Determine what to outsource and what to retain in-house. Evaluating core competencies helps with this task; however, once competencies and needs are determined, a strategy to address them is needed. Some common approaches to architecting the project for outsourcing are the following:

- Outsource the entire product systems, hardware, and software.
- Outsource the entire software life cycle and retain hardware and systems development and testing.
- Modularize functionality, so that some functions are done in-house and others are outsourced.
- Separate software development and verification teams (typically development is done in-house and verification is outsourced).
- Outsource software development and verification but use in-house experts to oversee and monitor the work.

When deciding what will be outsourced and what will be retained, consider intellectual property, key functionality that must work fast, interfaces between the various components, and areas highly visible to the customer. These items are typically best kept in-house.

Recommendation 7: Determine the best outsourcing approach. As mentioned in Recommendation #6, there are many possible approaches to outsourcing. One might outsource an entire system, or just have a supplier write and execute the tests. One might outsource to a local company, or to one on the other side of the globe. One might have a contractual arrangement with the supplier, or actually own the offshore subsidy. Mezak identifies six common strategies for software development [3]: (1) in-house engineers, (2) onshore contract outsourcing, (3) offshore contract outsourcing, (4) in-house and offshore blend, (5) offshore subsidiary, (6) build, operate, transfer (start with a contracted team with the option to transfer the engineers to a subsidiary in the future). Since the best approach varies, the options should be evaluated in order to determine the best fit for both short-term and long-term needs.

Recommendation 8: Select supplier wisely. This should be an obvious one; however, having seen several unwise selections, it's worth a reminder. Once the strengths, weaknesses, and needs of the customer's team are determined, look for a supplier that meets those needs and complements the customer's team. Apply the same scrutiny and expectations that would be applied when hiring a new full-time employee for a key position: check references and review resumes. Here are some specific areas to consider:

- Experience on safety-critical software projects
- Experience with DO-178C (and/or DO-178B)
- Similarity of experience to the planned tasks
- Language capabilities (written and spoken)
- Safety domain understanding
- Familiarity with the planned tool suite and environment
- Commitment to following established processes
- Stability and depth of staff
- Capability to support assigned tasks in the time allotted
- Ability to add resources if the project doesn't go according to plan
- Experience with the specific domain (e.g., navigation, communication, flight controls)
- Track record in the industry (check references, and not just those given by the supplier)
- Experience with the planned project size (such as small, medium, or large project)
- Overlap with the customer's team's work day
- Education (get resumes and have a means to ensure the team will be retained on the project)
- Location (safe, secure, easy to reach, etc.)

- Ability to support on-site, as needed (this may involve investigating the visa application process, travel options, extended stay lodging, etc.)
- Respect for intellectual property and proprietary data
- Probability of security vulnerabilities (some countries are perceived as having state-sponsored malware activities)
- Equipment (quality telecommunications system, consistent power, high-speed Internet, office space and equipment quality, video conferencing capabilities, lab space, etc.)
- Cultural compatibility between supplier and customer teams
- Established quality assurance
- Risks (such as government stability, security, conflicts of interest, power outages)

The *Outsourcing Handbook* identifies the following six common errors when selecting a supplier [1]:

1. Sacrificing needs analysis for a glamorous vendor
2. Evaluating a vendor with cost savings as the decisive factor
3. Poor risk assessment of the vendor
4. Rushing through the process of vendor selection
5. Lack of care in managing interactions between vendors
6. Failing to maintain a balance between using current and new vendors

Keep these in mind when evaluating and selecting an outsourcing partner. It cannot be emphasized enough: Don't choose based on price alone! It nearly always leads to regrets.

Recommendation 9: Insist on the same team for the duration. Some suppliers start with a solid, mature team, but those engineers are soon moved to another project after the project initiation. Be sure to get the team composition agreement in writing and to verify it throughout the project. Some staff turnover is to be expected and there may be some need to shift engineers to get the right fit. However, it is important to continually monitor the team composition.

Recommendation 10: Prepare the customer team. The customer's team will need some special attention. Some engineers may be opposed to outsourcing and especially offshoring. Make sure that they understand the reasons for the outsourcing and continually reinforce the fact that they are key to the project and company's success. Depending on the location of the supplier, it may be necessary to shift schedules to have overlap with the supplier's team. Also, some cultural sensitivity and diversity training may be needed for both the supplier and customer teams. Throughout the project, listen for feedback and take the input seriously. Most of the time, engineers have valid points that are worth considering. With time, it will become clear whose input can be relied upon and whose requires some filtering; however, everyone should be valued and given the opportunity to share freely.

Recommendation 11: Cross-pollinate teams. Depending on the size and nature of the project, it is often beneficial to have some of the customer personnel on-site at the supplier's facility and to have some of the supplier's personnel at the customer's site. Short-term stays seem to work best (3 months or less). The cross-pollination is valuable for both teams. It helps them to better understand and proactively address the project's issues.

Recommendation 12: Start small. It is usually best to start the outsourcing adventure with small, well-defined tasks. These tasks can help build the relationship, work out process issues, identify the leaders, and build expertise and confidence. Many companies start with a pilot project—typically a small, noncritical function (or a tool). This allows one to test the communication abilities, test the technical team's skills, and implement a small but needed function.

Regardless of whether the pilot project is performed or not, it's still wise to start small. Some companies outsource lower level software first (maybe level C or D). If that goes well, they may outsource more critical and larger projects.

Recommendation 13: Clearly specify expectations. The tasks to be performed and the data to be generated should be documented in the statement of work (SOW). The SOW should be complete, unambiguous, consistent, and measurable. The DO-178C artifacts should be included in the SOW. It's good to have a technical team and someone knowledgeable of DO-178C and certification review the SOW.

Recommendation 14: For staff augmentation, use common tools and methods. If the outsourced team will be augmenting the customer's staff, common tools and methods are recommended. The following tools should be common: requirements management and capture tool(s), design tool(s), source code tool(s), source code control tool(s), and problem reporting tool(s). It's also important to include detailed procedures, examples, and training. Sometimes procedures are clear to the customer's team who has used them for several years, but they may not be clear to an external supplier's team. Procedures to be used by the supplier may need to be enhanced.

When outsourcing an entire component or system, it may not be necessary to use the same tool set and processes. However, it will be important to (1) understand the supplier's processes, (2) ensure that the supplier either delivers or maintains the data and environment long term per the customer and certification authority requirements, and (3) plan and implement a transition strategy for transferring the product and data.

Recommendation 15: Develop a supplier management plan. FAA Order 8110.49 (Change 1) requires a documented supplier management plan [2].

The plan might be part of the Plan for Software Aspects of Certification (PSAC) or a system certification plan, or it may be an enterprise-wide plan. The supplier management plan needs to ensure that suppliers (including staff augmentation and offshored subsidiary partners) comply with all regulations, policy, guidance, agreements, and standards that apply to the project. The plan also explains how all suppliers will be managed. Specific areas of interest to certification authorities are compliance, integration of various artifacts (such as requirements, design, code), certification liaison, problem reporting and resolution, integration, verification, configuration management, compliance substantiation, and data retention [2]. The certification authorities want to know what is being done, by whom, and where. They also want to know how the outsourcing is being managed by the *applicant* (the entity applying for certification or authorization). Even if the offshore facility is owned by the customer or applicant or is a domestic team used as *staff augmentation* (as opposed to a *supplier*), the FAA wants to know how the effort is being managed.*

Recommendation 16: Coordinate the problem reporting process. Software developers and verifiers may not realize the impact of a low-level problem on the system or aircraft safety. Therefore, a common problem report categorization is recommended for all suppliers. Additionally, visibility into the supplier's problem reports is needed throughout the development and/or verification effort (it is too risky to wait until the end to look at problem reports). Any problem reports that are deferred for the initial certification will require justification and review by all stakeholders, including the certification authorities.

Recommendation 17: Train in key areas. At the beginning of the program it is advisable to identify training needs for the entire team (both the customer and the supplier teams). Some common areas of deficiency are DO-178B/C compliance, robustness testing, low-level requirements capture, documentation of repeatable analyses and procedures, integration, and data and control coupling analyses. For large teams, it may require creativity to get the right training to the right team members. A common strategy is to train the technical leaders first and then have them flow the knowledge down. Computer-based training with a variety of examples can also be valuable. Additionally, an online best practices guide that is regularly updated can be a great way to get the knowledge to the masses.†

Recommendation 18: Pay continual attention to communication. Poor communication is one of the main contributors to project failure; therefore, extra attention should be paid to the communication arrangement and its effectiveness. As Mezak writes: "Using outsourcing is like a marriage.

---

* EASA's Certification Memo CM-SWCEH-002 has similar guidance.
† Such a guide should be explained in the software plans and kept under configuration management.

It takes commitment from both sides to make the relationship work. Good communication is required" [3]. Here are some ways to enhance communication:

- Hold face-to-face meetings for team leads—especially early on and at key phases throughout the project.
- Arrange frequent teleconferences and/or virtual meetings between teams. If it is an offshore supplier, it may be a daily teleconference when the shifts overlap. Video conferencing and network-based document sharing can be helpful.
- Encourage the teams to get to know each other personally, not just professionally.
- Use graphics to communicate difficult concepts. A diagram can often help resolve communication issues.
- Implement some level of on-site presence (i.e., cross-pollinate teams).
- Encourage teams to immediately communicate issues and concerns rather than letting them get out of control.
- Develop an escalation method for when either company believes a significant issue (technical or programmatic) is not adequately addressed.
- Continually look for ways to improve communication, based on lessons learned.

Recommendation 19: Assess and mitigate risks. Software development has a number of risks. Outsourcing may increase some of those risks. As noted earlier, the supplier's qualifications should be carefully evaluated prior to selection. Any deficiencies should be identified as a risk. As the project evolves, the risks should be consistently reevaluated and risk mitigation established. It does little good to identify a risk and then do nothing about it. Unaddressed risks are like holes in the bottom of a boat. Without attention, they can sink the boat (or project). An ongoing risk assessment with concrete mitigation steps helps prevent sinking.

Recommendation 20: Consider undue burden. Closely related to the risk assessment is the undue burden mitigation. As noted earlier, an offshore supplier in a country without bilateral agreements could become an undue burden to the U.S. government. Therefore, the FAA needs to know that the supplier is being properly overseen and that risks are proactively addressed. Following are some of the common strategies to avoid undue burden:

- Limit offshore work to verification and perform testing for certification credit in the U.S.
- Involve the U.S. team (customer team) in the peer reviews of the data.

- Utilize FAA-authorized designees to perform on-site audits of the supplier's work to ensure it is compliant.
- Limit offshore work to lower software levels until the certification authorities have confidence in both the customer and supplier teams.

Recommendation 21: Manage it well. Many of the outsourcing fiascos that I've observed have been because of poor management. Outsourced projects should be managed by qualified and experienced personnel. Managers in this role ought to have the following skills and abilities: good decision-making ability in uncertain environments, able to embrace change, good relationship-building and communication skills, keen knowledge-management skills, and persistence [1]. Additionally, the technical manager should possess strong technical skills and be able to understand the technical concerns and risks.

A complete list of milestones, with detailed tracking and acceptance of these milestones, is an important tool for managing any software project, including an outsourced project. Dependencies between the teams and tasks should be identified, and the accuracy of status should be ensured. The schedule and milestones should be updated as needed, based on the actual project progress.

Recommendation 22: Be realistic. The technical expectations and schedule expectations should be realistic. It's not realistic to expect a team of new engineers to produce quality work quickly. In general, it's advisable to hope for the best but plan for the worst. Success-based planning doesn't do anyone any good. It might make everyone happy at some point in time, but if it isn't realistic, that happiness will not last. Good managers deal with reality! Johanna Rothman writes: "Plan for each project to take longer and cost more, especially at the beginning of an outsourcing relationship. My rule of thumb is to increase the estimated time by 30% for the first project. Then monitor the project to see if you need increase that estimate" [4].

Recommendation 23: Keep long-term goals in mind. Some companies get frustrated because their outsourced effort is not turning out as envisioned. It's important to keep the long-term goals in mind. If neither short-term nor long-term goals seem realistic, they may need to be revamped.

Recommendation 24: Plan for oversight. The certification authorities require insight into the oversight approach, including the management oversight, technical oversight, quality assurance oversight, security controls, and certification liaison oversight. Each of these areas should be detailed in the PSAC and/or some other plan accessible to the certification authorities.

Recommendation 25: Plan for postproject support. The outsourcing arrangement should consider who will perform the post project support. That is, who will maintain the data after the initial certification? This will be a factor when determining how roles and responsibilities are divided and what data go where. For example, if the supplier who developed test cases and procedures will not be utilized after the certification, it will be important to have all tools, data, equipment, etc. delivered so future updates to the test data can be made.

Whoever does the support should be sure to obtain all life cycle data, including open problem reports, Software Configuration Index, and Software Accomplishment Summary prior to making changes.

Recommendation 26: Implement continuous improvement. Every completed outsourcing effort provides numerous lessons learned. It's important to capture those lessons, to learn from them, and to modify processes accordingly.

26.5 Summary

Outsourcing is here to stay; therefore, it's important to learn how to implement and manage it effectively. This chapter has identified common issues and some recommendations to help prevent or resolve those issues. Please consult other resources and experts in the field. This is just one person's perspective. Good luck and enjoy the experience!

References

1. M. J. Powers, K. C. Desouza, and C. Bonifazi, *The Outsourcing Handbook: How to Implement a Successful Outsourcing Process* (Philadelphia, PA: Kogan Page, 2006).
2. Federal Aviation Administration, *Software Approval Guidelines*, Order 8110.49 (Washington, DC: Federal Aviation Administration, Change 1, September 2011).
3. S. Mezak, *Software Without Borders* (Los Altos, CA: Earthrise Press, 2006).
4. J. Rothman, 11 steps to successful outsourcing: A Contrarian's view, *Computer World* (article #84847, 2003). http://www.computerworld.com/s/article/84847/11_Steps_to_Successful_Outsourcing_A_Contrarian_s_View (accessed August 2011).

Appendix A: Example Transition Criteria

Acronyms

CM	configuration management
FDAL	function development assurance level
PR	problem report
PSAC	Plan for Software Aspects of Certification
SCI	Software Configuration Index
SDD	Software Design Description
SDP	Software Development Plan
SLECI	Software Life Cycle Environment Configuration Index
SQA	software quality assurance
SRD	System Requirements Document
SVP	Software Verification Plan
SWRD	Software Requirements Document

Overview

Tables A.1 and A.2 provide an example of transition criteria that might be included in a Software Development Plan (SDP) (Table A.1) and Software Verification Plan (SVP) (Table A.2).* The SDP criteria (Table A.1) only provide a high-level look at the reviews, because the SVP (Table A.2) provides detailed transition criteria for these reviews.

---

* These transition criteria are provided as an example and are not intended for certification use, since they only present part of the software life cycle being proposed in the SDP and SVP.

TABLE A.1

Example Transition Criteria for Software Development Plan

Phase	Entry Criteria	Activities	Exit Criteria
Develop and review software requirements	• System FDAL and software level are determined • System certification plan is released • PSAC is released • SDP is released • Software requirements standards are released • System requirements are mature enough to begin decomposition	• Document high-level software requirements (in SWRD) from the SRD for requirements allocated to software • Identify requirements that have a safety impact (tag with safety attribute) • Identify any derived software requirements (requirements that do not trace to SRD and that define implementation details) and provide rationale for why each one is needed • Develop bidirectional traceability between SRD and SWRD requirements • Include interfaces/interactions in SWRD as needed • Include reference to hardware data in SWRD, as applicable • Document any deviations from requirements standards and obtain SQA approval • Document robustness requirements (tag as robust) • Review SWRD using the company procedures defined in the SVP and make updates as needed • Validate derived requirements (as part of the review) to ensure they are the right requirements (correct and complete) • Involve safety personnel in the review of all derived requirements • Document the development environment in the SLECI	• SRD is released and under CM • SWRD is reviewed, updated, released, and under CM • Derived high-level software requirements are reviewed by appropriate safety personnel • SLECI is drafted and under CM

| Develop and review SDD | • PSAC is released
• SDP is released
• Software design standards are released
• SRD and SWRD are mature enough to begin design | • Document low-level requirements using SRD, SWRD, and other referenced requirements as input
• Identify interfaces/interactions in SDD as needed
• Identify any derived low-level requirements (requirements that do not trace to SWRD and that are implementation details) and provide rationale for why each requirement is needed
• Document robustness requirements (tag as robust)
• Validate derived requirements (as part of the review) to ensure they are the right requirements (that is, they are correct and complete) (see SVP for details on the review)
• Develop bidirectional traceability between high-level requirements and low-level requirements
• Develop software architecture to be consistent with the requirements
• Review the SDD using the process defined in the SVP and make needed updates
• Document any deviations from design standards and obtain SQA approval
• Involve safety personnel in the review of all derived requirements | • SDD (including low-level requirements and architecture) reviewed, updated, released, and under CM
• SRD and SWRD released and under CM
• Derived low-level requirements reviewed by appropriate system safety personnel |

(*continued*)

TABLE A.1 (continued)

Example Transition Criteria for Software Development Plan

Phase	Entry Criteria	Activities	Exit Criteria
Develop and review code	• PSAC is released • SDP is released • Code standards are released • SWRD and SDD are available and mature enough to begin coding	• Write code using the SWRD, SDD, SDP, and coding standards as input • Develop bidirectional traceability between the code and low-level requirements • Review code and update based on comments (see SVP for details of the review) • Document any deviations from code standards and obtain SQA approval • Develop and review SCI	• Code is reviewed, updated, released, and under CM • SRD, SWRD and SDD are released and under CM • SCI is reviewed and under CM
Integrate code	• Code available • SCI drafted • Compiler and linker options are identified in SLECI	• Develop build procedures (in SCI) and have an independent person run the procedures for repeatability • Build code, using the build procedures • Prior to formal testing, compile and link the released code, using the approved build procedures (in SCI) • Develop the load procedures (in SCI) and have an independent person run the procedures for repeatability • Update and review SCI (including build and load procedures) • Update (if needed) and review SLECI (prior to formal build of code)	• Code is released and under CM • SCI (including build procedures and load procedures) and SLECI are reviewed, released, and under CM • Released code is compiled and linked • Executable object code is released and under CM

TABLE A.2

Example Transition Criteria for Software Verification Plan

Review Type	Entry Criteria	Activities	Exit Criteria
Planning review(s)	• All five plans are under CM • Requirements, design, and coding standards are under CM	• Review all plans and standards using the planning checklists • Document review comments • Update plans and standards per review comments • Close or disposition action items and/or problem reports from the review • Release the updated plans and standards	• All plans and standards are reviewed, updated, released, and under CM • Action items and problem reports from the review are closed or dispositioned
SWRD review	• Plans and requirements standards are released • System requirements are released • SWRD is under CM (if the SWRD is reviewed in sections, each section needs to be mature and under CM prior to the review)	• Review SWRD against system requirements using the SWRD checklist in the SVP • Validate derived requirements • Ensure that requirements standards identified in the SDP were followed • Ensure bidirectional traceability between SWRD and SRD is correct and complete • Ensure that SWRD is properly updated based on review comments • Create action items (prior to baseline) or problem reports (after baseline) if any issues are found with related data that are under CM	• SWRD is updated based on review(s), released, and under CM • Action items and problem reports from the review are closed or dispositioned

(continued)

TABLE A.2 (continued)

Example Transition Criteria for Software Verification Plan

Review Type	Entry Criteria	Activities	Exit Criteria
		• Close or disposition action items and/or problem reports from the review • Ensure that appropriate system safety personnel have reviewed all derived requirements	
Software design review	• SRD and SWRD have been reviewed, updated, and released • SDD is under CM • Design standards are released • Traceability between the high-level and low-level requirements is under CM	• Review SDD using the design checklist from the SVP • Validate derived requirements • Ensure that design standards identified in the SDP were followed (for both low-level requirements and architecture) • Confirm accuracy of the bidirectional traceability between the high-level requirements and the low-level requirements • Ensure that SDD is properly updated based on review comments • Create action items (prior to baseline) or problem reports (after baseline) if any issues are found with related data that are under CM • Close or disposition action items and/or problem reports from the review • Ensure that appropriate system safety personnel have reviewed all derived requirements	• SDD is updated based on review(s), released, and put under CM • Action items and problem reports from the review are closed or dispositioned

Code review	• SRD, SWRD, and SDD have been reviewed, updated, and released • Source code is under CM • Coding standards are released • Traceability between the low-level requirements and source code is under CM	• Review source code using the coding checklist (from SVP) • Ensure that coding standards identified in the SDP were followed • Confirm accuracy of the bidirectional traceability between the low-level requirements and code • Ensure that code is properly updated based on review comments • Create action items (prior to baseline) or problem reports (after baseline) if any issues are found with related data that are under CM • Close or disposition action items and/or problem reports from the review	• Code is updated based on review(s), released, and put under CM • Action items and problem reports from the review are closed or dispositioned
Code integration review	• Code has been reviewed and released • SLECI is drafted • SCI is drafted (including build and load procedures)	• Review SLECI using checklist (from SVP) • Review SCI using checklist (from SVP) • Review any files (such as makefiles) used for building the software	• SCI and SLECI are updated based on review(s), and put under CM • Build files are reviewed, released, and under CM • Executable object code is built, released, and under CM • Action items and problem reports from the review are closed or dispositioned
Test cases and procedures development (high and low-level testing)	• SWRD and SDD are mature enough to begin developing tests • SLECI is under CM	• Develop test cases and procedures to exercise normal and abnormal (robust) conditions • Ensure all requirements are tested.	• SWRD, SDD, and code are released • Test cases and procedures are reviewed, updated, released, and under CM

(continued)

TABLE A.2 (continued)

Example Transition Criteria for Software Verification Plan

Review Type	Entry Criteria	Activities	Exit Criteria
		• Identify robustness test cases (if a requirement is written to be robust, document the rationale for why additional robustness testing is not needed) • Ensure that tools which require qualification used for the test execution have been qualified or are stable enough to begin test development • Develop bidirectional trace data between the test cases and requirements, and between test cases and test procedures • Document test build instructions and ensure that all tools needed to build or execute the tests have been documented in the SLECI • Add test information to the SCI • Review the test cases and procedures and update as needed (see "Test cases and procedures reviews" entry in this table)	• SLECI is released • SCI is updated with test procedure information and under CM

Activity	Entry Criteria	Tasks	Exit Criteria
Development of integration analyses (applies to all analyses)	• SVP is released • Data required for the analysis are mature enough for the analysis to begin	• Document analysis procedures using the criteria identified in the SVP or other applicable document • Review the procedures for repeatability and compliance with the identified criteria • Update the procedures based on review feedback • If any qualified tools are used to perform the analysis, ensure the tool qualification is completed • Perform *additional analyses* (such as link/load, worst-case execution timing, and memory map analyses) using released procedures • Document analysis results in the analysis report (see "Analysis results review" in this table) • Document any issues noted during the analysis in a PR	• Analysis procedures are reviewed, updated, and released • Analyses results are documented in an analysis report and under CM • Problem reports are generated for any issues noted during the analysis
Test cases and procedures reviews (high and low-level testing)	• SRD, SWRD, and SDD are released • Applicable test cases and procedures are under CM • SLECI and SCI are under CM • Traceability data between the test cases and requirements, and between test cases and test procedures are developed and under CM	• Review test cases and procedures using the checklist identified in the SVP • Verify the bidirectional traceability between the test cases and requirements, and between test cases and test procedures • Ensure that all requirements are tested • Ensure that expected results are identified, along with pass/fail criteria	• Test cases and procedures are updated based on review(s) and are released • Action items and problem reports from the review are closed or dispositioned

(*continued*)

TABLE A.2 (continued)

Example Transition Criteria for Software Verification Plan

Review Type	Entry Criteria	Activities	Exit Criteria
	• SVP is released • Test build instructions are under CM • Tests have been informally run by the test developer and results are available for review	• Run the test build instructions to ensure accuracy • Review informal test results to ensure that the tests are executable, tests pass, and the results are properly documented • Ensure that appropriate robustness exists (if robustness tests are not needed, ensure that rationale exists explaining why) • Close or disposition action items and/or problem reports from the review	
Test execution	• Test cases and procedures are released • Tool code and qualification data are released • Test build instructions are released • SLECI is released • All tests have been dry run	• SQA audits test station configuration to ensure all test stations are consistent with the configuration in the SLECI • Build test software using the released build procedures (SQA to witness the build) • Test execution plan has been identified	• Test results are under CM • Problem reports are generated for failed tests and for any redlines

Test readiness review	• Test readiness review has been performed and criteria satisfied	• Notify SQA and certification liaison personnel of intent to execute tests • Execute the tests • Analyze test results to determine pass or fail • Perform structural coverage analysis and document • Identify any failures and rerun if needed or document in a PR • Analyze any failures to determine if rework is needed • Document results in test reports • Review missing structural coverage and generate PRs as needed	
Test results review	• Test cases and procedures are released • Test cases and procedures have been executed • Test results are documented in test report • SCI has been updated with test results information • Test report is under CM	• Review test report for completeness and accuracy using the checklist in SVP • Verify that traceability between test procedures and test results is accurate and complete • Ensure that any test failures are explained and a PR has been opened • Ensure that redlines are documented and approved by SQA • Close or disposition action items and/or problem reports from the review • Review SCI for accuracy of test information and update as needed	• Test report is reviewed, updated, and released • Action items and problem reports from the review are closed or dispositioned • SCI is reviewed, updated, and released

(continued)

TABLE A.2 (continued)

Example Transition Criteria for Software Verification Plan

Review Type	Entry Criteria	Activities	Exit Criteria
Analysis results review (applies to all completed analyses)	• Data to be analyzed are released • Analysis procedure is released • Analysis results are in the analysis report and under CM • Tools used for the analysis requiring qualification have completed qualification • SCI has been updated with analysis information	• Review the analysis report to ensure analyses were performed per the procedures • Ensure that the analysis was performed with independence, when required • Confirm that the analysis results are consistent with the expected results • Ensure that the versions of the tools used are the same as the versions qualified (if applicable) • Review any resulting PRs and the justification to ensure they are acceptable • Review SCI for accuracy of analysis information and update as needed	• Analysis report is reviewed, updated, and released • Action items and problem reports from the review are closed or dispositioned • SCI is reviewed, updated, and released

Appendix B: Real-Time Operating System Areas of Concern

Acronyms

API	application program interface
C	dead or deactivated code
COTS	commerical off-the-shelf
CPU	central processing unit
D	data consistency
FAA	Federal Aviation Administration
I	interrupts and exceptions
I/O	input/output
M	memory and I/O device access
Q	queuing
RTOS	real-time operating system
S	scheduling
T	tasking
TCB	task control block

Overview

This appendix includes a table from the Federal Aviation Administration (FAA) Research Report DOT/FAA/AR-02/118, *Study of commercial off-the-shelf (COTS) real-time operating systems (RTOS) in aviation applications*. Table B.1 is publicly available at www.faa.gov; however, it is included for convenience. It identifies common areas of concerns for RTOSs. The following seven functional areas should be considered: data consistency (D), dead or deactivated code (C), tasking (T), scheduling (S), memory and input/output (I/O) device access (M), queuing (Q), and interrupts and exceptions (I). See Chapter 20 for additional information on RTOSs.

TABLE B.1

Typical RTOS Areas of Concern

Number	Functional Class	Concern	Description
D1	Data consistency	Data corruption or loss within the RTOS by the RTOS itself	Data, which are visible to the RTOS, are corrupted or "lost" by the RTOS.
D2	Data consistency	Input data corruption or loss by the RTOS	The RTOS incorrectly handles input data or loses it by storing it incorrectly, or incorrect data values are assigned to data variables or returned as results.
D3	Data consistency	Erroneous data or results caused by incorrect calculations or operations by the RTOS	Incorrect data values assigned to data variables or returned as results.
D4	Data consistency	Abnormal parameters	Calculations performed by the math library functions may return unpredictable small numbers if the values passed as parameters are abnormal.
C1	Inclusion of deactivated code or dead code	Inclusion of deactivated code	Unused functions may be loaded with the application even though they are never called. This activity can also be dependent on a linker or loader that is used to link the executable code into the executable image and/or load the image into the target computer memory. Unintended activation of this code may have unknown effects, typically leading to system failure.
C2	Inclusion of deactivated code or dead code	Generation of dead code	Additional software is generated by the compiler or linker, which is not verified during requirements-based testing or coverage analyses. This is especially a concern for Level A applications where the applicant needs to "account" for executable object code that is not traceable to source code; it can result in dead code, and compiler generated code can result in code that is not exercised during requirements-based test, nor is it included in structural coverage analysis which is typically performed at the source code level. Compiler/linker generated object code is not exempt from satisfying these objectives for compliance to requirements and robustness for Levels A–D, and for low-level requirements for Levels A–C.

TABLE B.1 (continued)

Typical RTOS Areas of Concern

Number	Functional Class	Concern	Description
T1	Tasking	Task terminates or is deleted	The task runs to completion or is deleted by another task. If the programming model requires a task to run forever, in a never-ending loop, then the API call to delete the task should be removed.
T2	Tasking	Kernel's storage area overflow	A central storage area in the kernel, which holds task control blocks and other kernel objects, may run out of space due to a malicious task that constantly allocates new kernel objects, which may in turn affect execution of other tasks. A quota system should be implemented to protect other tasks in the system.
T3	Tasking	Task stack size is exceeded	The task stack is overwritten leading to unpredictable system behavior and stack data corruption.
S1	Scheduling	Corrupted task control blocks (TCB)	TCBs may be corrupted, which compromises the scheduling operations of an RTOS. Scheduling information data should be protected from access from user software applications.
S2	Scheduling	Excessive task blocking through priority inversion	A user task of high priority may be excessively blocked by low-priority task because they share a common resource and an intermediate task pre-empts the low-priority task.
S3	Scheduling	Deadlock	If two tasks both require the same two resources but they are scheduled in an incorrect sequence, then they may cause a deadlock by blocking each other.
S4	Scheduling	Tasks spawn additional tasks that starve CPU resources	New tasks spawned by an existing task may affect the schedulability of all tasks in the system. User applications should not be allowed to spawn new tasks at their own will.
S5	Scheduling	Corruption in task priority assignment	Increasing or decreasing priorities of tasks in the system may lead to the task set not being schedulable or the system not responding in timely manner. Ability to change the priority of a task should be limited to special cases, such as to prevent the occurrence of priority inversion.

(continued)

TABLE B.1 (continued)

Typical RTOS Areas of Concern

Number	Functional Class	Concern	Description
S6	Scheduling	Service calls with unbounded execution times	Schedulability of tasks is impacted if there are kernel service calls that have unbounded execution time. The execution time of a task that makes such service calls may itself be affected, as well as accounting for the kernel's overhead while switching between tasks. Kernel service calls should have bounded execution time regardless of system load conditions.
M1	Memory and I/O device access	Fragmentation of heap memory space	Allocation, de-allocation, and the release of memory from the heap may lead to fragments of free memory, which complicates future allocations and may compromise timing analysis making it unpredictable. Dynamic memory allocation, de-allocation, and "garbage collection" should be very limited and controlled.
M2	Memory and I/O device access	An incorrect pointer referencing/de-referencing	An incorrect reference to an object, such as a semaphore, may be passed to the kernel via a service call, which can have disastrous results. The kernel should check validity of pointer references.
M3	Memory and I/O device access	Data overwrite	Data are written beyond their allocated boundaries and overwrite and corrupt adjacent data of other functions in memory.
M4	Memory and I/O device access	Compromised cache coherency	Increased access time occurs due to cache misses. This occurs when needed data are not available in cache and data must be accessed from other typically slower memory. Data loss due to missed memory updates.
M5	Memory and I/O device access	Memory may be locked or unavailable	The MMU page tables may be incorrectly configured or corrupted such that access to a region of memory is prevented.
M6	Memory and I/O device access	Unauthorized access to critical system devices	Unauthorized access to I/O devices may lead to improper functioning of the system. Kernel must implement mandatory access control to all critical devices.
M7	Memory and I/O device access	Resources not monitored	Proper allocations and usage of resources are to be monitored; otherwise resource could be deadlocked.

TABLE B.1 (continued)

Typical RTOS Areas of Concern

Number	Functional Class	Concern	Description
Q1	Queuing	Task queue overflow	May experience loss of information or change in scheduler performance. May result in missed schedule deadlines and incorrect task sequencing.
Q2	Queuing	Message queue overflow	Messages may be missed, lost, or delayed if the queue is not properly sized or messages are not consumed promptly, unless this is protected.
Q3	Queuing	Kernel work queue overflow	The work queue is used to queue kernel work that must be deferred because the kernel is already engaged by another request and the queue is full. Kernel work deferred to the work queue must originate from an interrupt service routine. The work queue may overflow if the interrupt rate is too high for the kernel to process tasks within the allotted time frame.
I1	Interrupts and exceptions	Interrupts during atomic operations	Certain operations that work on global data must complete before subsequent operations can be invoked by another task of execution. An interrupt arriving during this period may cause operations that modify or use a partially modified structure, or the interrupt may be lost if interrupts are masked during critical code execution.
I2	Interrupts and exceptions	No interrupt handler	No interrupt handler has been defined for an interrupt. A default interrupt handler should be provided by the RTOS if the user has specified none.
I3	Interrupts and exceptions	No exception handler	No exception handler has been defined for an exception raised by a task. A default exception handler should be provided to suspend the task and save the state of the task at the point of exception.
I4	Interrupts and exceptions	Signal is raised without a corresponding handler	A signal may be sent by a task to another task or by the hardware under defined exception conditions.
I5	Interrupts and exceptions	Improper protection of supervisor task	Supervisor task that is invoked due to an exception runs in an unprotected address space that may be corrupted.

Source: V. Halwan and J. Krodel, Study of commercial off-the-shelf (COTS) real-time operating systems (RTOS) in aviation applications, DOT/FAA/AR-02/118, Office of Aviation Research, Washington, DC, December 2002.

Reference

1. V. Halwan and J. Krodel, Study of commercial off-the-shelf (COTS) real-time operating systems (RTOS) in aviation applications, DOT/FAA/AR-02/118 (Washington, DC: Office of Aviation Research, December 2002).

Appendix C: Questions to Consider When Selecting a Real-Time Operating System for a Safety-Critical System

Chapter 20 discusses the typical characteristics, features, and issues with real-time operating systems (RTOSs). This appendix includes a list of questions to consider when choosing an RTOS for inclusion in a safety-critical system (as with other parts of this book, it is assumed that DO-178B or DO-178C is required). The questions are divided into three categories (general questions, RTOS functionality questions, and RTOS integration questions); however, the three categories are inter-related. In addition to the author's experience, two sources were used to compile this list of questions: (1) Federal Aviation Administration (FAA) Software Job Aid: *Conducting software reviews prior to certification* [1], and (2) FAA Research Report: *Commercial off-the-shelf real-time operating system and architectural considerations* [2].

C.1 General Questions

- Is an RTOS needed?
- Does the RTOS meet the objectives of DO-178C (or DO-178B) and is the supporting life cycle data available to show compliance?
- Will the necessary data be provided? If not, will the data be available to support certification and continued airworthiness? FAA Advisory Circular 20-148 lists typical data that are provided with a certifiable component.
- How complex is the RTOS?
- Are there sufficient data to support the required level of criticality?
- If the RTOS was reverse engineered, do the processes satisfy DO-178C objectives (see Chapter 25)?
- Have the technical and certification issues concerning RTOSs identified in Chapter 20, as well as any other project-specific issues, been addressed?
- Are the typical characteristics and features of an RTOS listed in Chapter 20 implemented? If not, does the missing characteristic or feature impact safety or compliance?

- What effect does the RTOS have on the development life cycle of the system that will use the RTOS?
- Is the RTOS compatible with other components being developed?
- Is this RTOS flexible? Or, does it put constraints on the system under development? If there are constraints, have they been addressed?
- In what language and which compiler was the RTOS developed? Does the compiler take advantage of processor attributes, such as out-of-order execution, cache, and pipelines? Is the compiler deterministic?
- What tool support does the RTOS offer?
- How will the possibility of hardware obsolescence impact the RTOS use?
- Will the RTOS supplier support the continued airworthiness requirements for the RTOS (e.g., support throughout the life of the aircraft that uses it)? What will happen if the RTOS supplier goes out of business? Usually, some kind of contract or legal agreement is needed to handle these situations.
- What kind of problem reporting system is in place for the RTOS product to support continued airworthiness?
- What assurance is there that the RTOS product will not be changed without the user's knowledge? If it is modified, what process will be used and how will the user be informed?
- Has an accurate User's Manual been produced? Does it address the needs of multiple users? Or, is it tailored for each user?
- What application program interface (API) is used? Does the API support portability? Does the RTOS offer comprehensive support for the API?
- What configuration management and software quality assurance processes were used in the RTOS development? Are these processes adequate for the system using the RTOS? Are these processes compatible with the configuration management and software quality assurance processes of the system using the RTOS?

C.2 RTOS Functionality Questions

- Does the RTOS scale when adding tasks?
- Is the run-time number of tasks limited? If so, is the upper limit adequate to support the user needs?
- What types of intertask communication are supported?

- Is the time slice adjustable?
- Does the RTOS prevent priority inversion?
- Does the RTOS support the selected synchronization mechanism?
- Does the RTOS allow the user to select the scheduling algorithm? Do the available scheduling approaches support safety?
- Are task switching times acceptable?
- How does the RTOS handle the following functions: (1) task handling; (2) memory management; and (3) interrupts? Does it do so in a deterministic manner?
- How does the RTOS use memory, such as internal and external cache?
- Have the following data consistency concerns been addressed: data corruption or loss, erroneous results, and abnormal parameters?
- Have common tasking concerns been addressed (including inadvertent termination or deletion, kernel storage overflow, and stack size exceeded)?
- Have common scheduling concerns been addressed (including corrupted task control blocks, excessive task blocking by priority inversion, deadlock, spawning tasks that starve central processing unit [CPU] resources, corruption in task priority assignment, service calls with unbounded execution times, and race conditions)?
- Have memory and input/output concerns been addressed (such as fragmentation of heap, incorrect pointer referencing, data overwrite, compromised cache coherency, memory locked, unauthorized access to devices, and unmonitored resources)?
- Have common queuing problems been addressed in the RTOS (including task queue overflow, message queue overflow, and kernel work queue overflow)?
- Have interrupt and exception concerns been addressed in the RTOS (such as interrupts during atomic operations, no interrupt handler, no exception handler, and improper protection of supervisor task)?
- If the RTOS is used to support partitioning/protection, have memory, input/output, and time partitioning/protection issues been addressed (see Chapter 21)?
- Has extraneous, dead, or deactivated code been addressed in the RTOS? In particular, how are unused parts of the RTOS being addressed in the requirements?
- Does the RTOS or system have a health monitor? Do the recovery mechanisms from problems meet the system safety needs?
- How has the data and control coupling analysis been addressed? Does it meet DO-178C Table A-7 objective 8?

- Is the kernel abstracted from the hardware in order to support portability onto other targets?
- If the RTOS is a commercial off-the-shelf (COTS) product, have the typical concerns of COTS software been addressed (see Chapter 24)?

C.3 RTOS Integration Questions

- Have the potential vulnerabilities of the RTOS been identified? Is the mitigation of any vulnerabilities that are not mitigated by RTOS design straightforward and feasible for the integrator/user?
- Is there a data sheet that summarizes the RTOS characteristics, limitations, available certification data, etc.? Is the approach for calculating the characteristics and limitations in the data sheet documented?
- Is the RTOS timing performance identified (such as latencies, thread switch jitter, scheduling schemes, and priority levels)?
- Who will be developing the board support package (BSP) and device drivers? Does the development team have the right information to perform this task?
- Are development kits available to assist the RTOS integration? Many RTOS vendors provide kits to help tailor the RTOS, develop drivers, develop BSPs, etc.
- Is the RTOS easily configured for specific use or does it require considerable customization?
- Was the RTOS developed for use with a specific processor or processor family?
- Are the RTOS and CPU(s) interfaces well understood?
- Is the RTOS effect on worst-case execution time effectively calculated?
- Is the source code available for scrutiny by higher criticality level application developers if needed to support low-level functionality?
- Which hardware resources does the RTOS affect?
- What effect does the software used to isolate the RTOS from the target computer (such as BSP or device drivers) have on the system certifiability?
- Which tools are available for analyzing the RTOS performance, and can these tools verify deterministic behavior?
- Are tools for remote diagnostics available? Many RTOS suppliers also provide tools that can help analyze a system's behavior and support analysis and testing activities.

- Does the RTOS have security capabilities that conflict with the system safety objectives? If so, how will these be addressed?
- Does the CPU have an internal memory management unit, and does it support partitioning?
- How is data and control coupling between the RTOS and the applications handled?

References

1. Federal Aviation Administration, *Conducting Software Reviews Prior to Certification*, Aircraft Certification Service (Rev. 1, January 2004).
2. J. Krodel, Commercial off-the-shelf real-time operating system and architectural considerations, DOT/FAA/AR-03/77 (Washington, DC: Office of Aviation Research, February 2004).

Appendix D: Software Service History Questions

Chapter 24 provided the four categories of questions to consider when evaluating software service history. This appendix includes the specific questions from the Federal Aviation Administration's *Software Service History Handbook* [1]. The handbook is available on the Federal Aviation Administration website (www.faa.gov); however, these questions are provided for convenience.

D.1 Questions Regarding Problem Reporting [1]

1. Are the software versions tracked during the service history duration?

2. Are problem reports tracked with respect to particular versions of software?

3. Are problem reports associated with the solutions/patches and an analysis of change impact?

4. Is revision/change history maintained for different versions of the software?

5. Have change impact analyses been performed for changes?

6. Were in-service problems reported?

7. Were all reported problems recorded?

8. Were these problem reports stored in a repository from which they can be retrieved?

9. Were in-service problems thoroughly analyzed and/or those analyses included or appropriately referenced in the problem reports?

10. Are problems within the problem report repository classified?

11. If the same type of problem was reported multiple times, were there multiple entries or a single entry for a specific problem?

12. If problems were found in the lab in executing copies of operational versions of software during the service history period, were these problems included in the problem reporting system?

13. Is each problem report tracked with the status of whether it is fixed or open?

14. If the problem was fixed, is there a record of how the problem was fixed (in requirements, design, code)?

15. Is there a record of a new version of software with new release after the problem was fixed?

16. Are there problems with no corresponding record of change in software version?

17. Does the change history show that the software is currently stable and mature?

18. Does the product have the property of exhibiting the error with a message to the user? (Some products may not have error trapping facilities, so they may just continue executing with wrong results and with no indication of failure.)

19. Has the vendor (or the problem report collecting agency) made it clear to all users that problems are being collected and corrected?

20. Are all problems within the problem report repository classified?

21. Are safety-related problems identified as such? Can safety-related problems be retrieved?

22. Is there a record of which safety problems are fixed and which problems are left open?

23. Is there enough data after the last fix of safety-related problems to assess that the problem has been corrected and no new safety-related problems have surfaced?

24. Do open problem reports have any safety impact?

25. Is there enough data after the last fix of safety-related problems to assess that the problem is solved and no new safety-related problems have surfaced?

26. Are the problem reports and their solutions classified to indicate how a fix was implemented?

27. Is it possible to trace particular patches with release versions and infer from design and code fixes that the new versions correspond to these fixes?

28. Is it possible to separate the problem reports that were fixed in the hardware or change of requirements?

29. Are problem reports associated with the solutions/patches and an analysis of change?

30. If the solutions indicated a change in the hardware or mode of usage or requirements, is there an analysis of whether these changes invalidate the service history data before that change?

31. Is there a fix to a problem with changes to software but with no record of the change in the software version?

32. Is the service period defined appropriate to the nature of software in question?

33. How many copies of the software are in use and being tracked for problems?

34. How many of these applications can be considered to be similar in operation and environment?

35. Are the input/output domains the same between the service duration and the proposed usage?

36. If the input/output domains are different, can they be amended using glue code?

37. Does the service period include normal and abnormal operating conditions?

38. Is there a record of the total number of service calls received during the period?

39. Were warnings and service interruptions a part of this problem-reporting system?

40. Were warnings analyzed to assure that they were or were not problems?

41. Was there a procedure used to log the problem reports as errors?

42. What was the reasoning behind the contents of the procedure?

43. Is there evidence that this procedure was enforced and used consistently throughout the service history period?

44. Does the history of warranty claims made on the product match with the kind of problems seen in the service history?

45. Have problem reports identified as a nonsafety problem in the original domain been reviewed to determine if they are safety related in the target domain?

D.2 Questions Regarding Operation [1]

1. Is the intended software operation similar to the usage during the service history (i.e., its interface with the external world, people, and procedures)?

2. Have the differences between service history usage and proposed usage been analyzed?

3. Are there differences in the operating modes in the new usage?

4. Are only some of the functions of the proposed application used in service usage?

5. Is there a gap analysis of functions that are needed in the proposed application but have not been used in the service duration?

6. Is the definition of normal operation and normal operation time appropriate to the product?

7. Does the service period include normal and abnormal operating conditions?

8. Is there a technology difference in the usage of the product from service history duration (manual vs automatic, user intercept of errors, used within a network vs standalone, etc.)?

9. Was operator training on procedures required in the use of the product during the recorded service history time period?

10. Is there a plan to provide similar training in the new operation?

11. Will the software level for the new system be the same as it was in the old system?

D.3 Questions Regarding Environment [1]

1. Are the hardware environment of service history and the target environment similar?

2. Have the resource differences between the two computers been analyzed (time, memory, accuracy, precision, communication services, built-in tests, fault tolerance, channels and ports, queuing modes, priorities, error recovery actions, etc.)?

3. Are safety requirements encountered by the product the same in both environments?

4. Are exceptions encountered by the product the same in both environments?

5. Is the data needed to analyze similarity of environment available? (Such data are not usually a part of problem data.)

6. Does the analysis show which portions of the service history data are applicable to the proposed use?

7. How much service history credit can be assigned to the product, as opposed to the fault-tolerant properties of the computer environment in the service history duration?

8. Is the product compatible with the target computer without making modifications to the product software?

9. If the hardware environments are different, have the differences been analyzed?

10. Were there hardware modifications during the service history period?

11. If there were hardware modifications, is it still appropriate to consider the service history duration before the modifications?

12. Are software requirements and design data needed to analyze whether the configuration control of any hardware changes noted in the service history are acceptable?

D.4 Questions Regarding Time [1]

1. What is the definition of service period?

2. Is the service period defined appropriate to the nature of software in question?

3. What is the definition of normal operation time?

4. Does normal operation time used in service period include normal and abnormal operating conditions?

5. Can contiguous operation time be derived from service history data?

6. Is "applicable-service" portion recognized from the total service history data availability?

7. What was the criterion for evaluating the service period duration?

8. How many copies of the software are in use and being tracked for problems?

9. What is the duration of applicable service?

10. Is applicable-service definition appropriate?

11. Is this the duration used for calculation of error rates?

12. How reliable was the means of measuring time?

13. How consistent was the means of measuring time throughout the service history duration?

14. Do you have a proposed accepted error rate justifiable and appropriate for the level of safety of proposed usage, before analyzing the service history data?

15. How do you propose that this error rate be calculated, before analyzing the service history data?

16. Is the error rate computation (total errors divided by time duration, by number of execution cycles, by number of events such as landing,

by flight hours, by flight distance, or by total population operating time) appropriate to the application in question? What was the total duration of time used for this computation? Has care been taken to consider only the appropriate durations?

17. What is the actual error rate computed after analyzing the service history data?

18. Is this error rate greater than the proposed acceptable error rate defined in PSAC [Plan for Software Aspects of Certification]?*

19. If the error rate is greater, was analysis conducted to reassess the error rates?

Reference

1. U.D. Ferrell and T.K. Ferrell, *Software Service History Handbook*, DOT/FAA/AR-01/116 (Washington, DC: Office of Aviation Research, January 2002).

---

* Brackets added for clarification.

Index

Printed in the United States
by Baker & Taylor Publisher Services